PMR Spectroscopy
in Medicinal
and Biological Chemistry

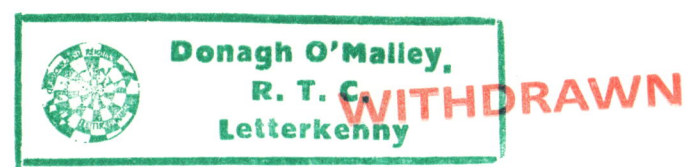
PMR Spectroscopy in Medicinal and Biological Chemistry

by

A. F. CASY

University of Alberta, Edmonton, Canada

1971

Academic Press: London and New York

ACADEMIC PRESS INC. (LONDON) LTD
24–28 OVAL ROAD,
LONDON NW1

U.S. Edition published by

ACADEMIC PRESS INC.
111 Fifth Avenue,
New York, New York 10003

A/574, 1'92

Library of Congress Catalog Card Number: 75-129784
ISBN: 0-12-164050-7

PRINTED IN GREAT BRITAIN BY
WILLIAM CLOWES AND SONS LIMITED
LONDON, COLCHESTER AND BECCLES

Preface

There is a good selection of books which give accounts of the fundamental principles of NMR spectroscopy appropriate to organic chemists (see selective bibliography, p. 403). Within the text of such books many examples of the applications of the technique to various problems are given, but descriptions of basic theory and data tabulations take precedence. I feel, therefore, that there is a need for an NMR text that, by assuming knowledge of essential principles, is left free to concentrate on applications, i.e. on what NMR can be used for and what problems it can solve.

The fields of medicinal and biological chemistry, embracing as they do the disciplines of organic chemistry, biology and pharmacology, present a rich variety of problems that are amenable to solution by NMR spectroscopy (and in some cases *only* by NMR spectroscopy). I came to appreciate this through personal research experience over the last 10 years and also through literature surveys that I carried out while preparing a review article on the topic for the *Journal of Pharmaceutical Sciences* [*J. Pharm. Sci.* (1967), **56**, 1049]. For this reason I decided that the NMR book I had in mind could be specially directed towards the interests of medicinal chemists (i.e. chemists in pharmaceutical industry, chemists teaching pharmaceutical and medicinal chemistry in schools of pharmacy plus graduate students in such departments), biochemists (two chapters deal exclusively with biochemical problems), and pharmacologists interested in drug-receptor interactions and structure-activity relationships. Since so much of the material described falls in the realm of organic chemistry, I hope that organic chemists not specifically working in a biological field will also find the book of value (e.g. for stereochemical problems).

The majority of NMR studies in fields connected directly and indirectly with life sciences concern protons and I have, therefore, restricted the book, with a few exceptions, to examples of proton magnetic resonance (PMR) as is reflected in the title. Rapid advances in studies of other nuclei (especially carbon-13), now in progress, will probably alter this balance during the next few years.

In the book's presentation, I have assumed an empirical knowledge (first order approach) of the fundamentals of NMR spectroscopy, and have adopted an explanatory rather than review style. I have attempted to give full accounts of all aspects discussed apart from basic principles. My choice of topics is largely a personal one and comprehensive accounts have not been attempted except in a few cases. Literature references up to 1970 have been included.

A few words about the content of the book are given below. Analytical applications of PMR spectroscopy are described in Chapter 1. These include procedures of a degree of accuracy appropriate to the assay of pharmaceuticals and also more approximate methods useful for the analysis of reaction product mixtures. Studies of alcohols and aromatic derivatives provide examples of the value of PMR spectroscopy as a tool in qualitative analysis. Chapter 2 is devoted to the NMR aspects of organic nitrogen derivatives, a topic included because of the wide-spread occurrence of nitrogen-containing biochemicals and drugs. The stereochemical utility of PMR data is described in Chapter 3. The fairly detailed account given is warranted by the increasing demand for knowledge of the specific molecular geometry of biologically active molecules, particularly in the solute condition. There is sufficient interest in the stereochemistry of cyclic bases to justify their separate discussion and this is provided in Chapter 4. Widespread observation of radical differences in the biological activities of enantiomorphic forms of chiral molecules has greatly stimulated study of optically active molecules over the past 10 to 20 years. The power of PMR spectroscopy in the resolution and configurational assignment of optically active molecules is described in Chapter 5. Specific examples of the use of the PMR technique in the study of compounds of pharmacological interest are given in Chapter 6 (narcotic analgesics, cholinergic agonists, histamine, and antihistamines) and Chapter 7 (tropanes, sympathomimetic amines, antibiotics and steroids). Some of these applications illustrate principles described in earlier chapters. The use of PMR spectroscopy in biochemistry is described in the last two chapters. Chapter 8 concerns proteins and their simpler components while studies of carbohydrates, nucleosides, nucleotides and nucleic acids are described in Chapter 9. Relaxation-time investigations, of particular importance to the study of protein conformation and the uptake of small molecules by macromolecules, are also included in Chapter 8. The book concludes with an appendix on some solvent and hydrogen bonding effects. The routine use of PMR spectroscopy in structure elucidation as encountered in synthetic and natural product work is not dealt with specifically in this book.

Of the many colleagues who have fostered my interest in PMR spectroscopy, I wish to thank Dr Alain Huitric (University of Washington) in particular. I also thank Drs H. Booth (University of Nottingham), R. U. Lemieux and G. Kotowycz (University of Alberta) for reading chapters of this book. I am indebted to the Medical Research Council of Canada for support of my own studies in this field and for the provision of an NMR spectrometer. Many of the spectra reproduced here were recorded on this machine and I acknowledge the able assistance of Mrs Sining Lee in this work. I owe much, in addition, to the willing co-operation of my graduate students and post-doctoral associates. Finally I wish to record my gratitude to the staff of the School of Chemical Sciences, University of East Anglia, for their kind hospitality during my completion of the manuscript.

A. F. CASY

University of Alberta
Edmonton, Alberta
October, 1971

Acknowledgements

The author is grateful to authors and publishers for permission to reproduce figures. The source is given in the legend and the full reference is listed in the reference list at the end of each chapter.

Foreword

The advent of proton magnetic resonance spectroscopy has, over the last 15 years, revolutionized organic chemistry. Initially an exciting new technique, which in the hands of forward-looking pioneer chemists provided elegant solutions to old problems, and enabled the solution of many new ones, NMR has now progressed to the single most useful and most used physical method in organic chemistry. It is now unthinkable for a strong research laboratory in organic chemistry to be without NMR facilities, or for an active organic chemist to be unskilled in its use.

The applications of NMR far outrange the boundaries of pure organic chemistry, and its potential in the medicinal and biological fields is enormous. Although such applications have in fact been numerous, the *routine* use of NMR by biologists, biochemists, pharmacologists and medicinal chemists is not yet general. Such routine use has been hindered by the lack of a suitable text book. It is to fill this gap that Professor Casy has prepared the present text. His book discusses a series of topics pertinent to the life sciences as he is well qualified to do.

It is to be hoped that this book will widely stimulate the application of NMR to chemical problems in the life science area, and it is commended to all workers in this area. In addition the book will be of interest to many pure organic chemists, particularly those concerned with nitrogen compounds and/or stereochemical problems. The work deserves wide dissemination.

A. R. KATRITZKY

University of East Anglia
October, 1971

Contents

Introduction

Nuclear magnetic resonance (NMR) spectroscopy is a physical technique which enables chemists to study the environment of certain atomic nuclei within molecules. Protons ($'H$) have been, and still are, the atomic species most commonly studied, and applications of proton magnetic resonance (PMR) form the specific topic of this book, except for a few cases in which reference is made to ^{13}C and ^{19}F spectra. PMR spectroscopy was first applied to organic compounds soon after 1950 and its use has grown enormously over the last 20 years. A general account of the basic principles of NMR spectroscopy is not included in this book; a reference list of texts providing this information is given in the bibliography at the end of the book (other books are mentioned throughout the text). Several elementary accounts of the use of physical methods in organic chemistry include useful chapters on NMR spectroscopy and these are also listed together with certain review series, journals and source books.

The interpretation of NMR spectra is based on three sets of parameters which characterize the absorption of radiofrequency radiation by atomic nuclei placed in a magnetic field (H):

1. The frequences of the absorbed radiation (v) expressed as the *chemical shift* relative to an arbitrarily standard absorption line;
2. The multiplicity of the lines originating from a given group of nuclei and described by the appropriate *coupling constants* ($\mathcal{J}$ values); and
3. The decay times characterizing the return of the nuclei excited by the absorption of radiation to a lower energy state, referred to as *relaxation times*.

Additional information is derived from the integration curve which enables the relative numbers of protons within each absorption band to be determined. Of the above parameters, the first two are the most widely applied to chemical problems, although more attention is now being given to relaxation times because of their value in the study of molecular interactions (Chapter 8).

NMR spectra are displayed as plots of a detector signal (ordinate) against the magnetic field strength (abscissa). Three scales are in current use for expressing $'H$ chemical shifts all of which are related to a standard (preferably internal) which is normally tetramethylsilane (TMS). The scales are:

1. Hertz (Hz) from TMS set at zero; use of Hz has now largely replaced that of the equivalent term cycles per second (cyc/sec).
2. Parts per million (ppm), obtained by multiplying the Hz value by the term 10^6/oscillator frequency in Hz. For most protons, values obtained fall in the range 0–10 ppm with TMS at zero. This scale is denoted by the Greek letter δ followed by the numerical value, e.g. δ 1·5 (the factor ppm is omitted since it is understood). The reverse form, e.g. 1·5 δ, is occasionally used in the literature.
3. The τ (tau or tor) scale where

$$\tau = 10 - \delta \text{ with TMS at 10, e.g., } \tau\ 5\cdot5\ (\equiv \delta\ 4\cdot5).$$

The last two scales are dimensionless, while the first depends on the operating frequency and must always be used in conjunction with the frequency value. Relative merits of the three systems have been discussed by Bible (1965).

Attempts are being made to encourage the presentation of NMR data in a uniform manner by the international Union of Pure and Applied Chemistry and other bodies, and it is probable that the δ scale will eventually gain the approval of the majority of NMR spectroscopists. Meanwhile, however, chemists employing the technique must be familiar with all three variants, and for this reason no attempt has been made to standardize chemical shift values in the examples discussed in this book. Coupling constants and the dimensions of a resonance peak (usually its width at half maximum height, W_H) are expressed in Hz (replacing cyc/sec). It is important to remember that an increase in the numerical value of a chemical shift expressed in Hz or on the δ scale implies a move to *lower* field (deshielding), while an increase on the τ scale means a move to *higher* field (shielding).

Most of the examples given in this book involve first order spectral analyses in which chemical shifts and coupling constants are read directly from spectra. Such an approach usually gives only an approximation of the true values but provides data which are adequate for the solution of many problems concerning molecular constitution and even stereochemistry. Garbisch (1968), in commenting upon this question, states that "the key to the solution may be simply recognition of the gross multiplicities of the bands (neglecting higher-order splittings which are often undetectable under the conditions which the spectra were determined) or the observation that a coupling constant is large, say 16 Hz, or small, say 3 Hz. The reliability of the solution of the problem will rarely rest on knowledge that the correct coupling constant is 16·87 or 3·44 Hz". Nevertheless, in the first order approach, care must be taken to guard against the possibility of the spectra being "deceptively simple", i.e. spectra which appear susceptible to first order treatment but which in fact are not (Becker, 1965). Guides to the second order analysis of complex NMR spectra have been provided by Garbisch (1968), Bible (1965) and Mathieson (1967).

Abbreviations

$CDCl_3$	deuterochloroform
D_2O	deuterium oxide
DMSO-d_6	deuterated dimethylsulphoxide
TMS	*tetram*ethyl*s*ilane
DSS	sodium 2,2-*d*imethyl-2-*s*ilapentane-5-*s*ulphonate
$J_{gem}(^3J)$	a geminal coupling constant
$J_{vic}(^2J)$	a vicinal coupling constant
W_H	width at half maximum height
Me	methyl
Et	ethyl
Ph	phenyl (and other common radical abbreviations)
ν	chemical shift
$\varDelta\nu$	difference in chemical shift

NOTE: In most representations of asymmetric molecules, only one of the two possible enantiomorphic forms is shown.

CHAPTER 1

Analytical Aspects

QUANTITATIVE ANALYSIS

The application of PMR to quantitative analysis depends on the fact that the area beneath a particular PMR signal is directly proportional to the number of protons from which the signal is derived. This area may be obtained from the corresponding integral trace (an integrator is now incorporated within most commercial instruments) or by means of a planimeter provided the signal is removed from nearby resonances which may distort or overlap the non-horizontal portion of the integral line. The accuracy of area measurements and the degree to which an area truly represents the number of protons giving rise to the integrated signal depends critically upon the instrumental operating conditions. Some of the factors affecting PMR integration accuracy are listed below (J. N. Shoolery, 1969, private communication).

(i) The signal-to-noise ratio must be as high as possible to ensure that the background noise contribution to the integration trace is negligible. A sensitive spectrometer should be used (modern instruments operated under optimal conditions provide signal-to-noise ratios near 20:1 with reference to a standard 1% ethylbenzene sample) and fairly concentrated ($\sim$10%) solutions analysed.

(ii) A non-saturating mode of operation must be employed. Saturation effects, i.e. the decrease in the population difference between nuclei in ground and excited states which occurs as a peak is being swept, lead to deceptively low areas, the extents of reduction varying from signal to signal. Corrections may be applied but require knowledge of the relaxation times T_1 and T_2 for each peak, and the avoidance of saturation is considered a more satisfactory general procedure (Paulsen and Cooke, 1964a). This can be achieved by the proper choice of r.f. power level (H_1) and sweep rate (dH/dt). Low H_1 and high dH/dt values minimize differences in integrated intensities due to saturation, but at the same time the integral is decreased and the signal-to-noise ratio degraded, so a compromise must be chosen. Paulsen and Cooke (1964a) identify the best conditions by measuring the ratio of areas of the sample and an internal standard at progressively increasing power rates (or decreasing sweep rates). The optimum area ratio is that obtained at the r.f. power level just below that at which a change in the ratio is detected.

(iii) The use of rapid sweeps allows a large number of independent measurements to be made per unit time and these can be averaged and statistically analysed to give an indication of the precision of the measurement.

(iv) The settings of both the baseline zero and the r.f. phase control are critical. Maladjustment of the latter leads to recording the integral of a mixture of absorption and dispersion mode signals. The dispersion mode is positive on one side of the centre and negative on the other, and this introduces baseline slopes near the integral which are of opposite sign (Fig. 1.1). On the other hand, correct phasing but incorrect detector zero introduces a constant slope on both sides of the integral. Correct setting of the controls gives a square step-function integral which can be measured accurately (Fig. 1.1).

(v) By careful adjustment of magnetic field homogeneity and use of high-precision sample tubes spinning at an optimal rate, it should be possible to reduce spinning side bands to a point where they may be neglected. If their area is significant, however, it should be added to that of the main peak to obtain the true area representative of the protons giving rise to the signal. Corrections for ^{13}C satellite lines are described on p. 13.

Further useful information on accuracy and optimum operating conditions in quantitative PMR work is available in papers by Alexander and Koch (1967) of the U.S. Food and Drug Administration.

In quantitative PMR procedures a particular integral value is related to a specific number of protons by the inclusion of an internal standard in the solution to be analysed. The reference and analytical peak should, if possible, have comparable areas to minimize errors caused by non-linearity of the electronic measuring circuits. The assay of meprobamate (Turczan and Kram, 1967), described below, illustrates the principles involved. The PMR spectrum of meprobamate (Fig. 1.2) displays sharp singlets for methylene (4 protons) and *t*-methyl protons; either could be used for analytical purposes but the former is preferable since it is well isolated, making impurity detection more easy (impurities give an erroneous integral

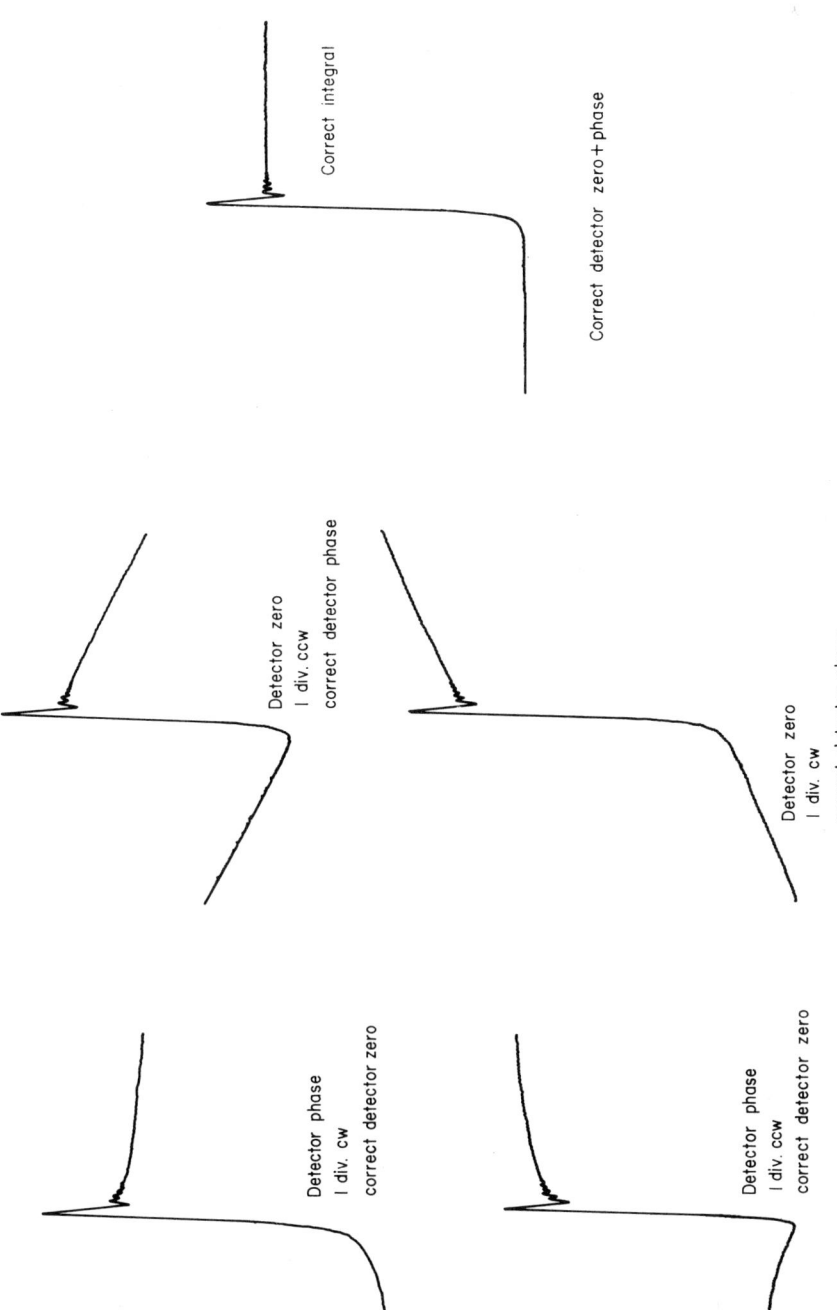

FIG. 1.1. Varian integrals. Integral traces recorded for 6% TMS in CHCl₃ under various settings of detector phase and zero (cw denotes clockwise and ccw counter clockwise).

value). The methylene signal of malonic acid falls near (but not too near) that of meprobamate and this acid is used as the standard.

Procedure

Twenty tablets are weighed and finely powdered. An accurate weight of this powder (equivalent to about 400 mg meprobamate) is placed in a glass-stoppered centrifuge tube together with pure malonic acid (400 mg) and acetone (5 ml), and the whole shaken and then centrifuged. Approximately 0·4 ml of the supernatant (the volume need not be accurate) is placed in an NMR tube and its spectrum recorded (care is taken to avoid spinning side bands between 3 and 4 ppm due to solvent) (Fig. 1.2). Peaks at 3·8 (meprobamate CH_2) and 3·3 ppm (malonic acid CH_2) are integrated several times in each direction.

Calculation

First, a particular integral value is equated to a precise amount of protons by use of data for the standard, malonic acid.

Thus 104 g malonic acid (M.W.) = 2 methylene protons. Therefore, 52 g malonic acid (M.W./2) = 1 proton.

If x is the weight (g) of standard taken and I_{mal} the corresponding integral

$$I_{mal} \equiv \frac{x}{52} \text{ protons}$$

(strictly, this integral only represents a fraction of x dependent upon the volume of acetone extract used. This point also applies to the unknown and may be ignored because the two proportionalities cancel. The ratio $I_{unknown}/I_{standard}$ is the significant quantity).

Now, 218·4 g meprobamate (M.W.) $\equiv$ 4 methylene protons. Therefore, 54·6 g (M.W./4) meprobamate $\equiv$ 1 proton. Thus if z g of meprobamate are present in the sample and I_{mep} is the integral due to the methylene signal of this material,

$$I_{mep} = \frac{z}{54 \cdot 6} \text{ protons}$$

and the ratio

$$\frac{I_{mal}}{I_{mep}} = \frac{x}{52} \times \frac{54 \cdot 6}{z}$$

gives the unknown quantity z (in the same weight units as those of x). This is contained in a known weight of sample and the value may then be related to tablet content, etc. A similar procedure has been described for

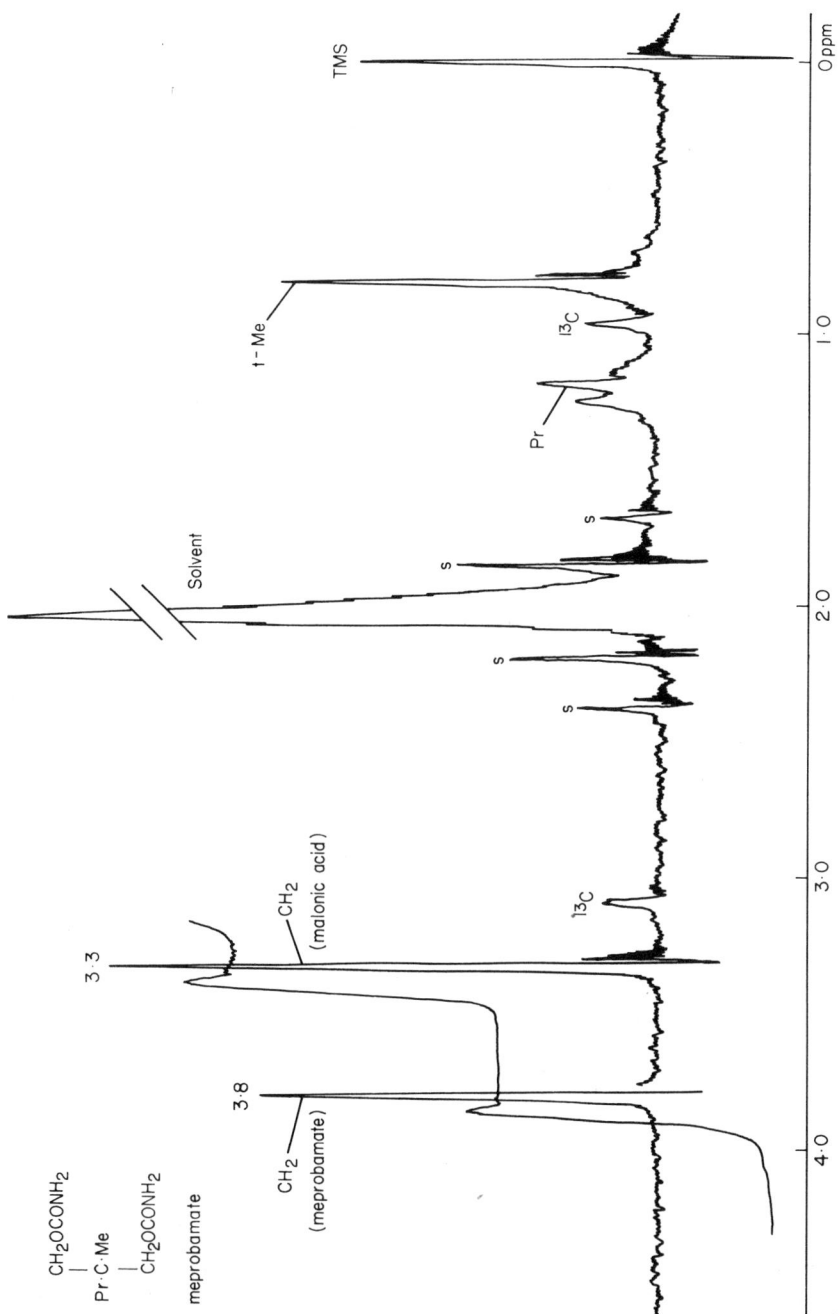

Fig. 1.2. 60-MHz-PMR spectrum of a mixture of meprobamate (451 mg) and malonic acid (415 mg) in acetone (4 ml). The singlets (S) are spinning side bands and ^{13}C satellite signals of the solvent methyl resonance.

the assay of dimethylsulphoxide (DMSO) in solutions and ointments (Kram and Turczan, 1968).

To be of value, an NMR assay procedure must be at least as accurate as other procedures and should have the advantage of rapidity and specificity. Regarding accuracy, meprobamate contents obtained by the described method agreed well with those found by the NF XII volumetric procedure in three out of four cases while the average deviation for seven standard preparations was 0·9%. Specificity is clearly an advantage of the NMR method (this is potentially true of all such assays because the spectrum of the unknown sample may be compared with that of authentic material), related agents being readily differentiated by signal characteristics in the methyl resonance region, e.g. carisoprodal **1** (CH$\underline{Me}_2$ doublet falls to low field of *t*-Me singlet) and mebutamate **2** (CH$\underline{Me}$ doublet overlaps *t*-Me singlet).

$$CH_2OCONHCHMe_2$$
$$|$$
$$PrCMe$$
$$|$$
$$CH_2OCONH_2$$

1

$$\underline{Me} \quad CH_2OCONH_2$$
$$| \qquad |$$
$$EtC—CMe$$
$$| \qquad |$$
$$H \quad CH_2OCONH_2$$

2

Mixtures

Integral data may be used for the analysis of mixtures when overlap between signals due to individual components does not occur (or may be corrected for), and PMR methods for the quantitative assay of mixtures of aspirin, phenacetin and caffeine (Hollis, 1963) and barbiturates (Philipsborn, 1964) have been described. The spectrum of an APC tablet in CDCl$_3$ is shown in Fig. 1.3 (binders such as starch and lactose are insoluble and do not interfere). The signal at τ7·7 is due to OCOMe of aspirin and its integral value (compared with that of one Me in a standard solution of caffeine) enables the aspirin content of the mixture to be determined. The signals at τ6·6 and 6·4 (due to N-1 and N-3 Me groups) give the caffeine content and the integral of the quartet centred near τ6·0 due to OC$\underline{H}_2$Me (and corrected for the overlapping N-7 Me signal of caffeine) gives the phenacetin content. The phenacetin triplet (OCH$_2\underline{Me}$) near τ8·7 was rejected as an analytical peak because commercial samples gave rise to impurity signals in this region. The method is claimed to be of an accuracy sufficient for quality control.

Maolate **3**, a muscle relaxant related to meprobamate, is commonly contaminated with its structural isomer **4**, and a PMR method for the assay of mixtures has been developed in which a peak height rather than integral ratio is employed (Slomp *et al.*, 1964). In the spectra of synthetic mixtures, it was found that the ratio of peak heights of signals at 245 and 225 Hz (characteristic solely of **3** and **4** respectively) were dependent only

$$ArOCH_2CHCH_2OCONH_2 \qquad\qquad ArOCH_2CHCH_2OH$$
$$\quad\quad\;| \qquad\qquad\qquad\qquad\qquad\qquad\qquad\;| $$
$$\quad\quad OH \qquad\qquad\qquad\qquad\qquad\qquad OCONH_2$$

3   **4**

$Ar = p\text{-}Cl.C_6H_4$

on percent composition and were independent of sample size and spectrometer performance factors. The value of this ratio in an unknown, together with a plot of peak height ratio versus composition, enabled analyses to be made and the error of the method is claimed to be less than 1%.

The treatment required for an analysis where signal overlap occurs for all analytical peaks is illustrated by the case of the phosphorus insecticide Systox (Babad *et al.*, 1968). This consists chiefly of two isomers, compound

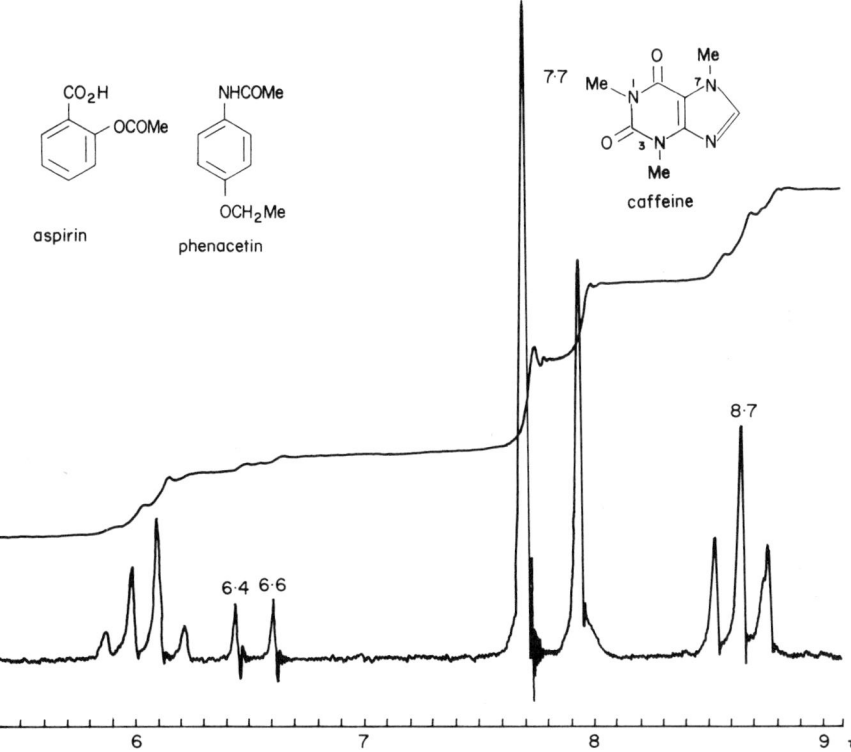

FIG. 1.3. 60-MHz PMR spectrum of a mixture of aspirin (220 mg), phenacetin (160 mg) and caffeine citrate (32 mg) in CDCl₃.

A (thiono) and B (thiol) plus 1–5% of compound C (Sulfotepp). Knowledge of the ratio of A to B is of interest because the two isomers differ in toxicity.

$$\begin{array}{c} S \\ \uparrow \\ (EtO)_2P\!-\!O\!-\!C_2H_4SEt \end{array}$$
A (thiono)

$3 \times \underline{OCH_2}$ (6 protons)
$2 \times \underline{SCH_2}$ (4)
$3 \times C\!-\!\underline{Me}$ (9)

$$\begin{array}{c} O \\ \uparrow \\ (EtO)_2P\!-\!S\!-\!C_2H_4SEt \end{array}$$
B (thiol)

$2 \times \underline{OCH_2}$ (4)
$3 \times \underline{SCH_2}$ (6)
$3 \times C\!-\!\underline{Me}$ (9)

$$\begin{array}{cc} O & O \\ \uparrow & \uparrow \\ (EtO)_2P\!-\!O\!-\!P(OEt)_2 \end{array}$$
C (Sulfotepp)

$4 \times \underline{OCH_2}$ (8)
$4 \times C\underline{Me}$ (12)

The PMR spectrum of a Systox sample (Fig. 1.4) shows three distinct multiplets, the low-field signal due to the $\underline{OCH_2}$ protons, the midfield to $\underline{SCH_2}$ and the high field to C-$\underline{Me}$, and the areas of each are measured by integration. To sort out the contributions of each compound to these signals, three simultaneous equations are constructed:

Let I_{OCH_2} and I_{SCH_2} and I_{Me} be the respective integrals and X_A, X_B and X_C the relative mole fractions of compounds A, B and C.

I_{OCH_2} will be made up of three contributions:

That due to compound $A = 6X_A$ ($3 \times \underline{OCH_2}$, see formulae); that due to compound $B = 4X_B$; that due to compound $C = 8X_C$.

Hence,

$$I_{OCH_2} = 6X_A + 4X_B + 8X_C$$

Similarly,

$$I_{SCH_2} = 4X_A + 6X_B$$

and

$$I_{Me} = 9X_A + 9X_B + 12X_C$$

Solving for X_A, X_B and X_C

$$X_A = \tfrac{3}{8}I_{OCH_2} + \tfrac{1}{8}I_{SCH_2} - \tfrac{1}{4}I_{Me}$$
$$X_B = \tfrac{1}{6}I_{Me} - \tfrac{1}{4}I_{OCH_2}$$
$$X_C = \tfrac{5}{48}I_{Me} - \tfrac{3}{32}I_{SCH_2} - \tfrac{3}{32}I_{OCH_2}$$

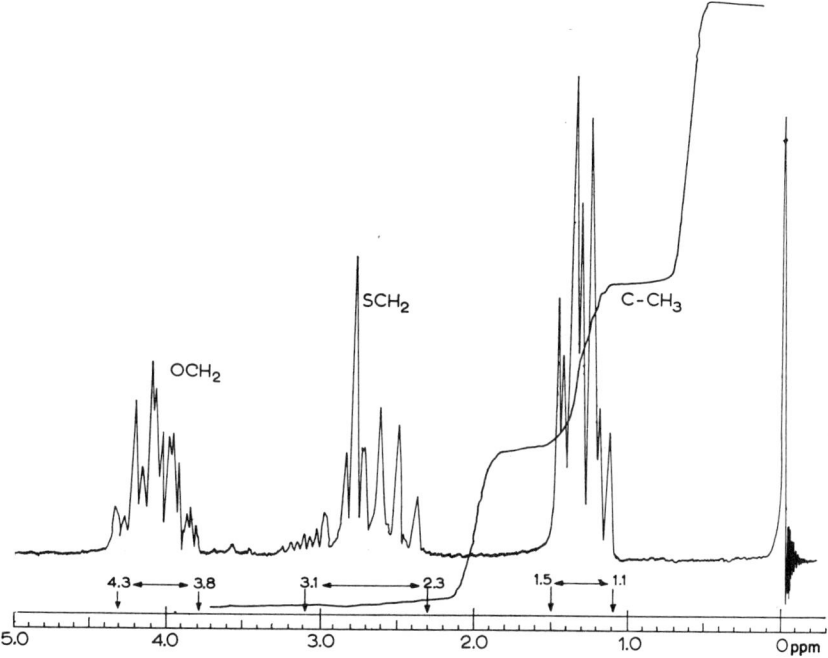

Fig. 1.4. 60-MHz PMR spectrum of 20% Systox in CCl₄. When the integration circuit was checked against the TMS peak the error of the procedure was ±1·5% (Babad *et al.* 1968).

Further, somewhat cumbersome mathematical manipulations allow the actual weights of A, B and C to be calculated. Independent knowledge of the concentration of C (possible by vapour phase chromatography) much simplifies the calculation.

Quantitative active hydrogen measurements by PMR are exemplified by the estimation of the number of hydroxyl groups in Bjorkman's lignin (Gagnaire and Robert, 1968). The lignin is dissolved in DMSO-d_6 containing a known weight of pyrazine as internal reference. Deuterated trifluoroacetic acid (CF_3CO_2D) is added when all OH functions (phenols, 1°, 2° and 3° alcohols) exchange their protons for deuterium. These protons contribute to a common proton signal which is low field and isolated from all other signals. Its integral, compared with that of the pyrazine protons enables the weight of OH in a given weight of lignin to be calculated. Allowance must be made for traces of OH impurities (e.g. water) in the solvent and reagents. In this example fast exchange between CD_3CO_2D and the hydroxyl protons is presumed so that the common signal accounts for all the active hydrogen. Paulson and Cooke (1964b) have described a treatment applicable when D/H exchange is slow. In such cases, e.g. exchange between acetone or phenylacetylene and D_2O catalysed by a base, separate HDO and active hydrogen signals arise and calculation of total active hydrogen requires knowledge of the H/D ratio in the water in addition to the proton content of the HDO signal. The calculation is as follows:

$$R—H + D_2O \ \rightleftharpoons \ R—D + DHO$$

In the above equilibrium, let A = total H in DHO signal; B = amount of active H derived from the active hydrogen site (this equals the amount of deuterium entering the site). B is obtained by correcting A for impurities (chiefly water) already present in the D_2O, and C = mols of deuterium added (from the weight of D_2O used).

Then, $C - B$ = mols of deuterium are left in the D_2O—H_2O, whence the H/D ratio in this mixture = $A/C - B$. It is assumed that the same ratio exists at the active hydrogen site. Hence, if there are B mols of deuterium at this site, there must be $B(A/C - B)$ mols of hydrogen.

Hence, total active hydrogen = $B + B(A/C - B) = B[1 + A/C - B]$.

Molecular Weight

A molecular weight procedure based on the integration of proton signals has been proposed (Barcza, 1963). Again, a certain integral is equated with a precise amount of protons by including a known weight of standard in the NMR sample of the unknown. A signal due to the latter must be sought where it is possible to make a reasonable guess as to the proton group from which it arises. If, for example, there are good grounds for believing one or more monosubstituted phenyl groups to be present, the aromatic signal (5, or a multiple of 5, protons) may be used; alternatively, a well-defined doublet near 1 ppm can be taken as a 3- (or multiple of 3) proton secondary methyl signal.

As before,

$$I_s \equiv \frac{W_s n_s}{M_s} \text{ protons}$$

where I_s = integral of standard, W_s = weight of standard, M_s = molecular weight of standard, and n_s = number of protons in standard peak.

Therefore,

$$I \equiv \frac{W_s n_s I}{M_s I_s} \text{ protons}$$

where I = integral of unknown signal, assumed to arise from n protons. The

$$\frac{W_s n_s I}{M_s I_s}$$

proton signal is due to weight W of unknown. Therefore, n protons correspond with

$$\frac{W n M_s I_s}{W_s n_s I} = \text{molecular weight of unknown}$$

The standard used must not overlap any signals of the unknown and hexamethylcyclotrisiloxane (5) (a very soluble inert solid with a high-field

$$\text{Me}_2\text{Si} \overset{O}{\diagup}\hspace{-0.3em}\diagdown \text{SiMe}_2$$

$$\text{O}\diagdown \underset{\text{Me}_2}{\text{Si}} \diagup \text{O}$$

5

chemical shift a few cycles downfield from TMS) is recommended. The author provides a series of examples demonstrating the accuracy of the method, but the procedure does not appear to have received much application in practice.

There are several reports, however, about the molecular-weight determination of polymers based upon end-group analysis by PMR spectroscopy (Page and Bresler, 1964; Liu, 1968). The OH resonance of the polyethylene glycol Carbowax 600 [$HO(CH_2CH_2O)_nH$], for example, may be integrated separately from the $—OCH_2CH_2—O$ signal and the number average molecular weight (NAMW) calculated as follows, taking X as the OH integral and Y as the OCH_2CH_2O integral: the X integral $\equiv 2$ protons; therefore, the Y integral $\equiv 2Y/X$ protons.

If the Y integral is due to 4 protons it is equivalent to a mass of 44 (CH_2CH_2O); therefore, $2Y/X$ protons = mass $22Y/X$, i.e.

$$\text{NAMW} = \frac{22Y}{X} + 18 \tag{a}$$

This method is not very reliable (the PMR value of 422 is much lower than the cryoscopic value of 598) because it depends on the accurate integration of a very small area (X) relative to a large area (Y) and upon the absence of water which would contribute to the OH integral. A better procedure is to employ the terminal methylene signal ($HO—CH_2$). This is not resolvable in the $CDCl_3$ spectrum but can be separately integrated when the spectrum is recorded in pyridine containing a trace of hydrogen chloride (to decouple $\underline{H}O$ from the adjacent $C\underline{H_2}$ signal) (Fig. 1.5). This signal forms the lower-field half of the $HOCH_2CH_2$-resonance (an A_2B_2 system), the upper half being obscured by the principal proton signal.

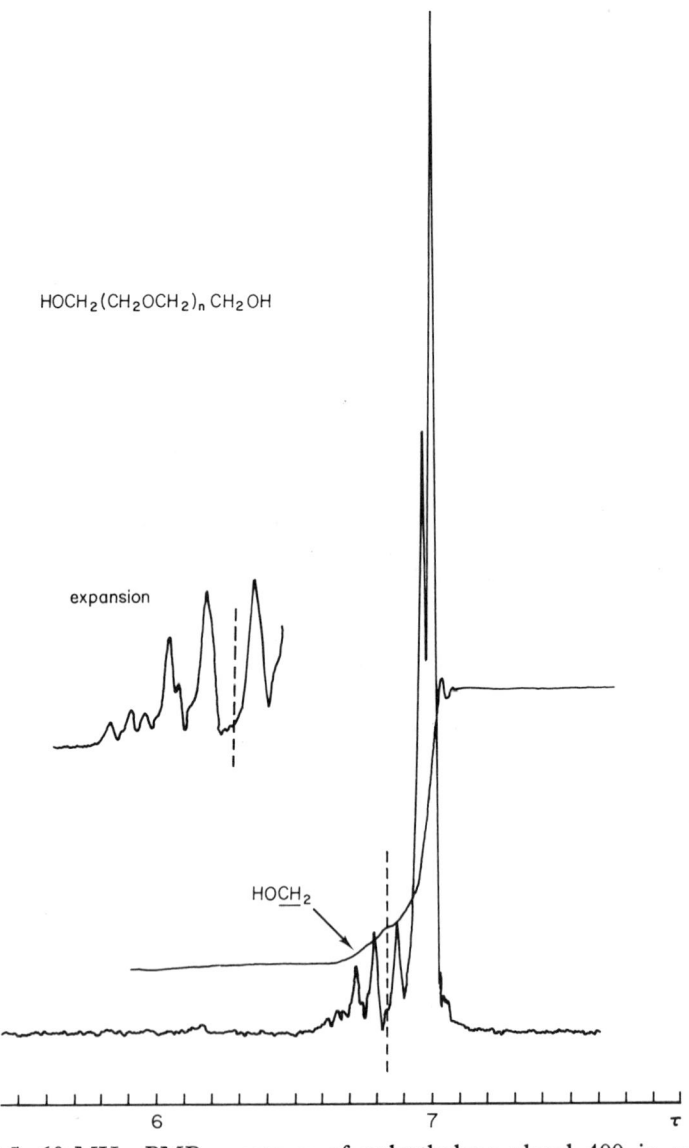

$HOCH_2(CH_2OCH_2)_n CH_2 OH$

expansion

$HOC\underline{H_2}$

6   7 τ

FIG. 1.5. 60-MHz PMR spectrum of polyethylene glycol 400 in pyridine–CF_3CO_2H.

The relevant formula:

$$\text{NAMW} = \frac{44\,Y}{X} + 62 \qquad\qquad (b)$$

where X is the HOC$\underline{\text{H}}_2$ and Y the remaining integral is derived in the same way as (a), noting that $2 \times$ HOCH$_2 = 62$ and X is equivalent to 4 protons. The value 592 obtained from this data agrees well with the cryoscopic result and is more accurate than the previous procedure because a larger integral is involved and water does not interfere. Modifications of this technique have been reported in which near-end group integrals are employed (Liu, 1968). The complete average structure of non-ionic surfactants (**6**) obtained by condensing ethylene oxide and/or propylene oxide with p-alkylphenols may be achieved by a similar PMR procedure; integration is simpler in this case because the signals of the various proton groups are more widely separated (Ludwig, 1968).

$$p\text{-RC}_6\text{H}_4\text{O(CH}_2\text{CH}_2\text{O)}_x\text{(CH}_2\text{CHO)}_y\text{H}$$

6

Johnson and Shoolery (1962) have described a PMR method for determining the degree of unsaturation and average molecular weight of natural fats, and their procedure provides a good example of correcting for ^{13}C satellite signals in quantitative work. The spectrum of safflower seed oil is shown in Fig. 1.6. Integrals of importance are:

X: a measure of the vinylic protons (signal A) plus the single methine proton of the glyceryl moiety.

$Y - X$: a measure of the glyceryl methylene protons (4 per gram molecule) (signal B).

Z: a measure of the total proton content.

If the area of signal B is taken to represent 4 protons, then to a close approximation:

$$\text{Area per proton} = \frac{Y - X}{4}$$

$$\text{Number of vinylic protons} = \frac{X - (Y - X/4)}{(Y - X)/4}$$

$$\text{Total number of protons} = \frac{Z}{(Y - X)/4}$$

These relationships are next corrected for ^{13}C satellite contributions. Each proton signal has satellites which arise as a result of the 1·108%

natural abundance of carbon-13 (spin number $\frac{1}{2}$); signals of protons adjacent to this isotope are split into doublets due to the large geminal ^{13}C-H coupling ($J > 100$ Hz). Thus, vinylic proton signals have ^{13}C satellite signals approximately 80 Hz each side of the corresponding ^{12}C-H signal, while protons on saturated carbon atoms have satellites plus or minus about 60 Hz from the main signals. Consequently, the high field ^{13}C satellite of signals in the area A will fall in area B, and the low field satellites of B in A.

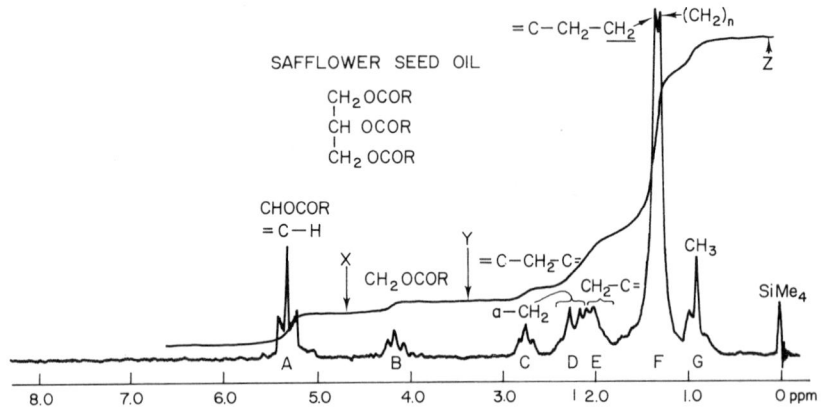

Fig. 1.6. 60-MHz PMR spectrum of safflower seed oil, 50% in CCl_4 (Johnson and Shoolery, 1962).

Likewise, the low field satellites of C are included in B. In all cases, the low field and high field satellites each amount to 0·55% of the corresponding main band. Thus, the area $Y - X$ represents 98·89% of the methylene protons and must be multiplied by 1·011 (100/98·89) to give the full value. The satellite contributions of signals A (0·0055X) and C must be deducted. Equations amended by significant correction terms are as follows:

$$\text{Area per proton} = [1 \cdot 011(Y - X) - 0 \cdot 0055X - 0 \cdot 0055(C)]/4$$
$$= 0 \cdot 253 Y - 0 \cdot 254X - 0 \cdot 0014(C) = A$$

$$\text{Number of vinylic protons} = \frac{1 \cdot 0055X - 0 \cdot 0055(Y - X) - A}{A}$$

$$= \frac{1 \cdot 265X - 0 \cdot 258 Y + 0 \cdot 0014(C)}{0 \cdot 253 Y - 0 \cdot 254X - 0 \cdot 0014(C)}$$

$$\text{Total number of protons} = Z/A$$
$$= Z/[0 \cdot 253 Y - 0 \cdot 254X - 0 \cdot 0014(C)]$$

The area C is approximately equal to B, and hence equals $Y - X$.

A general formula for natural fats may be written as follows:

$$CH_2OCO(CH_2)_a(CH\!=\!CH)_xMe$$
$$CH\!-\!OCO(CH_2)_b(CH\!=\!CH)_yMe$$
$$CH_2OCO(CH_2)_c(CH\!=\!CH)_zMe$$

The molecular weight using this formula is:

$$M.W. = 173 + 45 + 14(a + b + c) + 26(x + y + z) \qquad (1)$$

The first term, 173, is the formula weight of the $C_6H_5O_6$ glyceryl triester radical, while the second term, 45, is the weight of the three terminal methyl groups, 14 the CH_2 weight, and 26 the $CH\!=\!CH$ weight.

The total number of protons T is

$$T = 5 + 9 + 2(a + b + c) + 2(x + y + z) \qquad (2)$$

The number of vinylic protons V is

$$V = 2(x + y + z) \qquad (3)$$

Equation (2) may be solved for $(a + b + c)$ in terms of T and V.

$$(a + b + c) = (T - V - 14)/2 \qquad (4)$$

Likewise, equation (1) may be expressed in terms of T and V using equations (3) and (4).

$$M.W. = 218 + 7\cdot013(T - V - 14) + 13\cdot02V$$
$$= 120 + 7\cdot013T + 6\cdot01V \qquad (5)$$

T and V, obtained from integral data, thus enable the average molecular weight of the oil to be calculated. From this value, the degree of unsaturation of the oil may be expressed in a calculated iodine number.

$$\text{Iodine number} = \frac{126\cdot9\dagger}{M.W.} \times \text{number of vinylic protons} \times 100$$
$$= \frac{1269V}{M.W.}$$

The involved nature of this calculation, plus the operational demands required to record accurate integrals, are clearly deterrents to the routine use of this and other NMR techniques in quantitative analysis. A way of overcoming these obstacles, however, has been demonstrated by the application of a spectrometer controlled by a computer to the natural fat

† Atomic weight of iodine.

analysis just described. Shoolery and Smithson (1969) have programmed a small, general-purpose computer to control an NMR spectrometer, maintain adjustment of critical operating parameters, acquire data, numerically integrate appropriate spectral regions, calculate results and errors and print out a teletype record of results. The precision of the method may be judged by results upon a sample of pure trilinolein. Fifty individual analyses were made and the following results and standard deviations obtained. The instrument used was the A-60 with r.f. power level set at 0·05 milligauss, spectrum amplitude setting 2·5, and filter band width at 4 Hz.

Average molecular weight	875·4 ± 30·9	(Theory 879)
Iodine number	174·2 ± 1·8	(Theory 173·3)
Average number of vinylic groups	6·01 ± 0·24	(Theory 6·00)

An interesting modification of this assay involves the use of a 100-MHz spectrometer equipped with a wide-gap magnet; this enables the iodine value of oil present in a single corn kernel to be determined in a non-destructive manner, i.e. the kernel remains viable (Conway and Johnson, 1969). It is found that the lower the iodine value, the higher the oil content. Hence, kernels with low iodine values may be identified and then used for plant breeding purposes. The development of strains of corn yielding not only greater amounts of oil but also lower amounts of fatty acid is desirable because of correlations linking oil high in fatty acid with the onset of arteriosclerosis. The normal procedure of random selection of kernels from an ear, their destructive analysis and the planting of the remaining kernels if the test analyses are satisfactory, is open to error, because oil content varies widely among kernels in each ear.

NMR methods for the accurate determination of water in heavy water have been reported which may well find application in the pharmaceutical field. Although integrated intensities have been employed (Pople *et al.*, 1959), greater accuracy is achieved by measurement of ^1H resonance signal amplitudes (the $\underline{H_2O}$ signal is sharp and reproducible at 60 MHz) (Varian NMR at work No. 57); provided no change in nuclear spin relaxation times occur (p. 305) it may be assumed that the peak height is directly proportional to proton concentration. The method involves accurate dilutions of the unknown with small amounts of water. Extrapolation of the plot of signal amplitude against added H_2O to zero amplitude gives the amount of H_2O initially present (Fig. 1.7). A fluorine NMR method depends on the fact that the ^{19}F chemical shift of alkali fluoride solutions in $H_2O—D_2O$ of fixed concentration is sensitive to the isotopic composition of the solvent (Deverell and Schaumburg, 1967). On varying the solvent from H_2O to D_2O a progressive ^{19}F shift of 168 Hz to higher field is observed (56·4 MHz operating frequency), and, with the aid of a calibration graph, solutions of unknown concentration may be readily assayed, an accuracy of 1–2% being claimed.

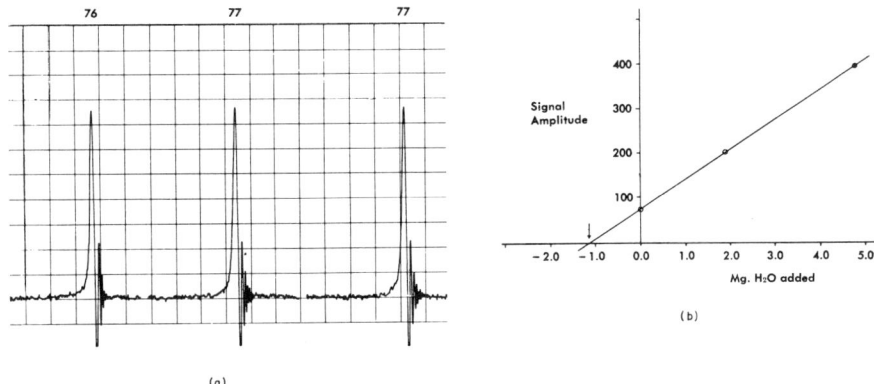

Fig. 1.7. Analysis of residual H_2O in D_2O: (a) trace of 60 MHz spectrum showing reproducibility of signal amplitude; (b) plot of signal amplitude against added H_2O (Varian Associates. NMR at work, No. 57).

Some Japanese analysts (Hatada *et al.*, 1969) have developed an assay of H_2O in D_2O by PMR in which water itself is used as the reference substance. This is possible by use of precision coaxial tubing (inner diameter: inner tube 1·3 mm, outer tube 4·2 mm). A mixture of water and acetic acid is placed in the central capillary and the sample of D_2O containing H_2O placed in the surrounding annulus. The signal caused by the water in the central capillary is shifted slightly downfield of that due to the water in the outer compartment (about 55 Hz at 100 MHz for 24·2 mole % of acetic acid). The intensity ratio of the two signals can be determined very precisely because the positions of the two peaks can be adjusted by the concentration of acetic acid so as to be close together without overlapping which is a special feature of the technique. After the additions of known amounts of water to the D_2O the intensity ratios are measured. The plot of intensity ratios against amount of added water is linear and may be used as a calibration-curve for the analysis of unknown mixtures.

Saur *et al.* (1968) have reported a combined NMR–mass spectrometry isotopic analysis of ethyl acetate ($CH_xD_{3-x}CH_yD_{2-y}OCOMe$) derived from the mixture of alcohols obtained by fermentation of glucose in D_2O by *Saccharomyces cerevisiae*. Accurate data upon the components of such a mixture provide valuable information about the kinetics and mechanism of the enzymic reactions that are involved in this fermentation. The PMR signals of the methyl and methylene resonances in the spectrum of the partially deuterated ethyl acetates are both broadened as a result of deuterium coupling (deuterium has a spin number, I, of unity) but become well resolved when the sample is irradiated with a radio-frequency in the region of the deuterium resonance absorption (Fig. 1.8). Under deuterium decoupling the undefined methylene resonance is resolved into six lines, one of them showing up as a shoulder on peak 3. This shoulder is recognized as the centre part of a triplet (1,3,6) with a proton coupling constant of 7·1 Hz, the same as found in ordinary ethyl acetate. Peaks 2 and 5 are parts of a doublet, while peak 4 is a singlet. The methyl groups present in the partially deuterated components of the mixture must therefore be CH_2D, CHD_2 and CD_3.

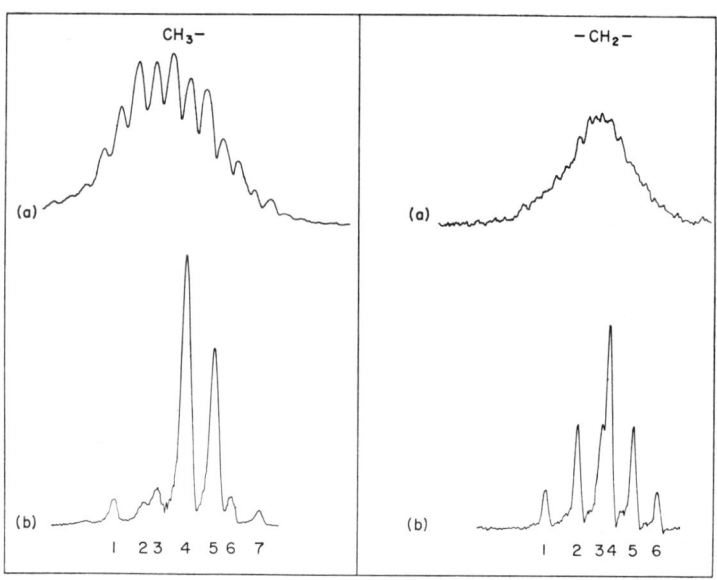

Fig. 1.8. 100-MHz PMR spectra of neat samples of partially deuterated ethyl acetates; (a) undecoupled; (b) deuterium decoupled. The methyl resonances range from 97 to 137 Hz (maximum at 117 Hz), the methylene resonances from 390 to 414 Hz (maximum at 402 Hz), both downfield from TMS (Saur *et al.*, 1968).

Information about the methylene groups is given by analysis of the decoupled methyl resonances. The doublets (1,6) and (2,7) indicate two components with —CHD— group. The singlet 3, 4 and 5 are due to three different species with —CD$_2$— group.

Combination of the data for the two groups of resonances leads to the following compounds as components of the mixture:

$$CH_3CD_2R \qquad CH_2DCHDR$$
$$CH_2DCD_2R \qquad CHD_2CHDR$$
$$CHD_2CD_2R \qquad CD_3CHDR$$

Absent from the mixture are ordinary (no deuterium present) ethyl acetate as well as CH_3CHD—R and all possible species with no deuterium present in the methylene group. Since CD_3CD_2OAc does not show any resonances in the methyl and methylene region, its presence can only be inferred by a comparison of the relative areas of methyl, methylene and acetyl resonances.

Quantitative determinations of the components in the mixture were first based on integration of the PMR spectra only. By normalizing the multiplet areas of methylene and methyl protons relative to the three acetyl protons, the composition of the mixture can be calculated. The amount of CD_3CD_2OAc is determined as the difference of the total and the partially deuterated components. The sensitivity of the acetyl signal to saturation led to its area being deceptively low, and a more accurate analysis was obtained after applying corrections from mass spectrometry data which allow direct measurement of the fully deuterated component. The data are summarized and compared with mass spectral values on the same mixture in Table 1.1, columns 3 and 4.

TABLE 1.1

Quantitative analysis of ethyl acetate isotopic mixtures: comparison between PMR[a] data, mass spectral data and values corrected by combination of the two methods

1	2	3	4	5
Compound	Empirical formula	Composition from PMR (mole %)	Composition from mass spectra (mole %)	Composition from PMR corrected from mass spectra (mole %)
CH_3CD_2OAc	$C_4H_6D_2O_2$	2·3 } 6·2	} 4·5	5·2 { 1·8
$CH_2DCHDOAc$	$C_4H_6D_2O_2$	3·9 }		{ 3·4
CH_2DCD_2OAc	$C_4H_5D_3O_2$	17·2 } 24·3	} 27·6	28·0 { 23·2
$CHD_2CHDOAc$	$C_4H_5D_3O_2$	7·1 }		{ 4·8
CHD_2CD_2OAc	$C_4H_4D_4O_2$	26·6 } 30·9	} 34·6	34·6 { 28·8
$CD_3CHDOAc$	$C_4H_4D_4O_2$	4·3 }		{ 4·8
CD_3CD_2OAc	$C_4H_3D_5O_2$	38·6	33·3	33·3

[a] PMR data were based on planimeter measurements of the methyl and methylene resonance peak areas and electronic integration scans of the acetyl peak.

REACTION MIXTURES

Chemical reactions yielding mixtures of products—particularly structural isomers—may be conveniently analysed on a semi-quantitative basis by an NMR procedure. Some examples are given below.

Alkylations

Reactions between 2-chloro-1-aminopropane and a nucleophile usually proceed via an ethyleniminium ion and may result in binary mixtures of alkylated products, as shown below for the alkylation of the diphenyl-acetonitrile anion with 2-chloro-1-dimethylaminopropane (**7**). The isomeric

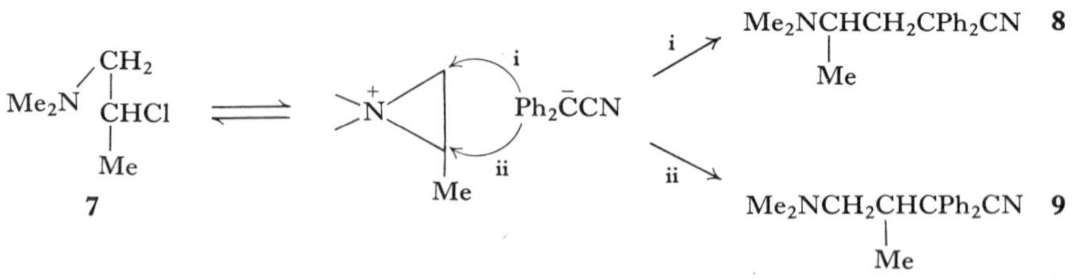

Ethyleniminium alkylation reactions

aminocyanides **8** and **9** differ in respect of their N—Me and secondary (*sec*) Me PMR signals and both compounds may be detected in mixtures by these signals as seen in Fig. 1.9. Since the two *sec* Me signals do not overlap their integrals may be measured separately allowing a fair estimate of the isomer ratio to be made; in addition, isomeric separations may readily be monitored by observing the disappearance of one set of isomer signals as purification proceeds.

The fact that a *total* reaction product may be analysed (allowance being made for signals due to possible impurities such as by-products and starting materials) makes the NMR procedure a good tool for the study of structural influences upon reactions of the above type. For example, the total product spectrum of Fig. 1.10 shows that steric factors appear to govern the reaction course between the chloroamine (**7**) and the 2-benzyl-benzimidazole anion since the more hindered isomer Y is formed only in minor amounts. This result also shows that the alkylation reaction is a practical route to the isomer X.

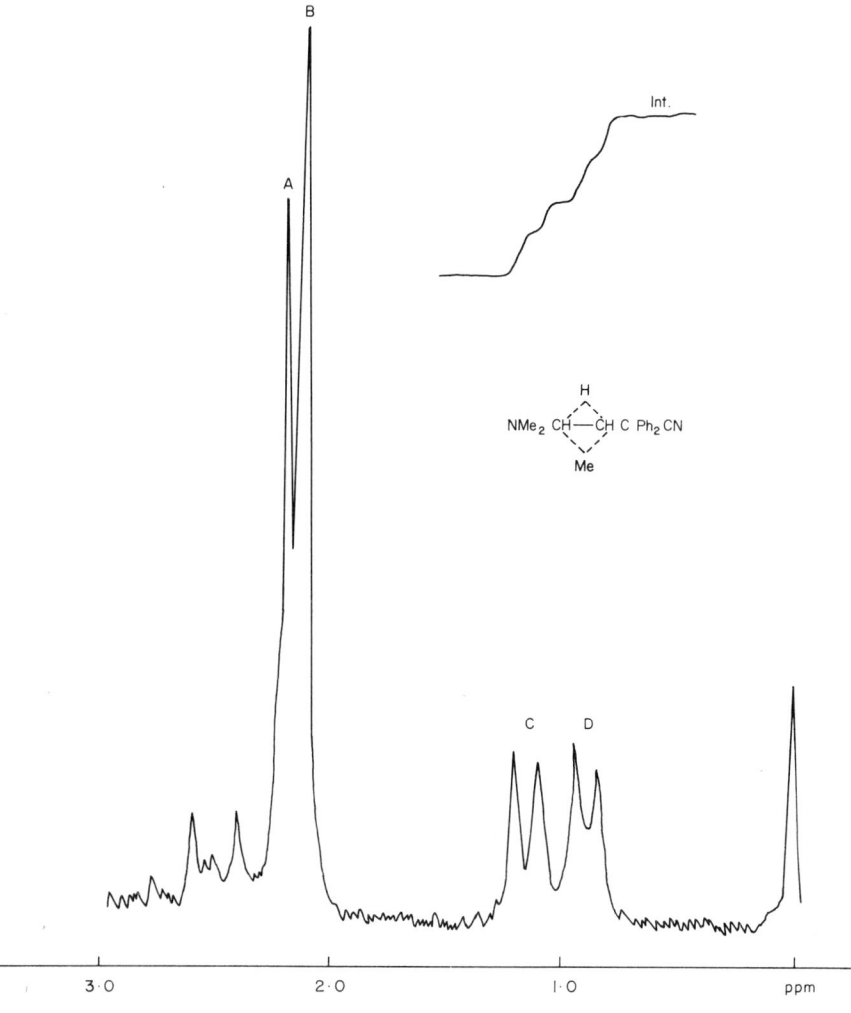

Fig. 1.9. Part of the 60-MHz PMR spectrum of mixed methadone–isomethadone cyanides in $CDCl_3$. A (N—Me) and C (*sec*-Me) are due to $NMe_2CH_2CHMeCPh_2CN$ (iso) and B (N—Me) and D (*sec*-Me) to $NMe_2CHMeCH_2CPh_2CN$ (Wright, 1965).

Diels-Alder Isomer Ratios

Addition of cyclopentadiene to substituted ethylene gives mixtures of *endo* and *exo* 2-substituted norbornenes. Moen and Makowski (1967) have used the vinylic PMR characteristics of such mixtures for the determination

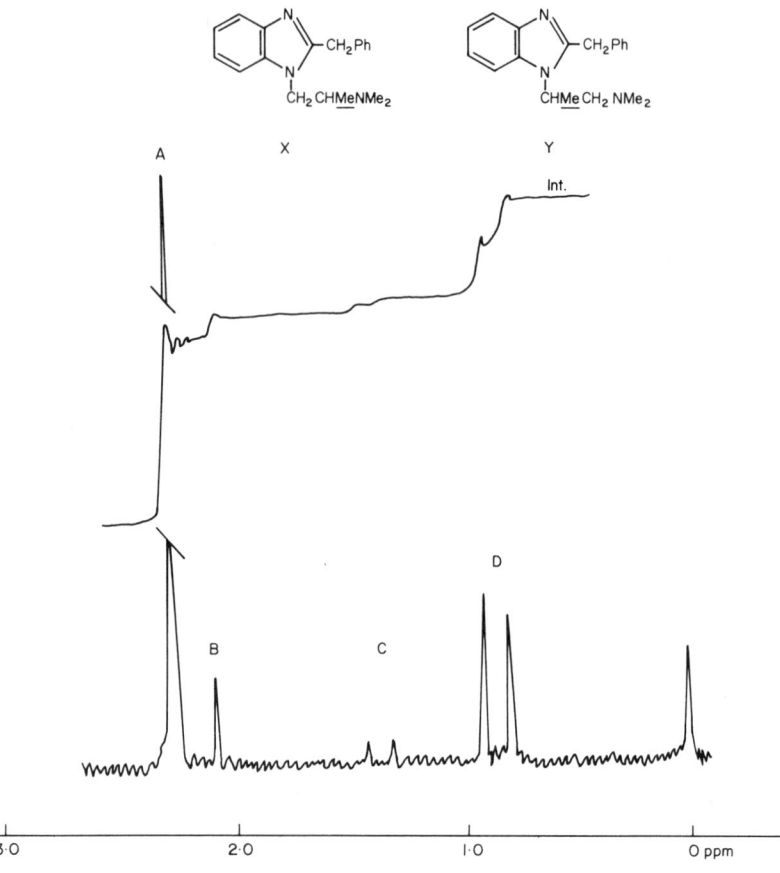

Fig. 1.10. Part of the 60-MHz PMR spectrum of the product of alkylating 2-benzylbenzimidazole with 2-chloro-1-dimethylaminopropane (solvent $CDCl_3$). A (NMe_2) and D (*sec*-Me) are due to isomer X, and B (NMe_2) and C (*sec*-Me) are due to isomer Y (Casy and Wright, 1966a).

of epimeric ratios. Typically, the C-5 and C-6 vinylic protons of each isomer give rise to eight peaks (four per proton—part of an ABX system). The entire vinylic signal is removed from other signals and in several cases four of the eight *endo* peaks, due to one of the vinylic protons, can be isolated (Fig. 1.11). These are higher field of the dividing line; peaks to low field are due to the other *endo* proton which overlaps the two (more closely placed) *exo*-vinylic signals. Areas to left (A) and right (B) of the median were measured (using a planimeter rather than the integrator) and the ratio calculated from the expression $endo/exo = 2B/A - B$. Agreement between

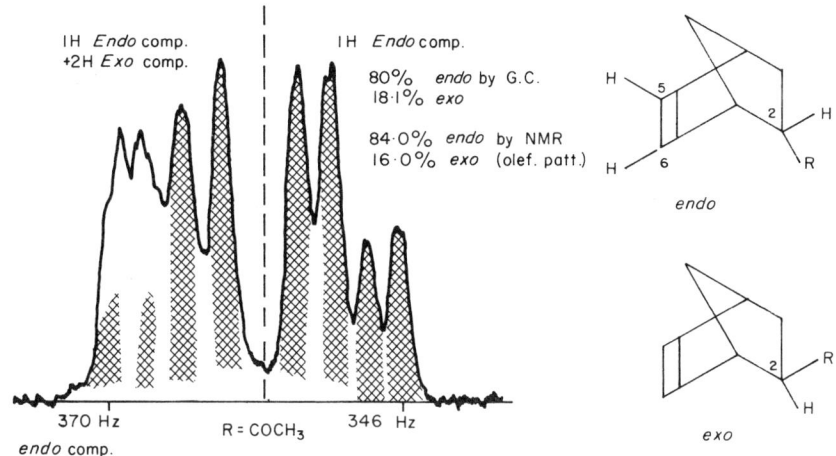

Fig. 1.11. 60-MHz vinylic proton pattern of H-5 and H-6 of 2-acetyl-5-nor-bornene. Neat sample used (Moen and Makowski, 1967).

NMR and gas chromatograph results were good. Similar principles have been used by other workers (Noland *et al.*, 1967).

Isomeric Tetrahydropyridines

Mixtures of structurally isomeric alkenes are also amenable to assay by the integration technique. Figure 1.12 shows the spectrum of the product of dehydration of 3-methyl-4-phenyl-4-piperidinol. The 3-methyltetra-hydropyridine content of the mixture is obtained from the 3-methyl integral (broad singlet at 92 Hz) and the 5-methyl isomer from integrals of the secondary methyl (doublet at 59 Hz) and vinylic proton (triplet at 357 Hz) signals (Casy *et al.*, 1965). By this analytical procedure, a study of the relative stabilities of isomeric 3- and 5-methyltetrahydropyridines, and of the influence of structure and stereochemistry upon the direction of elimination of 4-aryl-3-methyl-4-piperidinols has been made (Casy *et al.*, 1967). Thus the extent of isomerization of the pure 5-methyltetrahydropyridine *a* (see Table 1.2) to the 3-methyl isomer *b* after treatment with a hot acetic–hydrochloric acid mixture, could readily be followed by observing the development of the 3-methyl PMR signal in the total mixture. After 96 hours the 3-methyl isomer was the major component (Table 1.2, Nos 1–4). In contrast, the PMR spectrum of the pure 3-methyl isomer *b* showed little evidence of vinylic and secondary Me signals after the same treatment (Table 1.2, Nos 5–7). The fact that the 3-methyl alkene *a* was unchanged

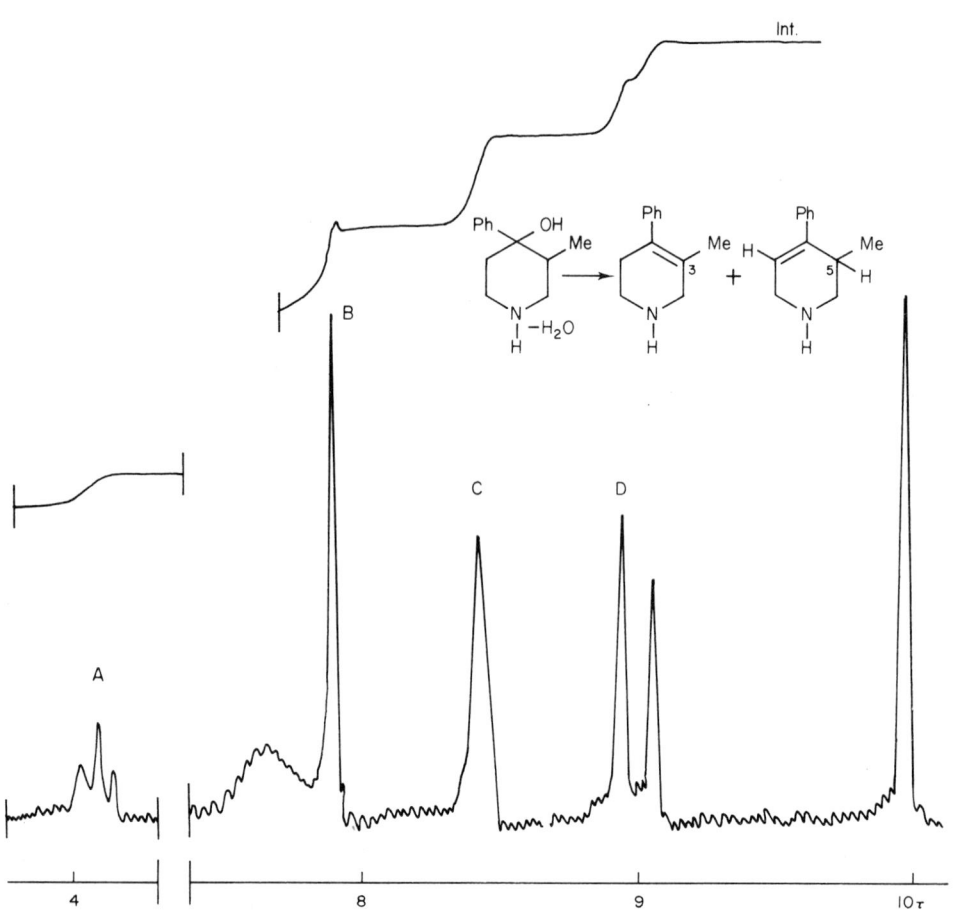

Fig. 1.12. Part of the 60-MHz PMR spectrum of the dehydration product of 3-methyl-4-phenyl-4-piperidinol in CDCl$_3$. C (*tert*-Me) is due to 3-methyl-4-phenyl-1,2,5,6-tetrahydropyridine, and A (vinylic) and D (*sec*-Me) to the 5-methyl isomer; B (N—H) is due to both isomers.

TABLE 1.2

Integral data for mixtures of tetrahydropyridines and related compounds

No.	Sub-strate	Heating period (hours)	PMR integrals			Ratio 3-Me (*b*):5-Me (*a*)
			vinylic	s-Me	t-Me	
1	*a*	6	4	11	2·5	1:4·8
2	*a*	24	1·5	7	5	1:1·2
3	*a*	48	v. small	5·5	13·5	2·5:1
4	*a*	96·	v. small	2·5	15	6:1
5	*b*	6	v. small	v. small	†	7:1
6	*b*	24	not seen	v. small	†	8:1
7	*b*	7 days	not seen	v. small	†	8:1
8	*c*	6	2·5	8	13	1·6:1
9	*cis d*	24‡	5	17	—	5-Me sole elimination product

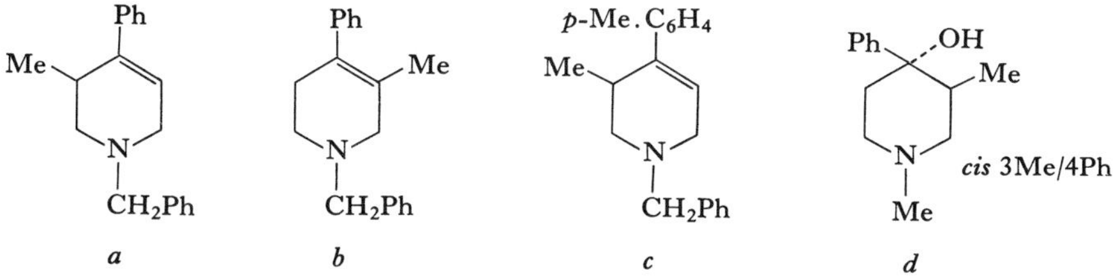

a   *b*   *c*   *d*

† Ratio obtained by comparing the 3-Me integral with the calculated 100% value (based upon the aromatic integral).

‡ Solvent, conc. HCl (2 ml) and water (2 ml), 50–52°.

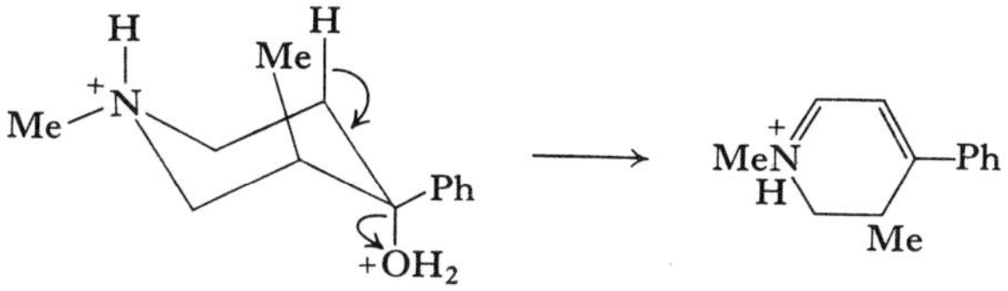

after 24 hours in boiling toluene (again made evident from PMR data), showed that these equilibrations probably proceed via addition products such as the original *t*-alcohol which subsequently eliminate. A para methyl substituent in *a* significantly increases the equilibration rate (Table 1.2, Nos 1 and 8). Stereochemical data follows from experiments upon diastereoisomeric 3-methyl-4-phenylpiperidinols. Under mild conditions, the *cis* isomeric 4-piperidinol yields a 5-methyltetrahydropyridine exclusively, while the *trans* isomer is unaffected (Table 1.2, No. 9) and this result indicates a preference for a *trans* elimination pathway.

The NMR technique may, in principle, be applied to any study which requires reasonably accurate knowledge of isomer or reaction product proportions. Other techniques, such as gas chromatography, may well be applicable, but the advantages of the NMR method lies in its simplicity and the fact that, in most cases, a positive association of spectral characteristics and product structure may be made (in other words, one knows what one is measuring).

NMR spectroscopy has great potential in clinical analysis, e.g. determination of the structure and amounts of urinary and plasma constituents (Dean, 1968), but its development in this field is hampered by the fact that the relatively large samples required for a reasonable NMR spectrum are generally not available from clinical media. The wider availability of signal enhancement equipment (see Hall, 1968, for a recent review on this topic) will, it is hoped, overcome this obstacle.

Two examples of its use in drug metabolism studies are available, however. These concern the analgesic pentazocine, described in Chapter 6 (p. 197), and isomeric 2-tolylcyclohexanols (Galpin *et al.*, 1969). In the latter study, a total of 1 g of racemic I (see Scheme) was given to six rats (20 mg doses) and their urine collected and pooled over a period of 7 days.

Racemic I

Racemic *trans*-2-
(*p*-carboxyphenyl)cyclohexanol   **A**

(1S, 2R)-(+)-*trans*-2-*p*-Tolylcyclohexanol   **B**

Racemic II

(+)-*trans*-2-*o*-Tolyl-*trans*-
5-hydroxycyclohexanol   **C**

Metabolism scheme

The urine was acidified with concentrated hydrochloric acid, boiled to hydrolyse any conjugates, and then extracted with ether. The residue from the extract was separated into a neutral and an acidic fraction. The PMR spectrum of the neutral product (100 mg) in $CDCl_3$ was readily seen to be identical with that of the starting product even though it was less clearly resolved; it differed, however, in being optically active (**B**). The spectrum of the acidic fraction (100 mg) in D_2O—Na_2CO_3 had an A_2B_2 aromatic signal typical of *p*-disubstituted benzenes (see p. 39) and this fact identified the product as the *p*-carboxylic acid **A**. The *o*-tolyl isomer II (1 g) yielded 125 mg of a mixture of neutral metabolites. The major component was identified as the axially hydroxylated derivative **C** by comparison of its PMR spectrum with that of authentic material.

The PMR spectra of urinary constituents are almost bound to be less well resolved than those of pure materials because of the limited amounts available and the presence of impurities. However, as the above work demonstrates, provided distinctive signals can be recognized (such as the A_2B_2 aromatic signal referred to above) and reference spectra are available, PMR spectroscopy can powerfully assist the analysis of clinical and metabolic specimens.

Some Qualitative Aspects

Although the NMR technique lacks the power of infra-red spectroscopy in regard to the detection of functional groups, it is, nevertheless, a valuable tool for the detailed study of established classes of compounds. As examples, compounds with hydroxyl and aromatic substituents are chosen to illustrate this point in the present Chapter, while compounds with nitrogen functions are described in Chapter 2.

Detection and Characterization of Hydroxyl Functions

Protons attached to oxygen generally give rise to sharp singlets which have chemical shifts in the range $\tau 9\cdot5$–$6\cdot0$ sensitive to temperature, solvent and solute concentration through the influence of hydrogen bonding upon the OH magnetic environment (see Appendix). Thus, the movement of a singlet induced by a small temperature rise is good evidence for its assignment to OH, while disappearance of the signal after addition of D_2O provides confirmation (RO$\underline{H}$, by exchange with D_2O, forms ROD, the proton signal falling in the residual water peak of the heavy water near $\tau 5\cdot4$ in $CDCl_3$). When an alcohol is examined in DMSO, however, the OH signal has a markedly low field position due to strong hydrogen bonding between OH and solvent and its position is much less sensitive to temperature and concentration changes (Chapman and King, 1964; Rader, 1966).

The OH resonances lie below the low-field ^{13}C side band of the solvent signal, thus ordinary DMSO may be used rather than the more expensive deuterated solvent. A further consequence of solvent–OH bonding is that the proton exchange rate is sufficiently reduced to allow observation of spin–spin coupling between OH and protons on the carbon atom adjacent to oxygen (α-protons), normally only possible when there is an extremely low concentration of acids or bases which catalyse OH proton exchange. The various classes of alcohols may then be differentiated by OH signal multiplicities, primary giving triplets, secondary doublets and tertiary singlets. The methanol signal is a quartet (Fig. 1.13).

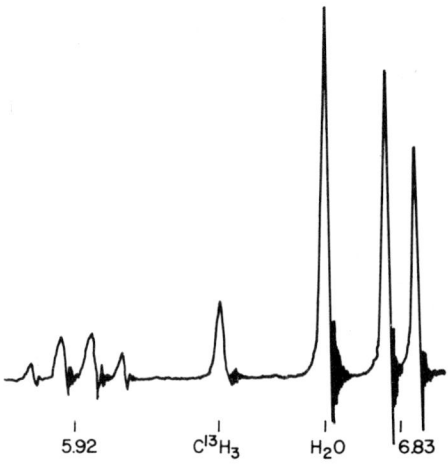

5.92   C^{13}H$_3$ H$_2$O   6.83

FIG. 1.13. PMR spectrum of methanol (5 mole %) and water (5 mole %) in dimethylsulphoxide showing the independence of the methanol and water hydroxyl proton resonances and the splitting of the methanol hydroxyl proton. The lower field ^{13}C side band of dimethylsulphoxide (221 Hz at 60 MHz) is also shown (Chapman and King, 1964).

The method has been shown to be valid for a variety of examples but is unreliable for alcohols containing strong electron withdrawing substituents close to OH, e.g. the OH signals of both 2-cyanoethanol and 2,2,2-trichloroethanol are singlets, not triplets (Traynham and Knesel, 1965). The presence of a basic centre within the molecule also eliminates HCOH coupling; thus, the OH signals of 3-hydroxymethylpyridine and isomeric 1,3-dimethyl-4-piperidinols are singlets in DMSO (Bass and Sewell, 1969; Casy and Jeffery, unpublished results). The OH signal of corresponding methiodides display coupling but those of the piperidinol hydrochlorides remain as singlets.

Hydroxy signals in DMSO may also provide stereochemical information as is seen from some data on epimeric cyclohexanols (Rader, 1966, 1969). In several pairs the axial OH signal was found at a higher field (0·21–0·27 ppm) than its equatorial counterpart, differences consistent with the latter forming stronger hydrogen bonds due to greater steric accessibility (see p. 386). In addition, the coupling constants between OH and the α-protons were greater by 1·0–1·5 Hz in the equatorial epimer (see below for example), a difference interpreted in terms of probably rotamers about the O—C bond as shown (**10–12**). The population of **10** is probably only significant in

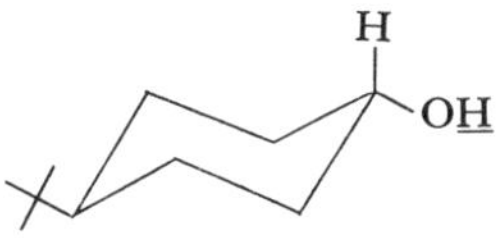

τ5·65
J 4·5 Hz

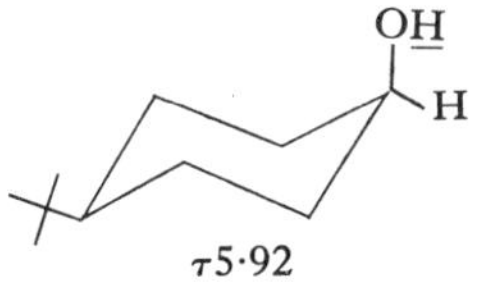

τ5·92
J 3·0 Hz

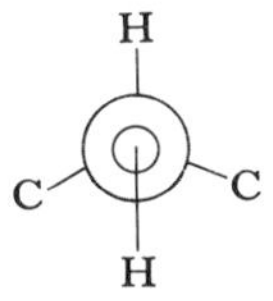

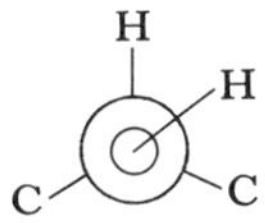

11

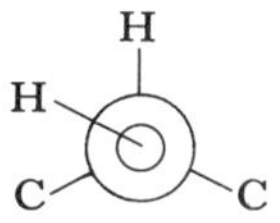

12

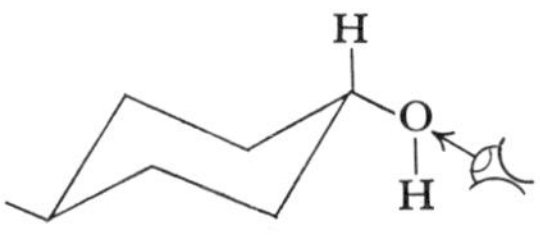

10

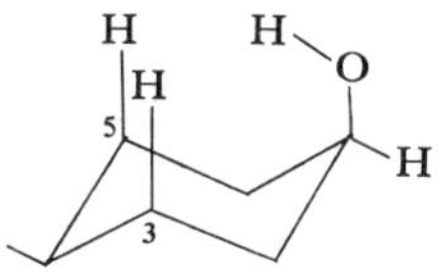

13

equatorial alcohols, since it entails 1,3-diaxial interactions between OH and the 3- and 5-axial hydrogen atoms in the axial epimer (see **13**). The HO—CH coupling constant $J_{H/H}$ (dihedral angle 180°) in **10** will be greater than that in **11** and **12** (dihedral angles 60°) on the basis of Karplus's expression (see p. 88), hence J_e (OH equatorial) should be greater than J_a (OH axial) because only the former receives a contribution from **10**. A number of examples of alcohols in which the HCOH dihedral angle is fixed by intramolecular bonding are now available (Bauld and Rim, 1968; Stolow and Gallo, 1968; Kiefer et al., 1968; Fraser et al., 1969) and the values obtained parallel the Karplus relationship for J_{HCCH} (see examples below).

J_{HCOH} 12·5 Hz
ϕ 180°

11·3 Hz   in solvent CCl_4
167 ± 2°

A particularly interesting case is that of the primary alcohol (Fig. 1.14), prepared by the methoxymercuration of methallyl alcohol. The conformation of this alcohol is fixed by mercury–oxygen chelation. The OH signal is a quartet in $CDCl_3$ with J values of 11 and 2·8 Hz corresponding with *trans* and *gauche* coupling respectively. Elimination of the mercury atom by iodode-mercuration restores free rotation and the OH proton appears as a broad singlet.

Further information about the OH environment may be obtained from alcohol derivatives, notably esters. The greater deshielding of α-protons of the alcohol moiety of esters compared with the same protons in the free alcohols is a well known phenomenon in NMR spectroscopy and is termed the "acylation shift". The degree of shift differs for secondary (1·0–1·15 ppm) and primary alcohols (0·45–0·60 ppm) and this fact is the basis of several characterization procedures. Culvenor (1966) has placed this shift upon a more rational basis by considering the anisotropic effects of the ester system upon the α-protons. Preferred conformations for secondary esters are the staggered forms **14** and **15** and the eclipsed conformer **16**. For primary esters preferred conformers are **17** (staggered) and **20** and **21** (eclipsed) although the staggered forms **18** and **19** must also be considered since these are equivalent to **14** and **15**. Considering staggered and eclipsed forms separately, it is seen that a 2:1 ratio between secondary and primary acylation shifts agrees very closely with number of favoured eclipsed conformers in the two alcohol types. Averaging of **20** and **21** will cause each α-proton of primary esters to suffer half the deshielding experienced in the secondary ester **16**. It is argued that contributions from the primary and secondary staggered forms will not seriously upset the ratio since they might be expected to cancel. In methanol, three α-protons share positions in the deshielding zone and the acylation shift for methanol (0·23–0·36 ppm) is, in fact, about one-half to two-thirds the primary shift, a result which supports the above interpretations.

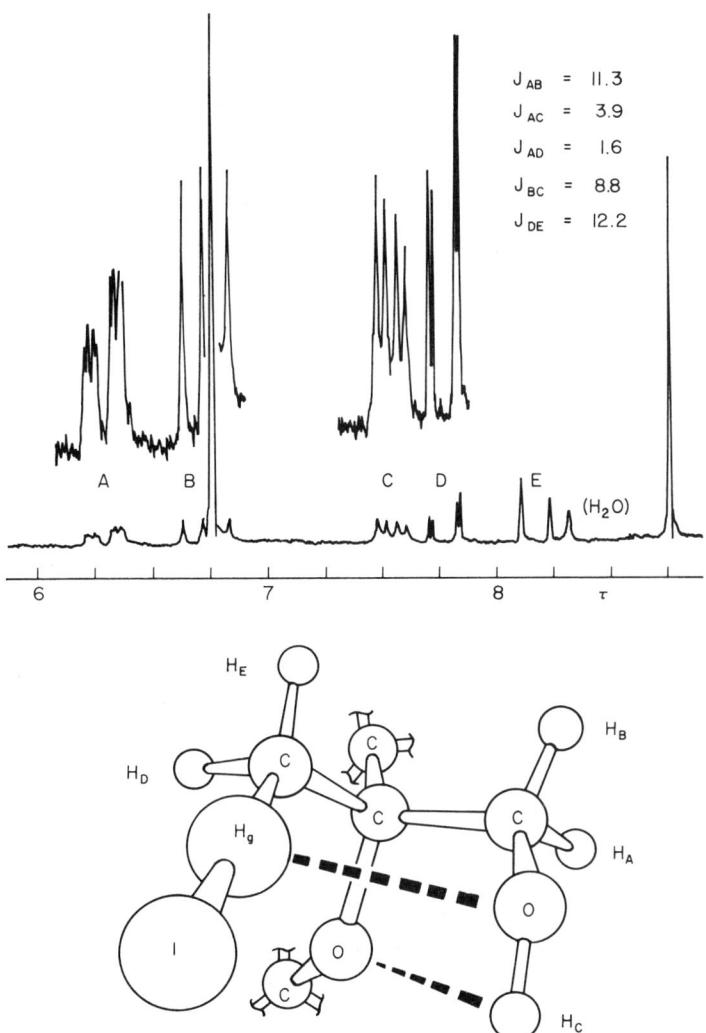

$$J_{AB} = 11.3$$
$$J_{AC} = 3.9$$
$$J_{AD} = 1.6$$
$$J_{BC} = 8.8$$
$$J_{DE} = 12.2$$

Fig. 1.14. The 100-MHz PMR spectrum in $CDCl_3$ and preferred conformation of 3-iodomercuri-2-methoxy-2-methyl-1-propanol (Kiefer *et al.*, 1968). J_{HCOH} values quoted on p. 30 are the limiting values calculated using data from spectra recorded over a temperature range.

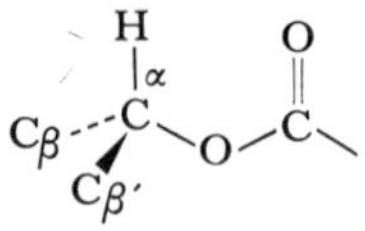

secondary 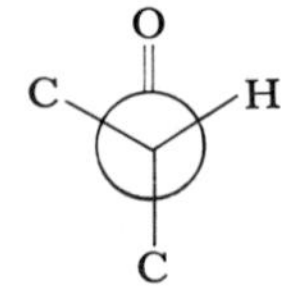   **14** 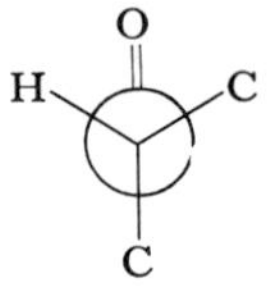   **15** 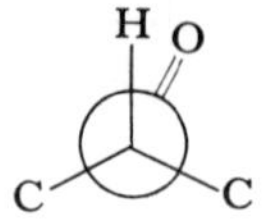   **16**

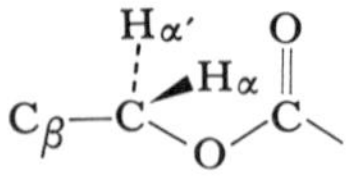

primary 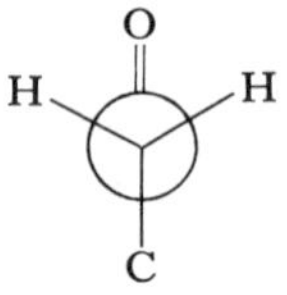   **17** 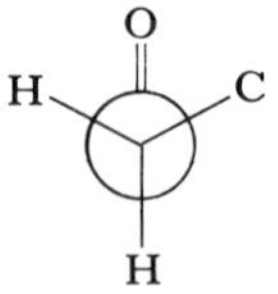   **18** 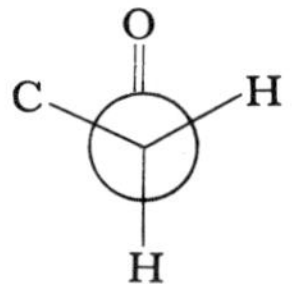   **19**

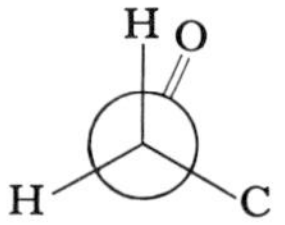

20 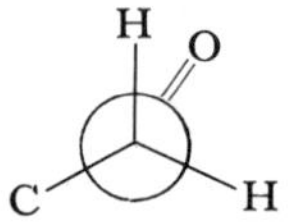   **21**

In addition to shift magnitudes, the integral of the α-proton signal provides confirmation of primary and secondary character, but measurement may be hampered by overlap with other signals.

In some esters the PMR signal due to the acid moiety may serve for characterization purposes. Thus, the proton resonance of the dichloroacetyl group $CHCl_2$ forms a sharp line removed from most alkyl and aryl signals, the order of shielding being $3° > 2° > 1°$; signal separation is better in DMSO than in $CDCl_3$ (Babiec *et al.*, 1968). The number of alcoholic components in a mixture may, in principle, be determined by counting the $CHCl_2$ singlets while tertiary may be differentiated from primary and secondary species from relative chemical shift positions (ranges for the last two classes are not so well separated) (see Fig. 1.15). A similar approach may be followed by utilizing the ^{19}F NMR signals of trifluoroacetyl esters which provide larger chemical shift differences between 1°, 2° and 3° resonance, the shielding order being the same as that for $CHCl_2$ signals (Manatt, 1966).

Trichloracetyl isocyanate (TAI) reacts rapidly with phenols and all classes of alcohols, and the use of resultant carbamate PMR signals have

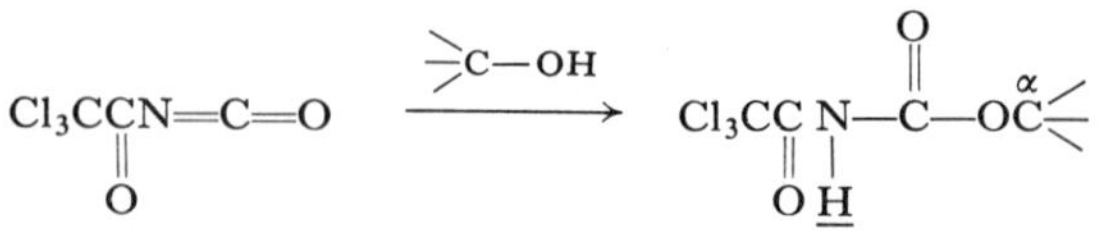

$$Cl_3CCN{=}C{=}O \quad \xrightarrow{\geq C-OH} \quad Cl_3CC\,N-C-OC\leq$$

TAI

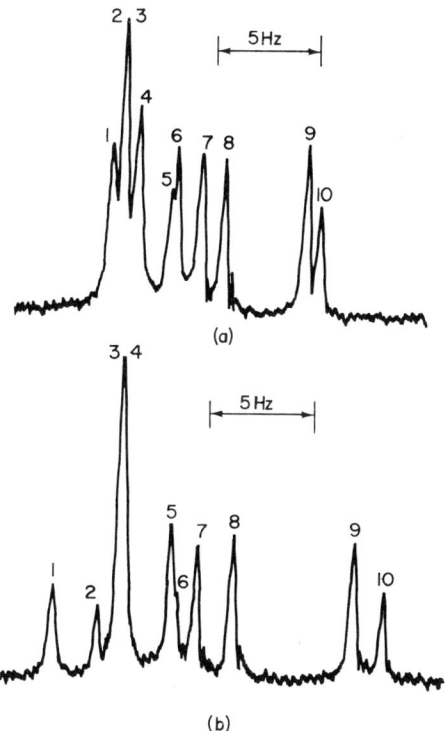

FIG. 1.15. 60-MHz PMR spectra of the dichloroacetyl resonance of some hydroxyl compounds; chemical shifts given in parentheses from TMS. Increasing field to right; (a) Mixture in $CDCl_3$; 1, $C_6H_5CH_2OH$ (6·032); 2 and 3, $HOCH_2CH_2CH_2OH$ and $CH_3CHOHCH_2OH$ (6·026); 4, CH_3OH (6·011); 5, $HOCH_2CHOHCH_3$ (5·988); 6, $CH_3CH_2O\overline{H}$ (5·981); 7, $CH_3CHOHCH_2CH_3$ (5·961); 8, $CH_3CH\overline{OH}CH_3$ (5·941); 9, $(CH_3)_2(CH_3CH_2)COH$ (5·875); 10, $(CH_3)_3COH$ (5·860); (b) Mixture in DMSO-d_6; 1, $C_6H_5CH_2OH$ (6·878); 2, $CH_3CHOHCH_2\underline{OH}$ (6·841); 3 and 4, $HOCH_2CH_2CH_2OH$ and CH_3OH (6·818); 5, CH_3CH_2OH ($\overline{6·784}$); 6, $CH_3CHO\underline{H}$ $CH_2\underline{OH}$ (6·777); 7, $CH_3CHOHCH_2CH_3$ (6·761); 8, $CH_3CHOHCH_3$ (6·730)$\overline{;}$ 9, $(CH_3)_2(CH_3CH_2)COH$ (6·633); 10, $(CH_3)_3COH$ (6·608) (Babiec *et al.*, 1968).

been proposed for alcohol characterizations (Goodlett, 1965). Down-field shifts of α-protons (again in the ratio of about 1:2 for 1° and 2° alcohols) provide structural data and use may also be made of the NH signals. These are fairly sharp singlets which fall at very low field near 10 ppm in acetone, and their number and area provide the total OH groups present per molecule or sample regardless of the type of OH substrate. Water does not interfere because the NH_2 peak of trichloroacetamide (the reaction product of TAI and water) occurs near 7·8 ppm. Application of this method to the classification of sterols has been examined because of the common occurrence of hydroxylation in steroid metabolism (Trehan and Monder, 1968). Thus, addition of TAI to 4-pregnen-11β, 21-diol-3,20-dione

converts the broad OH signals in the τ5–8 region to 2 one-proton singlets at τ0·7 and 1·15 and shifts the carbinol protons down field by the characteristic amounts. The use of chloral alcoholates is similarly recommended for the characterization of sterols (McClenaghan and Sykes, 1968). Chloral reacts rapidly at room temperature with certain hydroxy steroids to give a

mixture which is epimeric about the starred carbon atom. The signal of the methine proton attached to this atom appears as 2 singlets of varying intensity ratio between τ4·7 and 5·2 (the adducts are not isolated, spectra being recorded before and after the addition of chloral) and their separation (Δ) is characteristic of the steroid hydroxy position (Δ 0·3 ppm for 3-OH, 0·06 ppm for 17-OH). The vinylic signal in a 3β-hydroxy-Δ^4-steroid moved downfield by 0·1 ppm after reaction with chloral, whereas that of the Δ^5 isomer was unaffected after this change.

Finally the use of ethers may be mentioned. In several secondary alcoholic sterols it has been found that O-methylation results in the α-proton moving *upfield* by about 0·6 units and this change combined with the downfield acylation shift serves to identify this particular proton (Narayanan and Iyer, 1965). No data is provided regarding the order of the methylation shift in primary alcohols or in epimeric pairs so its application to alcohol characterization remains to be seen.

Aromatic Compounds

Aromatic protons are strongly deshielded and generally absorb in the range τ2·0–3·5 for benzenoid, and τ1·0–3·5 for heterocyclic derivatives.

Hence, provided allowance is made for the presence of other highly de-shielded protons such as

$$\overset{+}{\underset{/}{\overset{\diagdown}{N}}}\underline{H} \quad \text{and} \quad H_2NCOR$$

observation of signals in this spectral region allows the rapid identification of aromatic features in an unknown compound. The creation or removal of an aromatic feature in a reaction may similarly be monitored by PMR spectroscopy. For example, completion of the reduction of 2,4-lutidine

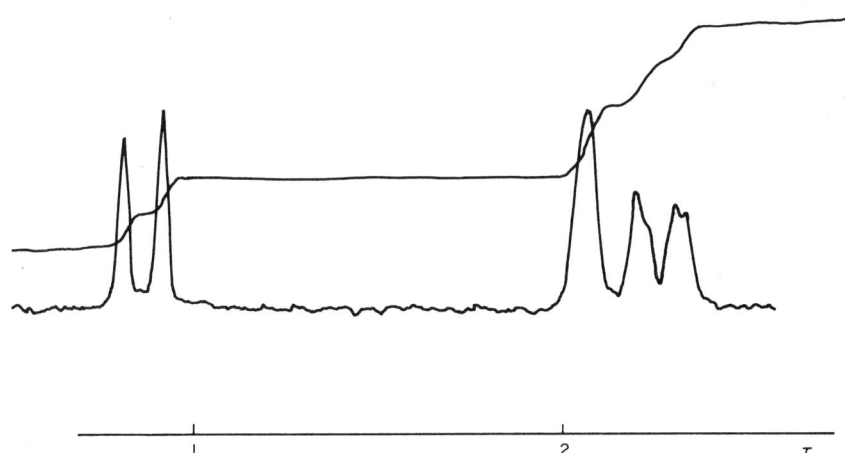

methiodide by sodium borohydride is readily made evident by the complete disappearance of aromatic signals near $\tau 2 \cdot 2$ and $0 \cdot 8$ (Fig. 1.16); in the spectrum of the reaction product, a signal near $\tau 4 \cdot 7$ appears due to the vinylic protons of the tetrahydropyridine product. PMR spectroscopy also lends itself to the identification of the positional isomers of aromatic compounds, and this aspect forms the main part of this section.

Fig. 1.16. The 60-MHz aromatic protons signal of 2,4-lutidine methiodide in $CDCl_3$.

The position taken by an aromatic substituent can often be identified by PMR spectroscopy from considerations based upon a combination of chemical shift, spin–spin couplings (*ortho* 5-9, *meta* 2-3 and *para* 0-1 Hz coupling for substituted benzenes) and integration data. Simple benzene derivatives are discussed first.

Derivatives with Two Identical Substituents

o-   *m-*   *p-*

Of the three isomers above, the *p*-member is generally readily identified because its aromatic protons (all identical) give rise to a sharp 4-proton singlet, in contrast to the more complex *o*- and *m*-aromatic signals. A distinguishing feature of *o*-isomers (these usually constitute $A_2'B_2'$ systems with A′ coupled unequally with the two B′ protons), is an aromatic signal which, although complex, is approximately symmetrical about its mid-point. Signal expansion is often necessary to establish this fact as seen in the case of catechol (Fig. 1.17). When X is a group, which lacks pronounced electronic effects, the two halves of the A_2B_2 spectrum approach each other so closely as to appear a singlet similar to that of the corresponding *p*-isomer and in such cases (e.g. the isomeric xylenes, X = Me) only the *meta*-isomer provides a distinctively complex aromatic signal. Some *meta* derivatives form aromatic patterns that approximate to those expected for AMX_2 systems. Thus, the anticipated triplet A (2 *meta* couplings) and quartets X (*o/m* coupling) and M (2 *ortho* couplings) may be distinguished within the aromatic signal of *m*-dinitrobenzene (acetone solvent) even though all signals are much deformed [Fig. 1.18(a)]. Spectral examination at a higher field strength generally provides a more nearly first order pattern because peak separation is greater with increasing field strength, whereas coupling constants do not change (example given later). Another analytical aid is a change of solvent which may materially alter the appearance of a spectrum through the influence of the diamagnetic susceptibility of the medium upon resonance frequencies.† Thus, the chemical shift separations

† The use of tris(dipivalomethanato) europium [Eu(DPM)₃] and related compounds containing a paramagnetic ion is the most recent means of achieving spectral simplification (Sanders and Williams, 1971). These so-called shift reagents complex reversibly with lone pair bearing organic compounds whence protons in the vicinity of the bonding centre suffer pseudocontact shifts which may be of the order of several ppm. The aromatic examples quinoline and 1,2-diphenylethanol are described in the reference quoted.

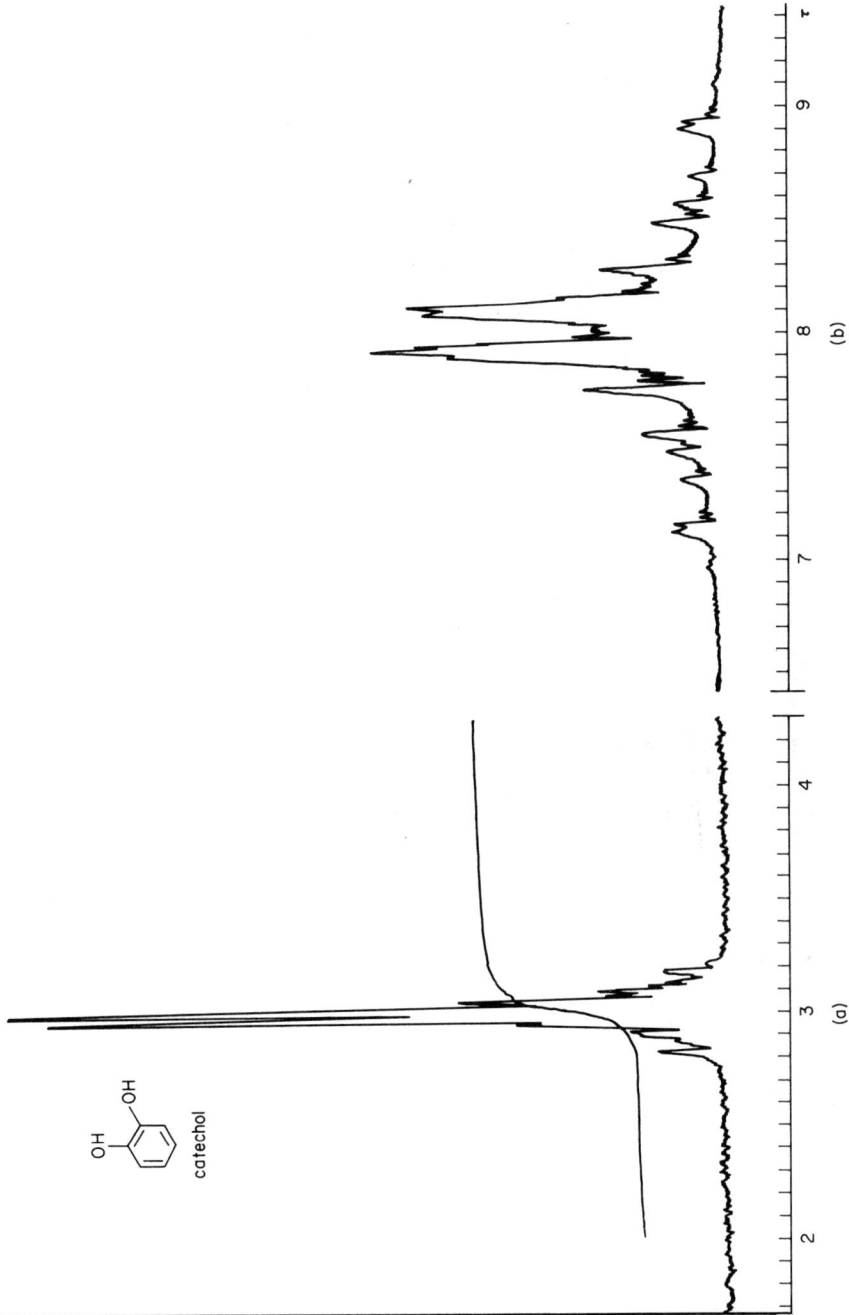

Fig. 1.17. 60-MHz PMR spectrum of catechol aryl protons in D₂O at sweep widths of 500 (a) and 100 Hz (b). The expanded signal is offset 405 Hz.

of the aromatic protons of *m*-dinitrobenzene are much larger in benzene than in acetone so that the spectrum becomes almost first order in the former solvent [Fig. 1.18(b)].

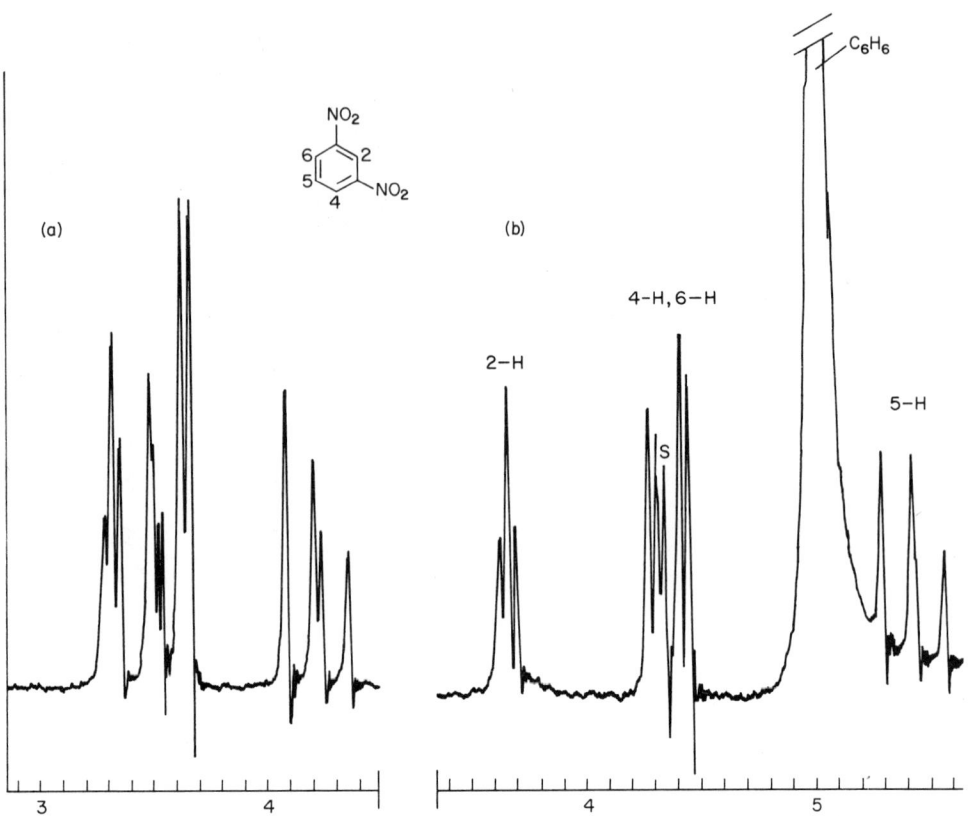

FIG. 1.18. 60-MHz PMR spectrum of *m*-dinitrobenzene in acetone (a) and benzene (b).

Two Non-Identical Substituents

o-   *m-*   *p-*

Again, of the three isomers, the aryl signal of the *p*-member is most easy to recognize through its symmetrical $A_2'B_2'-A_2'X_2'$ nature. For example, the aromatic signals of the chlormethylation product of phenetole and the nitration product of α-prodine both establish the substituents to have a *para* orientation (Fig. 1.19). In contrast, *ortho* and *meta* di-derivatives provide complex aromatic signals that lack any element of symmetry (e.g. *o*-methoxybenzoic acid and *m*-cresol) (Fig. 1.20).

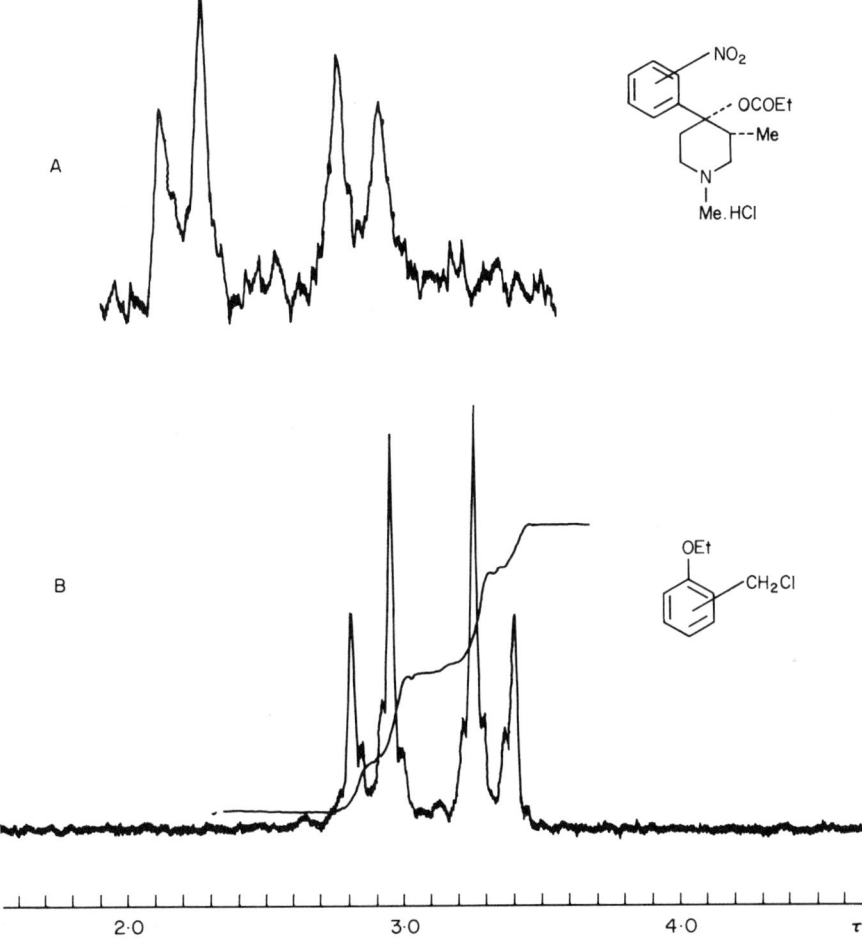

FIG. 1.19. 60-MHz PMR aromatic protons signal of A, the nitration product of α-prodine in D₂O (Casy and Armstrong, 1965), and B, the chloromethylation product of phenetole in CCl₄. Signal A is offset 50 Hz.

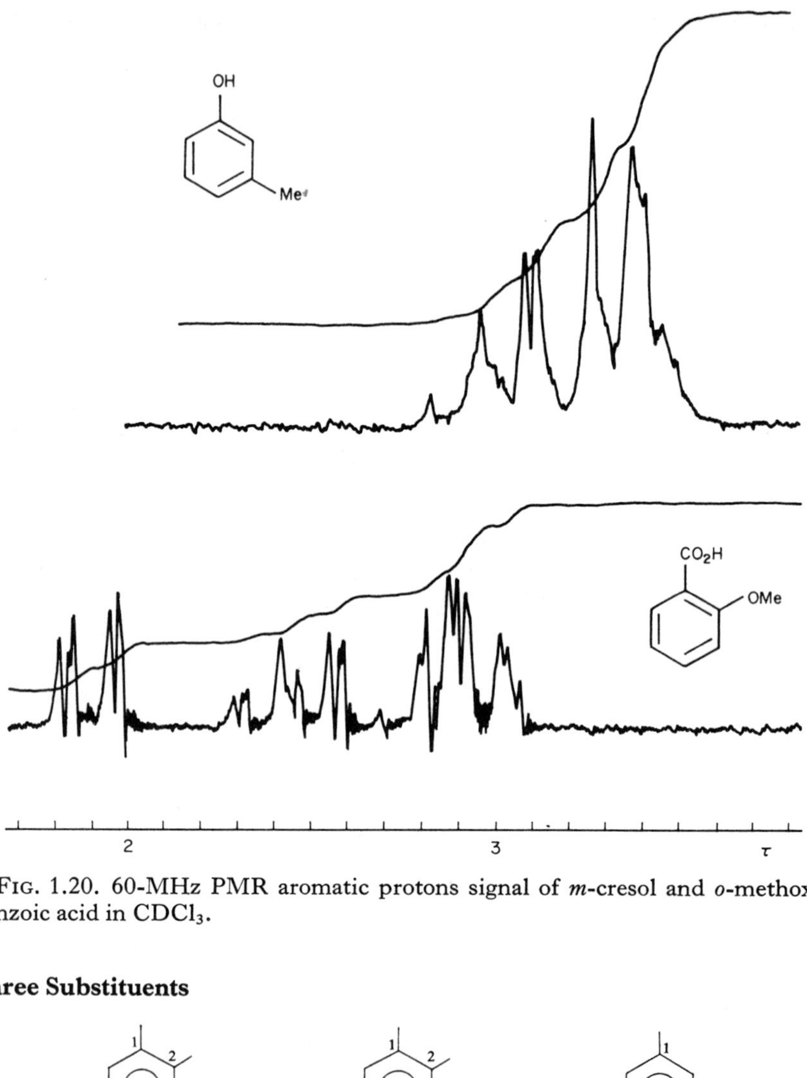

OH

Me

CO₂H

OMe

2   3   τ

FIG. 1.20. 60-MHz PMR aromatic protons signal of *m*-cresol and *o*-methoxy-benzoic acid in CDCl₃.

Three Substituents

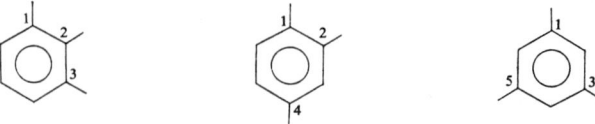

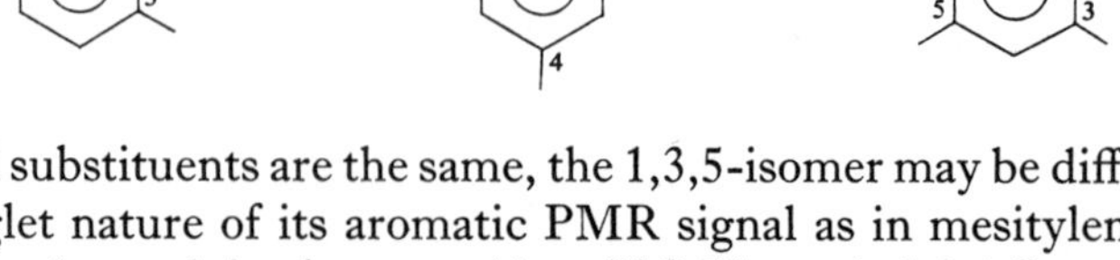

When all substituents are the same, the 1,3,5-isomer may be differentiated by the singlet nature of its aromatic PMR signal as in mesitylene (Ar—H 6·78 ppm) and *sym*trinitrobenzene (Ar—H 9·55 ppm). 1,2,4-Isomers generally form ABC systems and given complex signals that are, however, often amenable to analysis especially at high frequencies. A good example is

1,2,4-trichlorobenzene—its aromatic signal is complex at 60 MHz but near first order at 100 MHz (Fig. 1.21).

Three coupling constants may be abstracted from this spectrum and their orders show that one *ortho* (9 Hz), one *meta* (2·5 Hz) and one *para* coupling (0·5 Hz) are operating, consistent with the 1,2,4-isomer (requires

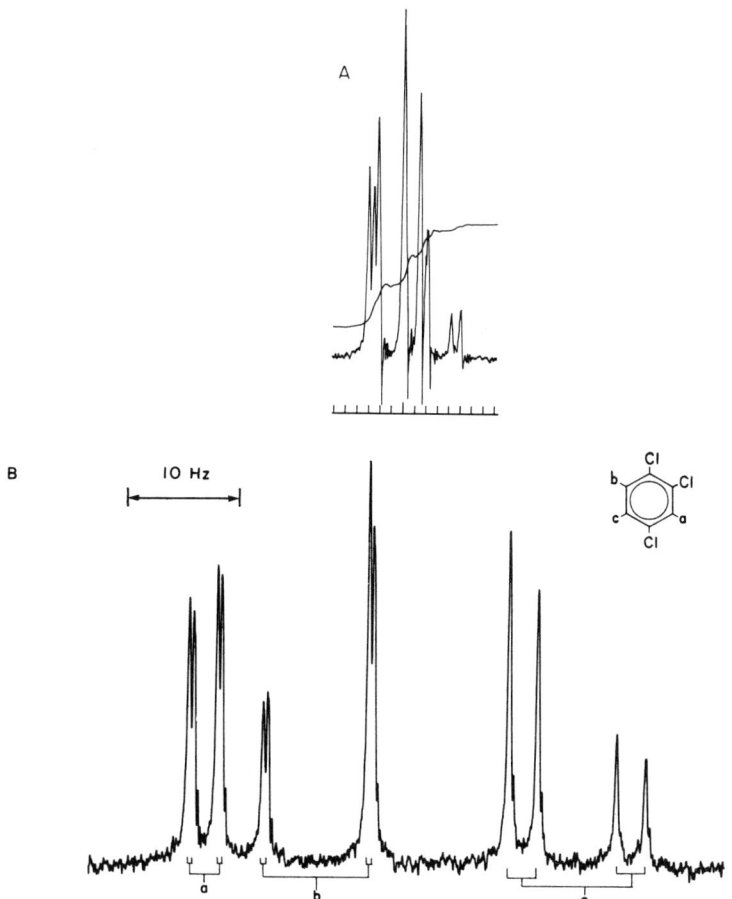

Fig. 1.21. PMR spectra of 1,2,4-trichlorobenzene, at 60 MHz in $CDCl_3$ (A), and at 100 MHz in CCl_4 (B) (Reed, 1967).

m-p, *o-p* and *o-m* signals) but not with the 1,2,3-member which requires an *o-m* and *o-o* signal. Sometimes approximate first order (AMX) spectra are obtained for 1,2,4-derivatives at 60 MHz giving a pattern composed (in varying order) of an *o*-doublet, a *m*-doublet and an *o-m* doublet of doublets (*p*-coupling is not usually apparent at lower frequencies).

Trisubstituted 1,2,3-isomers of the type **22** constitute AB_2 systems. Such spectra consist of nine lines, four for the A proton and four for the

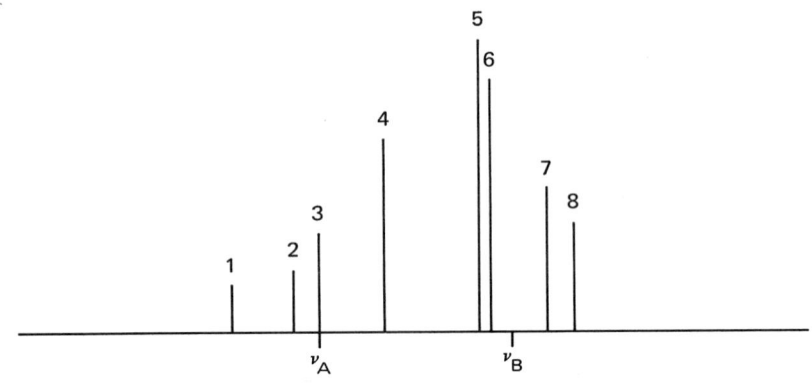

22

B proton plus a weak combination line that is usually not seen. If the peaks are numbered from left to right, the chemical shift of the A proton is given

Calculated AB_2 spectrum for the case $J_{AB}/\varDelta\nu = 0{\cdot}4$.

by line 3 and that of B by the average of lines 5 and 7. J_{AB} may then be calculated and the theoretical spectrum (computed from ν_A and ν_B and J_{AB}) compared with that actually recorded. If the two spectra are similar, support is provided for assignment of orientation to the trisubstituted benzene. The spectrum of pyrogallol provides an example of this type (Fig. 1.22).

5-Membered heterocyclic derivatives of the type **23** also have 3 adjacent aryl protons which sometimes give rise to PMR signals that may be analysed

X = NH, O, or S

23

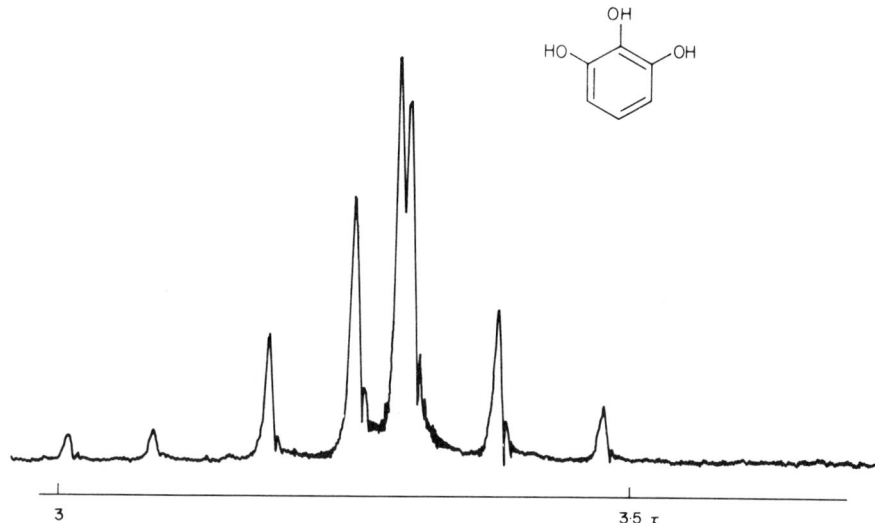

FIG. 1.22. 60-MHz PMR aromatic protons signal of pyrogallol in D_2O (sweep width 100 Hz).

as AMX systems. Furan-2-aldehyde, for example (Fig. 1.23) displays three well separated doublet of doublets that involve (moving up-field), o/m, o/m and o/o coupling respectively. The highest field signal is clearly due to H-4, while the lowest field signal must arise from H-5, since this proton (being adjacent to oxygen) should be the most deshielded. Resolvable signals are less likely for 3-substituted derivatives of **23** because the shielding influence of the hetero-atom is not then largely confined to one proton.

It is clear from the above that the ability to recognize aromatic patterns enables orientational assignments to be made even when deviations from first order systems occur. Added complications arise in polycyclic aromatic compounds where proton signals due to the ring in question may overlap those due to other aryl protons. However, provided suitable reference compounds are available, orientation problems may still be solved as illustrated by the next two examples.

Analysis by the Aid of Reference Compounds

Acetylation of 9,10-dimethylphenanthrene and reduction of the product gives a monoethyl derivative (**24**) of unknown orientation (Bavin *et al.*, 1965). The aryl PMR signal of this compound will be made up of two parts: (i) that due to the unsubstituted ring (B)—resembling the spectrum of 9,10-dimethylphenanthrene, and (ii) that due to the substituted ring (A)—

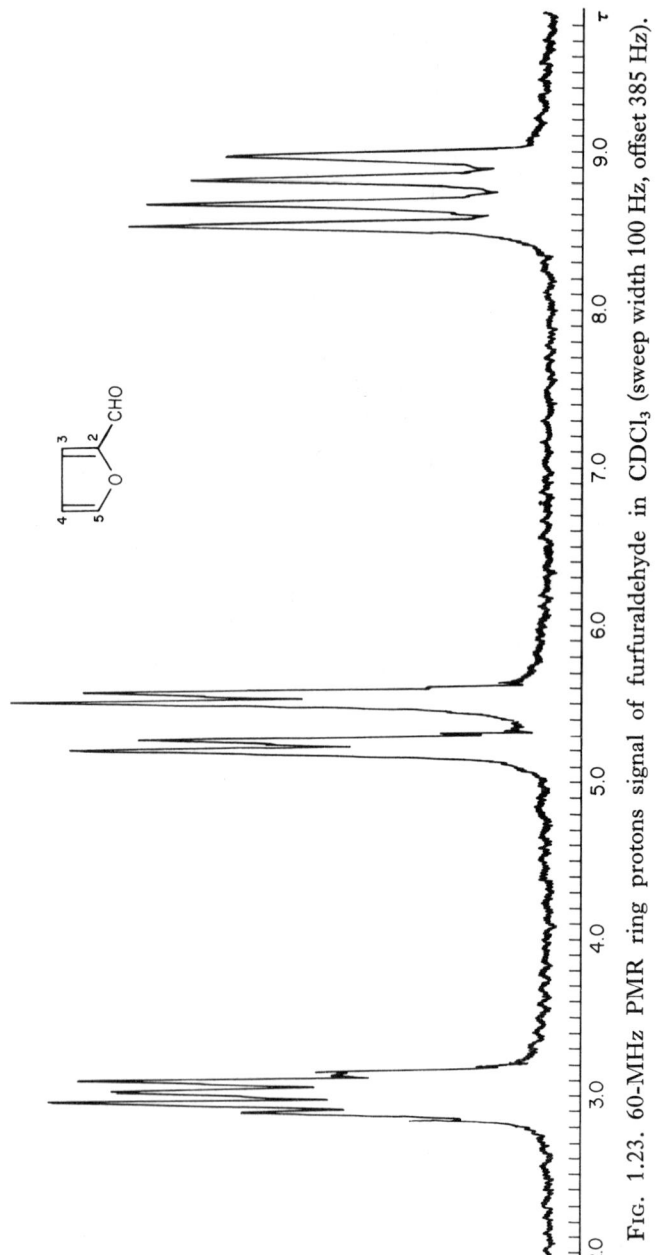

Fig. 1.23. 60-MHz PMR ring protons signal of furfuraldehyde in CDCl₃ (sweep width 100 Hz, offset 385 Hz).

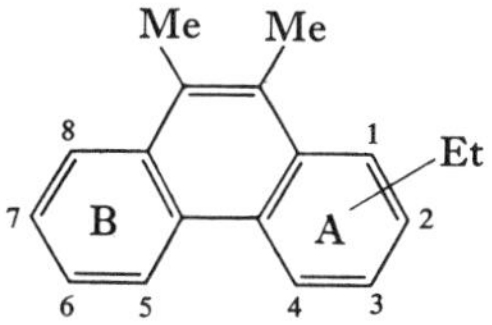

24 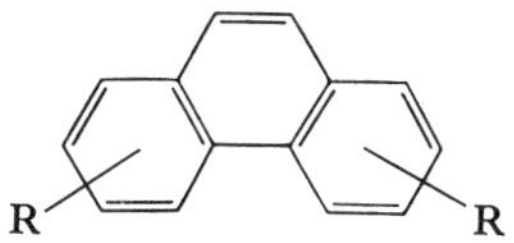   25

resembling the spectrum of *symmetrically* disubstituted analogues of type
25. Provided the (i) and (ii) signals are sufficiently different and removed
from one another the possibility exists of picking both patterns out from
the aromatic signal of **24** and thereby solving the orientation problem.

This is the case in the present example as is seen by comparison of
spectra A, B and C (Fig. 1.24).

Thus patterns due to both the unsubstituted ring (analysed as an ABMX
system) and the substituted ring of 3,6-dialkyl derivatives (analysed as
ABX: H_4 forms a singlet broadened by meta coupling, H_1 half of an AB
system showing *ortho* coupling and H_2 the other half showing additional
meta coupling); hence, the unknown must be the 3-ethyl derivative, since
the pattern due to the substituted ring would be markedly different if the
ethyl group were placed elsewhere, e.g. the H-4 signal would be a doublet
(*J* near 7 Hz) in the 2- and absent altogether in the 4-ethyl-isomer.

The same principles may be used to identify the dinitro-product obtained
from 2-benzylbenzimidazole (Fig. 1.25). The A_2B_2-4 proton quartet of the
3'-nitro derivative A and signals due to protons of the benzimidazole nucleus
of the 5-nitro isomer B may be discerned (in spite of signal overlap) in the
spectrum of the dinitro derivative C. In the 5-nitro isomer assignment of the
lowest field doublet (511 Hz from TMS) to the 4-proton is confirmed by the
fact that it exhibits a coupling constant (2 Hz) typical of *meta* placed protons.
The doublet of doublets (centre 490 Hz *J* 2 and 9) must arise from the
6-proton since this is *meta* coupled to the 4- and *ortho* coupled to the
7-proton, the latter giving a signal of the expected multiplicity and *J* value
(doublet at 464·5 Hz *J* 9 Hz).

Chemical Shift Predictions

Another approach to aromatic orientation problems is to compare a
particular chemical shift with that calculated for the various possibilities
by means of substituent effect data (sometimes used in the form of an
empirical equation). This procedure has been much used, principally by
Zürcher (1963), to identify angular methyl group positions in steroids and
depends upon the essentially additive nature of frequency shifts caused by
various shielding factors. Some aromatic examples follow.

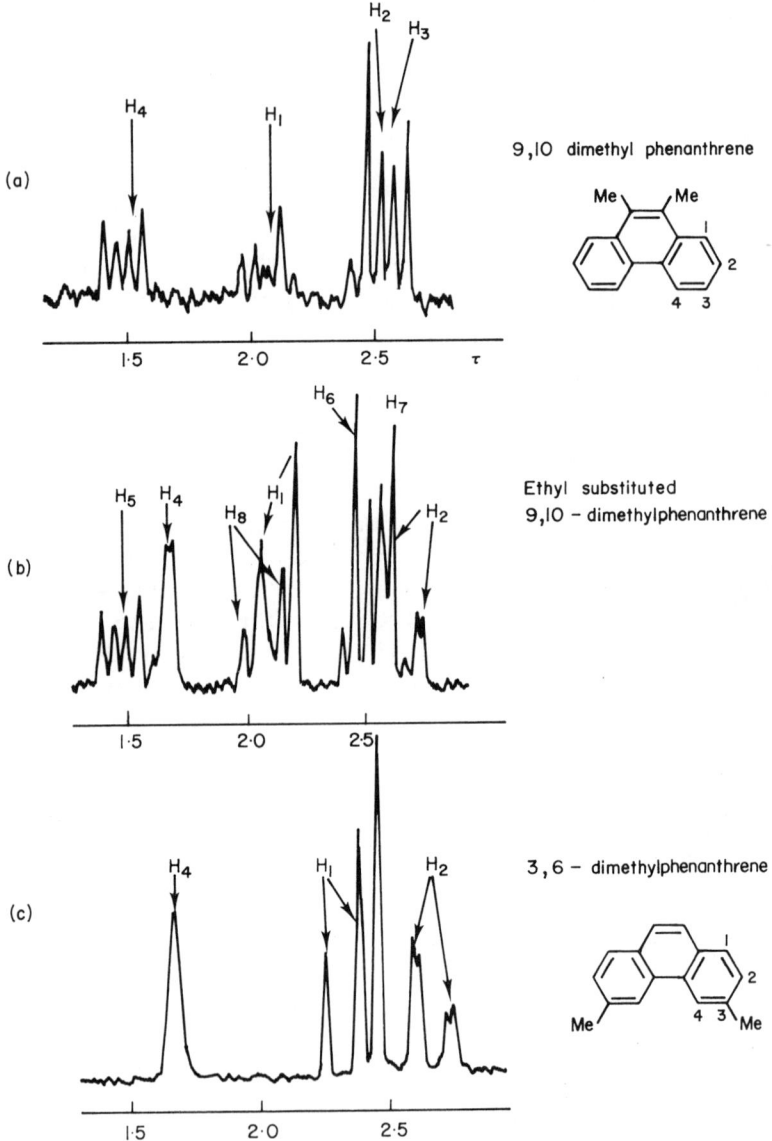

Fig. 1.24. 60-MHz PMR aromatic protons signals of (a), 9,10-dimethylphenanthrene; (b), ethyl substituted 9,10-dimethylphenanthrene; (c), 3,6-dimethylphenanthrene (solvent $CDCl_3$) (Bavin *et al.*, 1965).

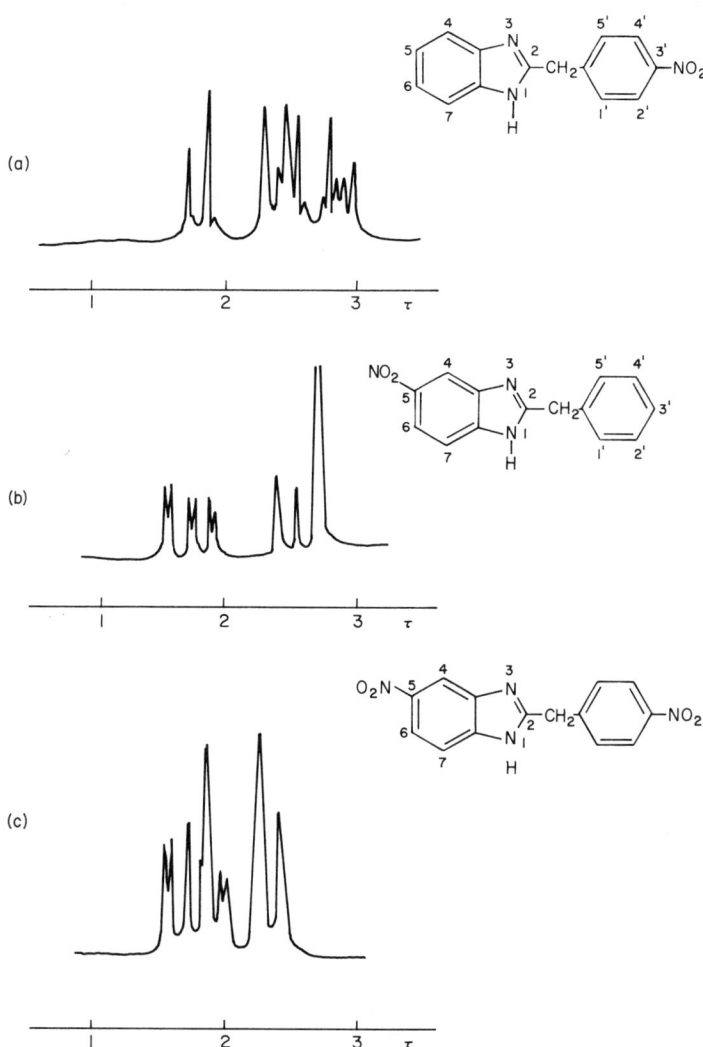

Fɪɢ. 1.25. 60-MHz PMR aromatic protons signals of some nitro-2-benzyl-benzimidazoles in DMSO: (a), 3′-nitro; (b), 5-nitro; (c), 3′,5-dinitro (Casy and Wright, 1966b).

A petroleum research group (Yew *et al.*, 1964) studied the NMR spectra of 24 polymethylnaphthalenes (2–5% in carbon tetrachloride) and found the resonances of α- and β-methyl groups to fall in the ranges $\tau 7\cdot36$–$7\cdot50$ and $7\cdot5$–$7\cdot7$ respectively. 1,8- and 4,5-derivatives (di-α-type) had

abnormally low chemical shifts ($\tau 7\cdot14$–$7\cdot26$), a result attributed to steric effects (see below). In general a methyl group becomes more shielded as further groups are introduced into the nucleus, the order of effect being adjacent Me > non-adjacent Me (same ring) > Me in opposite ring. When a substituent leads to a sterically hindered molecule (as in 1,8- and 4,5-derivatives) down-field shifts result through the distortion of electron clouds in the hindered groups (Pople, 1958). The equation

$$\delta_{\alpha,\,\beta} = \delta^0_{\alpha,\,\beta} + \Delta\delta_1 + \Delta\delta_2 + \Delta\delta_3 + \Delta\delta_4$$

where

$\delta_\alpha{}^0 = \tau 7\cdot36$ for 1-methylnaphthalene;

$\delta_\beta{}^0 = \tau 7\cdot52$ for 2-methylnaphthalene;

$\Delta\delta_1 = +0\cdot10$ ppm, contribution to chemical shift by an adjacent Me substituent;

$\Delta\delta_2 = +0\cdot05$ ppm, contribution for each non-adjacent Me substituent (same ring);

$\Delta\delta_3 = +0\cdot01$ ppm, contribution by each Me substituent in opposite ring (except for 1,8- and 4,5-cases);

$\Delta\delta_4 = -0\cdot2$ ppm, contribution to chemical shift of C-1 methyl by substitution of methyl at C-8, and vice versa

was developed from these data. Differences between experimental and calculated values were less than 0·04 ppm except for cases where ring distortions are likely such as 1,2,3- and 1,2,3,4-methylnaphthalenes. Ouellette and van Leuwen (1969) have made a similar analysis of methyl-arenes and related the methyl chemical shifts of compounds containing 3 or fewer rings to the summation of the inverse cubes of the distances ($\sum R^{-3}$) which separate the methyl group and the centres of the aromatic rings.

Ballantine and Pillinger (1967) have compiled a table of substituent shielding values (S) embracing 14 main classes of substituent from statistical analysis of data upon 450 phenols in $CDCl_3$ and CCl_4. A few examples are given below (figures in brackets show the number of examples used to arrive at each value). A positive sign denotes an up-field shift, and a negative a down-field shift.

Substituent	S *ortho*	S *meta*	S *para*
OH	+0·45 (78)	+0·1 (69)	+0·40 (45)
OCOR	+0·2 (60)	−0·10 (52)	+0·20 (41)
CH₃	+0·15 (80)	+0·10 (36)	+0·10 (14)
COR	−0·70 (54)	−0·25 (129)	−0·10 (10)
Cl	−0·10 (16)	0·00 (44)	0·00 (2)

An example showing the way these values are used to calculate the chemical shifts of aromatic protons is shown below. The value $\tau 2·7$ is the aromatic chemical shift of a 10% solution of benzene in $CDCl_3$.

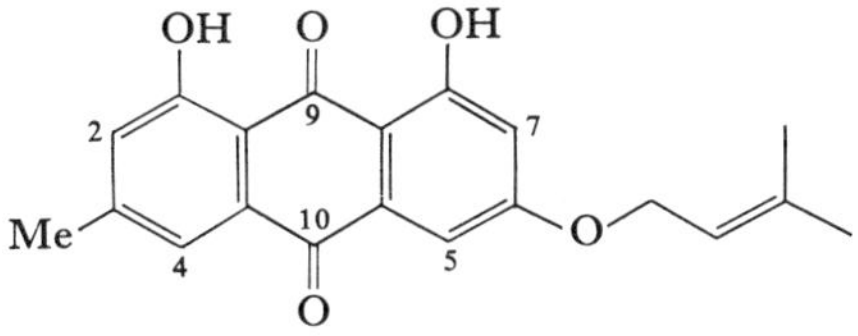

OH Me 9CO 10CO

Calc. τ for 2H = 2·70 + 0·45 + 0·15 − 0·25 − 0·10 = 2·95. Found 3·00
Calc. τ for 4H = 2·70 + 0·40 + 0·15 − 0·25 − 0·70 = 2·30. Found 2·50

OH Me 9CO 10CO

Calc. τ for 7H = 2·70 + 0·45 + 0·45 − 0·25 − 0·10 = 3·25. Found 3·40
Calc. τ for 5H = 2·70 + 0·40 + 0·45 − 0·25 − 0·70 = 2·60. Found 2·70

In many cases shielding values may be very roughly correlated with electronic influences. Thus, the order of OH shielding ($o \simeq p > m$) is probably linked to the much more efficient electron release by this group to *o*- and *p*-positions (by resonance) than to the *meta* position. However, theoretical prediction of these values is not at present possible because many factors (resonance, inductive, steric and magnetic anisotropic effects) contribute to the net value. The shielding values do not give an exact result in many cases, but the authors calculate that there is a greater than 70% probability of calculated values lying within ±0·2 ppm of the experimental result. Major deviations were noted in cases where long range effects were to be expected such as where the protons arrowed are subject to the magnetic anisotropic influence of the carbonyl group.

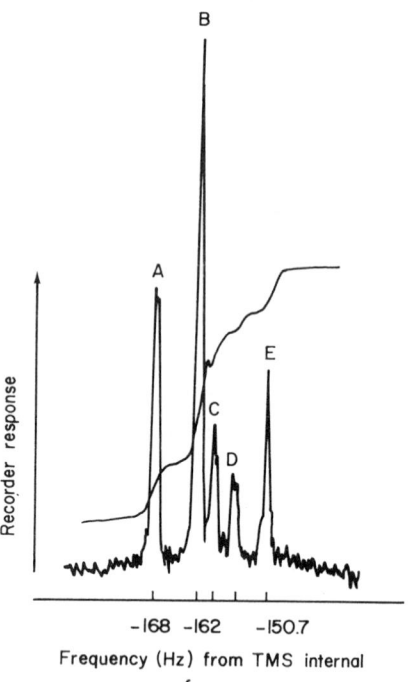

A similar, although more sophisticated, approach has been taken by Reed (1967) for the prediction of aryl proton chemical shifts in derivatives with OH, Cl and Me substituents.

Fig. 1.26. 60-MHz PMR methyl protons signals of a synthetic mixture of TNT isomers in $(CD_3)_2CO$ (Gehring and Reddy, 1968).

Peak	Isomer
A	2,4,5-TNT
B	2,4,6-TNT
C	2,3,4-TNT
D	2,3,5-TNT
E	2,3,6-TNT

An interesting example of the use of Ar-methyl chemical shifts is the determination of di- and trinitrotoluene isomers in 2,4,6-TNT (Gehring and Reddy, 1968). The Ar-methyl signals of these isomers are sufficiently separated at 60 MHz to permit quantitative integration of the individual components of an isomeric mixture (Fig. 1.26). The analytical technique involves a crystallization procedure which removes the bulk of the 2,4,6-TNT from the sample (thereby concentrating the impurities) and use of calibration curves set up from data upon samples of known composition. The methyl resonances show long range coupling with aryl protons of the order 0·75 (*ortho*), 0·35 (*meta*) and 0·6 (*para*) ppm, a fact used to aid isomer identification. Thus, an unknown resonance observed at 151 Hz was resolved into a quartet with inner and outer spacings of 0·35 and 0·60 Hz respectively showing it to arise from the 2,3,6-isomer.

References

Alexander, T. G. and Koch, S. A. (1967). *Appl. Spectrosc.* **21**, 182; *J. Ass. Offic. Anal. Chem.* **50**, 676.

Babad, H., Taylor, T. N. and Goldberg, M. C. (1968). *Anal. Chim. Acta* **40**, 387.

Babiec, J. S., Barrante, J. R. and Vickers, G. D. (1968). *Anal. Chem.* **40**, 610.

Ballantine, J. A. and Pillinger, C. T. (1967). *Tetrahedron* **23**, 1691.

Barcza, S. (1963). *J. Org. Chem.* **28**, 1914.

Bass, R. J. and Sewell, M. J. (1969). *Tetrahedron Lett.* 1941.

Bauld, N. L. and Rim, Y. S. (1968). *J. Org. Chem.* **33**, 1303.

Bavin, P. M. G., Bartle, K. D. and Smith, J. A. S. (1965). *Tetrahedron* **21**, 1087.

Casy, A. F. and Armstrong, N. A. (1965). *J. Med. Chem.* **8**, 57.

Casy, A. F. and Wright, J. (1966a). *J. Chem. Soc.* (C), 1167.

Casy, A. F. and Wright, J. (1966b). *J. Chem. Soc.* (C), 1511.

Casy, A. F., Beckett, A. H. and Iorio, M. A. (1967). *Tetrahedron* **23**, 1405.

Casy, A. F., Beckett, A. H., Iorio, M. A. and Youssef, H. Z. (1965). *Tetrahedron* **21**, 3387.

Chapman, O. L. and King, R. W. (1964). *J. Amer. Chem. Soc.* **86**, 1256.

Conway, T. F. and Johnson, L. F. (1969). *Science* **164**, 827.

Culvenor, C. C. J. (1966). *Tetrahedron Lett.* 1091.

Dean, J. A. (1968). *Clin. Chem.* **14**, 326.

Deverell, C. and Schaumberg, K. (1967). *Anal. Chem.* **39**, 1879.

Fraser, R. R., Kaufman, M., Morand, P. and Govil, G. (1969). *Can. J. Chem.* **47**, 403.

Gagnaire, D. and Robert, D. (1968). *Bull. Soc. Chim. Fr.* 781.

Galpin, D. R., Cochran, T. G. and Huitric, A. C. (1969). *Biochem. Pharmacol.* **18**, 979.
Gehring, D. G. and Reddy, G. S. (1968). *Anal. Chem.* **40**, 792.
Goodlett, V. W. (1965). *Anal. Chem.* **37**, 431.
Hall, G. E. (1968). *In* "Annual Review of NMR Spectroscopy" (E. F. Mooney, ed.), Vol. 1, p. 227. Academic Press, London and New York.
Hatada, K., Terawaki, Y., Okuda, H., Nagata, K. and Yuki, H. (1969). *Anal. Chem.* **41**, 1518.
Hollis, D. P. (1963). *Anal. Chem.* **35**, 1682.
Johnson, L. F. and Shoolery, J. N. (1962). *Anal. Chem.* **36**, 1136.
Kiefer, E. F., Gericke, W. and Aminoto, T. (1968). *J. Amer. Chem. Soc.* **90**, 6246.
Kram, T. C. and Turczan, J. W. (1968). *J. Pharm. Sci.* **57**, 651.
Liu, K-J. (1968). *Makromol. Chem.* **116**, 146.
Ludwig, F. J. (1968). *Anal. Chem.* **40**, 1620.
McClenaghan, I. and Sykes, P. J. (1968). *Chem. Commun.* 800.
Manatt, S. L. (1966). *J. Amer. Chem. Soc.* **88**, 1323.
Moen, R. V. and Makowski, H. S. (1967). *Anal. Chem.* **39**, 1860.
Narayanan, C. R. and Iyer, K. N. (1965). *Tetrahedron Lett.* 3741.
Noland, W. E., Langager, B. A., Manthey, J. W., Zacchei, A. G., Petrak, D. L. and Eian, G. L. (1967). *Can. J. Chem.* **45**, 2969.
Ouellette, R. J. and van Leuwen, B. G. (1969). *J. Org. Chem.* **34**, 62.
Page, T. F. and Bresler, W. E. (1964). *Anal. Chem.* **36**, 1981.
Paulsen, P. J. and Cooke, W. D. (1964a). *Anal. Chem.* **36**, 1713.
Paulson, P. J. and Cooke, W. D. (1964b). *Anal. Chem.* **36**, 1721.
Philipsborn, W. D. V. (1964). *Arch. Pharm.* **34**, 58.
Pople, J. A. (1958). *Mol. Phys.* **1**, 199.
Pople, J. A., Schneider, W. G. and Bernstein, H. J. (1959). "High Resolution Nuclear Magnetic Resonance". McGraw-Hill, New York.
Rader, C. P. (1966). *J. Amer. Chem. Soc.* **88**, 1713.
Rader, C. P. (1969). *J. Amer. Chem. Soc.* **91**, 3248.
Reed, J. J. R. (1967). *Anal. Chem.* **39**, 1586.
Sanders, J. K. M. and Williams, D. H. (1971). *J. Amer. Chem. Soc.* **93**, 641.
Saur, W., Crespi, H. L., Harkness, L., Norman, G. and Katz, J. J. (1968). *Anal. Biochem.* **22**, 424.
Schoolery, J. N. and Smithson, L. H. (1970). *J. Amer. Oil. Chem. Soc.* **47**, 153.
Slomp, G., Baker, R. H. and MacKellar, F. A. (1964). *Anal. Chem.* **36**, 375.
Stolow, R. D. and Gallo, A. A. (1968). *Tetrahedron Lett.* 3331.
Traynham, J. G. and Knesel, G. A. (1965). *J. Amer. Chem. Soc.* **87**, 4220.
Trehan, I. R. and Monder, C. (1968). *Tetrahedron Lett.* 67.
Turczan, J. W. and Kram, T. C. (1967). *J. Pharm. Sci.* **56**, 1643.
Varian Associates. NMR at work, No. 57.
Wright, J. (1965). Ph.D. Thesis, University of London.
Yew, F. F., Kurland, R. J. and Mair, B. J. (1964). *Anal. Chem.* **36**, 843.
Zürcher, R. F. (1961). *Helv. Chim. Acta* **44**, 1350.
Zürcher, R. F. (1963). *Helv. Chim. Acta* **46**, 2054.

CHAPTER 2

NMR Spectral Features of Nitrogen-Containing Organic Compounds

The introduction of nitrogen into an organic molecule results in the deshielding of nearby protons and, to a first approximation, the effect of this atom in various structural environments is consistent and additive. Collections of spectral data enabling the chemical shifts of protons in many environments to be estimated to a fair degree of accuracy are available. The two catalogues compiled by Bhacca and others (Varian Associates, 1962, 1963) are the best known collections and their use requires the application of a coding scheme followed by inspection of a functional group index. Nitrogen functions are included in these volumes

$$\text{e.g.} \quad \begin{array}{c} R \\ R \end{array}\!\!>\!\!NH, -NO_2, -C\!\equiv\!N$$

but are dealt with more extensively by Slomp and Lindberg (1967) who have correlated and charted chemical shift data on 2306 protons influenced by nitrogen in a variety of ways. The extended form of the Varian coding scheme which is used requires careful study and will not be outlined here. The most recent spectra collection (Bovey, 1967) lists over 4000 examples which are arranged according to their empirical formula ($B_3Cl_3H_2N_3 \rightarrow C_{18}H_{17}BrO_3$) in the same order as that adopted in *Chemical Abstracts*. These tables contain many examples of nitrogen-containing organic compounds but are designed to be of value in searching for specific compounds rather than for examples of a particular nitrogen environment. The NMR spectra of nitrogen heterocyclic compounds, chiefly those with aromatic properties, have been reviewed (White, 1963). Coupling constants found in pyridine derivatives and other heterocyclic molecules are generally smaller than those of benzenoid derivatives (see below) (electronegative

J_o 7–10 Hz
Jm 2–3 Hz
Jp 1 or less

J_{23} 5·5, J_{34} 7·5
J_{24} 1·9, J_{35} 1·6, J_{26} 0·4
J_{25} 0·9   (all Hz)

substituents decrease the coupling constant magnitude), and this fact aids the differentiation of pyridyl and phenyl aromatic signals.

In this chapter attention is focused upon the magnetic resonance properties of protons closely linked to the chief isotope of nitrogen (^{14}N).

N—H

The N—H protons of most primary and secondary amines in the solute condition undergo rapid exchange with each other and with other labile

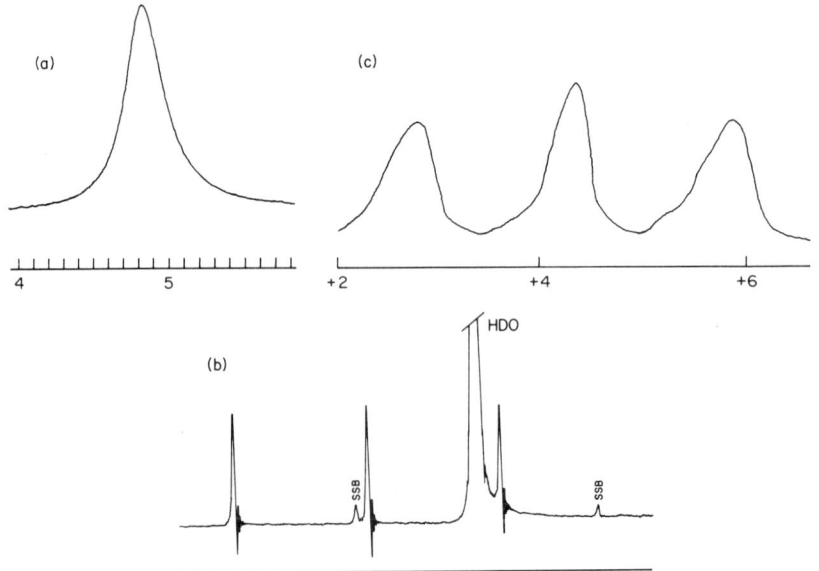

Fig. 2.1. PMR spectra of: (a) NH_4Cl—H_2O; (b) NH_4Cl—HNO_3—H_2O (a and b recorded at 60 MHz, sweep width 500 Hz; $J_{14N-H} = 52.5$ Hz); (c) intensively dried liquid NH_3 recorded at 30 MHz (Ogg, 1954); $J_{14N-H} = 46$ Hz.

protons with the result that their PMR signals form sharp singlets typical of such behaviour (cf. alcohols, p. 27). When other proton groups capable of exchange are present a common signal is obtained which represents the average environment of all species concerned. Thus, the spectrum of a NH_4NO_3—NH_3—H_2O solution consists of one broad band [Fig. 2.1(a)]. N—H protons, like OH, engage in hydrogen bonding, hence their PMR signal positions are sensitive to temperature and concentration variations. For this reason chemical shift ranges quoted for NH signals are wide, aromatic amines (τ 5·3–7·4) absorbing lower field than aliphatic types (τ 7·8–9·7). Signals due to the protons of amine salts ($\geq\overset{+}{N}H$) have much

lower values and usually fall to low field of the aromatic region. The rate
of NH exchange processes decreases as the hydrogen ion concentration
increases but it is usually only slower than the NMR experiment in strongly
acidic solutions (pH <1·0). Under these conditions fine structure may
become apparent in the NH signals. In theory, the NH signal should be a
triplet due to spin–spin coupling with the ^{14}N nucleus which, having a
nuclear spin number (I) of unity has its magnetic energy levels split in

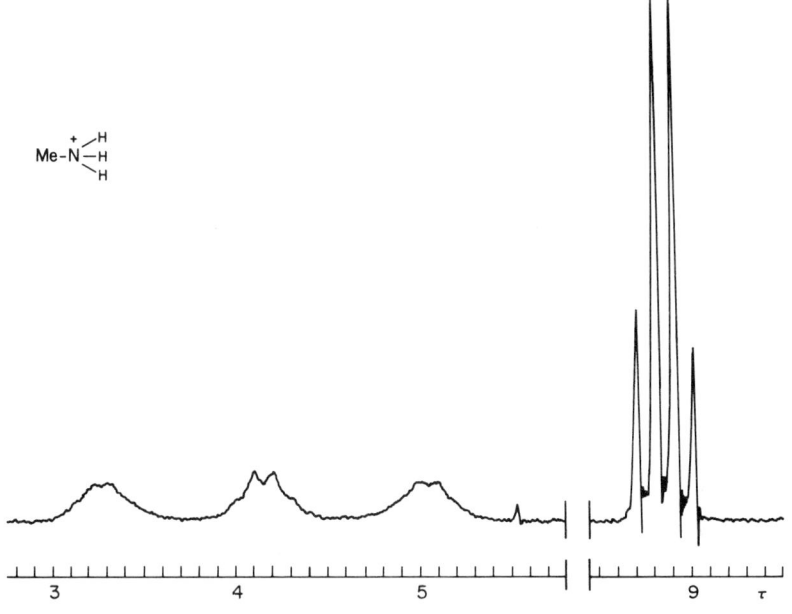

FIG. 2.2. 60-MHz PMR spectrum of methylamine–H_2O, pH 1·0 ($\overset{+}{H_3NMe}$);
$J_{N-H} \sim 53$ Hz, $J_{HNCH} \sim 6$ Hz (resolution is best in central component of the N—H
triplet).

three; magnetic quantum numbers (m) for ^{14}N are +1, 0 and −1 giving a
multiplicity ($2n\,I + 1$) of 3 for the N$\underline{\text{H}}$ resonance. Such signals are in fact
obtained, that of sodium-dried ammonia and the ammonium ion (NH$_3$—
H$_2$O, acid pH) being well defined 1:1:1 triplets (Ogg, 1954) with each
component of equal intensity in contrast with the more common 1:2:1
triplet which arises from the system CH$_2$C$\underline{\text{H}}$ [Fig. 2.1(b)]. Many amines,
e.g. ammonia [Fig. 2.1(c)], methylamine (Fig. 2.2) and pyrrolidine, give
similar NH triplet signals when examined in acid solution, with J_{NH} values
between 50 and 60 Hz (Roberts, 1956). In most cases the triplet components
are broadened due to (a) spin–spin coupling between the N—H and any
vicinally placed protons [this coupling is evident, for example, in the

quartet nature of each component of the NH triplet and methyl signal of methylamine ($\overset{+}{\text{MeNH}_3}$, Fig. 2.2)] and (b) electric quadrupole effects of the ^{14}N nucleus which arise through interaction of the nuclear charge with the surrounding cloud of valency electrons when this is non-symmetrical (Roberts, 1959). Quadrupole effects are seen in all nuclei with I values of unity—these behave as if their nuclear charge were distributed over an ellipsoidal surface, in contrast with nuclei having $I = \frac{1}{2}$ (such as

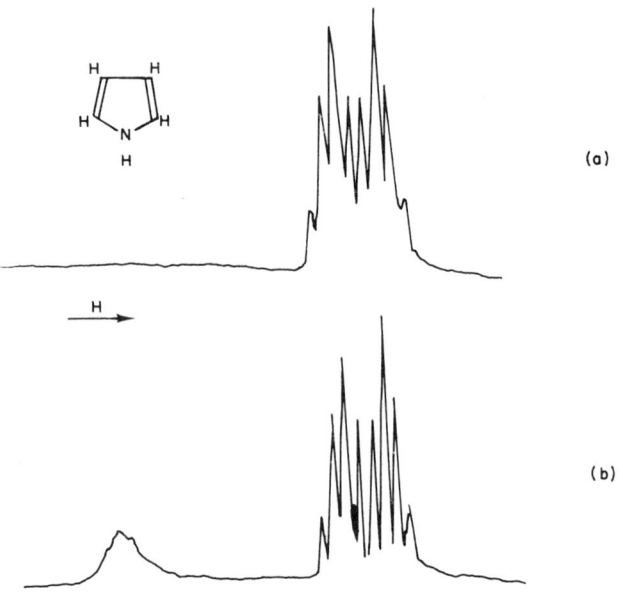

Fig. 2.3. (a) The 40-MHz PMR spectrum of pyrrole; (b) as in (a) but with strong irradiation of the ^{14}N nucleus (Varian Associates, *NMR at Work*, No. 50).

protons) that are regarded as spherical bodies of uniform charge distribution. These interactions provide an additional relaxation mechanism for the ^{14}N nuclear magnets and render differences between the ^{14}N magnetic energy levels less distinct as far as their influence upon the N—H resonance signal is concerned. The efficiency of this process determines the form of the N—H signal (this may be a singlet, a sharp triplet or a broad shallow band corresponding with high, low and intermediate efficiencies respectively) and is temperature dependent through its relationship to the rate of molecular tumbling. The real nature of ^{14}N quadrupole effects has been confirmed by double resonance experiments. When pyrrole is irradiated at the ^{14}N frequency (thereby eliminating all nitrogen magnetic influences) the N—H resonance is converted from the very broad (scarcely apparent) band of the normal spectrum to a distinct near-triplet through spin–spin coupling with the two adjacent C—H protons (Fig. 2.3). In symmetrical molecules such as the ammonium ion, quadrupole effects do not operate and the NH signal forms a sharp triplet at low pH values [Fig. 2.1(b)].

Exchange processes are much slower in amides than in amines so broad N—H signals (τ1·5–5·0) are more commonly encountered in the former class since both coupling (with nitrogen and α-protons) and quadrupole effects may operate. The N—H signal of methacrylamide (Fig. 2.4) is a good example.

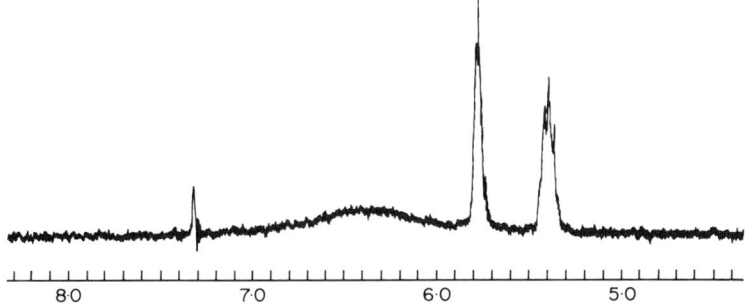

Fig. 2.4. Vinylic and NH_2 (broad band centre 6·38 ppm) 60-MHz PMR signals of methylacrylamide in $CDCl_3$ (Varian Associates, 1962).

N-METHYL

The N-methyl resonance of a tertiary organic base forms a sharp 3 (or multiple of 3) proton singlet since it is usually not subject to ^{14}N coupling because of nitrogen quadrupole effects. When these effects do not operate, as in symmetrical molecules, N—CH coupling may be seen, J values being commonly less than 1 Hz. Thus, the Me signal of tetramethylammonium bromide (Fig. 2.5) is a 1:1:1 triplet, $J \sim 0·55$ Hz (Grunwald *et al.*, 1957);

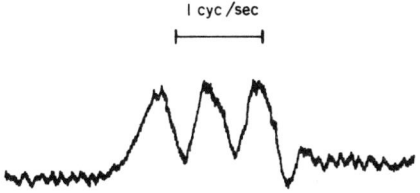

Fig. 2.5. 31·65-MHz PMR signal of the Me_4N^+ ion in H_2O (Grunwald *et al.*, 1957).

3-bond NCCH coupling is larger (1·5–2·0 Hz) (see p. 217). The angular dependence of vicinal ^{1}H–^{14}N spin coupling (cf. Chapter 3, p. 87) has been established by work on rigid bicyclic compounds using proton double and triple resonance techniques to measure J values (Terui et al., 1968). The N—Me chemical shift range of τ 7·6–8·0, based on 34 examples of amines, (Slomp and Lindberg, 1967), shows that nitrogen has a deshielding influence

$$(cf. \quad \diagdown\!\!\diagup\!\!C\!-\!\underline{Me} \quad \tau\ 9\cdot1)$$

which is similar in extent to that of phenyl (Ph—$\underline{Me}$ τ 7·66) and less than that of oxygen (RO$\underline{Me}$ τ 6·7). This influence is greatly enhanced when nitrogen bears a positive charge as in acid salts and quaternary compounds, down-field shifts of close to one tau unit often resulting, the tau range being 6·87–7·3 for 25 $^+$NMe examples (Slomp and Lindberg, 1967). The N—Me salt-base chemical shift difference is a valuable diagnostic aid (as will be explained in more detail below) since a singlet that suffers a pronounced down-field shift when the solvent is acidified may confidently be identified as due to N-methyl and be differentiated from Ar—Me and O—Me signals which are little affected when the pH is reduced (see Fig. 2.6). Ideally, N—Me protonation shifts should be observed by comparing base and salt spectra in the same solvent, whereby chemical shifts variations due to solvent change do not arise. This procedure may be inconvenient, however, because it requires the separate preparation of a salt and also a common solvent for both forms; CDCl$_3$ or DMSO-d$_6$ often, but not always, meet these requirements. A more convenient method is to examine the base first in CDCl$_3$ and then in an acid which either lacks, or contains few, protons such as trifluoroacetic (TFA), perdeuteroacetic (PDA) or sulphuric acid. This method naturally introduces solvent effects but is useful provided interest is centred chiefly upon N-methyl protonation shifts. Ma and Warnhoff (1965) have made an extensive study of amines by this technique and their findings warrant discussion in detail.

In an equilibrium system of base and protonated base, the magnitude of the N-methyl down-field shift would be expected to be related to the proportion of the protonated species present and it has in fact been established that the methyl chemical shifts of mono-, di- and trimethylamine in aqueous acid are linear functions of the concentration of protonated amine (Grunwald et al., 1957). A differentiation between weak and relatively strong bases may be made upon this principle by the use as solvent of the two acids TFA (pK_a 0·26 in water) and PDA (pK_a 4·76, acetic acid in water) of widely differing acidities, since only the stronger bases will be extensively protonated in the weaker acid. Thus, the N—Me protons of many aliphatic

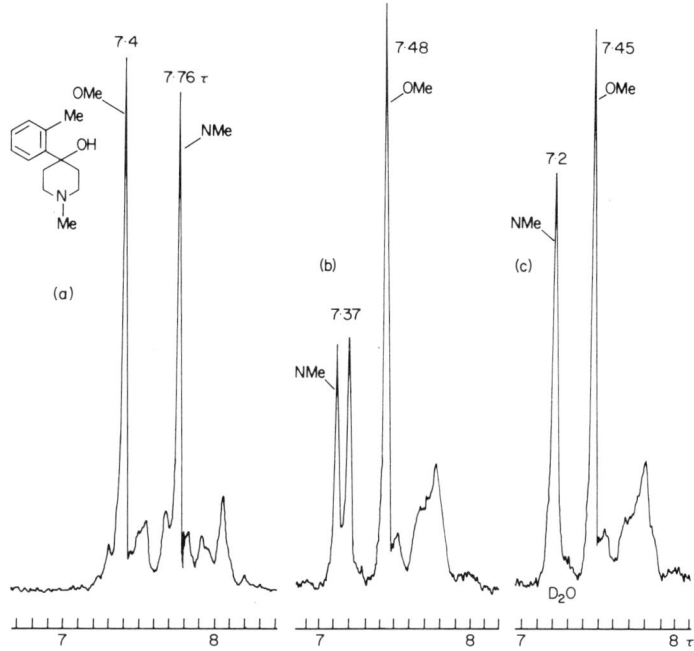

FIG. 2.6. Part of the 60-MHz PMR spectra of 1-methyl-4-*o*-tolyl-4-piperidinol:
(a) in CDCl₃; (b) in CDCl₃—CF₃CO₂H; (c) in CDCl₃—CF₃CO₂H—D₂O.

tertiary and secondary amines (i.e. reasonably strong bases) show down-field
shifts in PDA that are at least two-thirds the value of those observed in
TFA; in contrast, shifts for most aromatic amines (very weak bases) in
PDA are substantially less than those seen in the stronger acid (Table 2.1).

TABLE 2.1

NMe chemical shift differences (ppm)

Class	$\Delta_{PDA-CDCl_3}$	$\Delta_{TFA-CDCl_3}$
3° aliphatic [e.g. $Me(CH_2)_3NMe_2$]	+0·40 to +0·81 +0·64	+0·53 to +0·92 +0·75
2° aliphatic [e.g. $Me(CH_2)_3NHMe$]	+0·26 to +0·54 +0·33	+0·43 to +0·67 +0·53
Aromatic and N-heteroaromatic [e.g. $C_6H_5NMe_2$]	+0·01 to +0·54 +0·16	+0·34 to +0·82 +0·61

It is seen from Table 2.1 that the average protonation shift for a tertiary amine is significantly greater than that for secondary derivatives (two closely related amines are included in the table) and this difference, together with spin–spin coupling effects (see later), provides a means of differentiating 2° and 3° amines of related structure. The reason for the greater deshielding of tertiary amines is not clear. It cannot be due to different extents of protonation because secondary amines are usually the stronger bases. The positive charge on nitrogen appears to be more dispersed and hence partially neutralized in 2° amines—this could arise through solvation (this amine class is better solvated than 3° amines, a fact accounted for by the provision of more opportunities for hydrogen bonding between solvent and N—H and/or less steric hindrance to the approach of solvent molecules) or ion-pair phenomena. Ion-pair formation is highly probable in solvents of low polarity such as substituted acetic acids (the dielectric constant of acetic acid is 6·15) and the approach of the anion to the $\overset{+}{N}H$ centre should also be dependent upon steric factors. NMR evidence for ion pair formation has been obtained in quaternary salts of aromatic amines in DMSO (Fraenkel, 1963; Fraenkel and Kim, 1966). It was found that whereas the aromatic protons of the iodide (a) were markedly descreened by the N⁺

$$Cl-\left\langle\bigcirc\right\rangle-\overset{+}{N}Me_3\ \overset{-}{X}$$

X = (a) I, (b) Cl, (c) BF₃

centre, those of the chloride (b) had values close to protons in chlorobenzene. The interpretation was that the N⁺ influence is counteracted by the chloride anion but not by the larger iodide anion which cannot approach this centre so closely for steric reasons and does not therefore form an ion-pair. In support, the aromatic chemical shift of the boron trifluoride adduct (c), in which the charged centres are known to be close, was near the chloride value. The generally lower field position of N-methyl signals in quaternary salts compared with the N-methyl chemical shift of corresponding hydro-halide salts (as in the 1,3-dimethylpiperidine salts below) may be accounted for in the same way. Modification of the deshielding influence of charged

ν_{N-Me} in CDCl₃ τ 7·16

τ 6·42 and 6·56

nitrogen by the solvent is well illustrated by data upon 1-methylpyridinium iodide (Reynolds and Priller, 1968). All chemical shifts (measured at infinite dilution) of the salt moved upfield with increasing dielectric constant of the solvent (Fig. 2.7), the latter property being a measure of solvating ability. The authors consider the most probable mode of solvent–cation interaction to be direct solvation of the positive nitrogen centre; this would decrease the effect of the positive charge and consequently cause high field shifts relative to chemical shifts in non-polar solvents. The solvents chosen were those for which extents of ion-pair formation should be low.

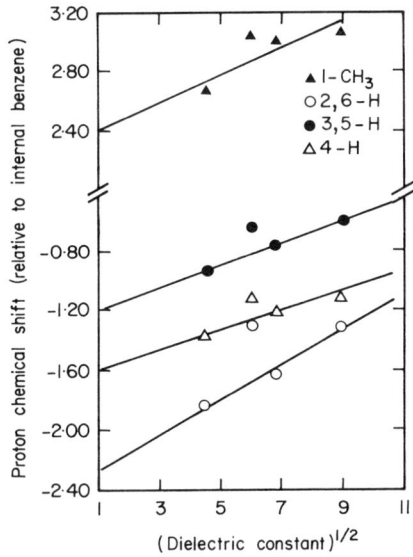

FIG. 2.7. Plot of infinite dilution proton chemical shifts of the 1-methylpyridinium ion in different solvents versus (dielectric constant)$^{\frac{1}{2}}$ for the solvent at 27·5°C (Reynolds and Priller, 1968).

Steric influences upon N—Me protonation shifts are also reflected in data upon some dimethylaminocholestanes (**1**). Of the five isomers examined

1

those with axial NMe_2 groups (2β, 7α, 6β) gave larger $\Delta_{CD_3CO_2D-CDCl_3}$ shifts ($+0.67$, $+0.69$ and $+0.81$ ppm respectively) than those with equatorial groups (2α, $+0.58$; 6α, $+0.64$ ppm) while the largest shift was recorded for the 6β-isomer, which would be expected to have the most hindered NMe_2 group.

Amides and imides showed negligible N—Me protonation shifts in PDA as expected from their very weakly basic nature although shifts near

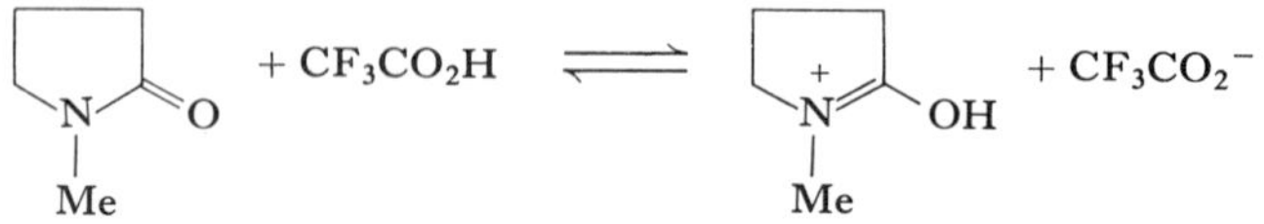

N-Methylpyrrolidone

0.2 ppm occur in TFA as a probable result of the formation of O-protonated forms as shown for N-methylpyrrolidone. Charged species of this nature

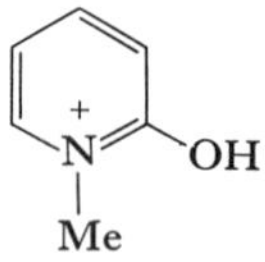

are more stable in N-heterocyclic analogues, N-methyl-2-pyridone for example showing a much larger protonation shift (0.6 ppm) in TFA (NMR evidence for the protonation structure shown is given later).

Signals due to C-methyl attached to aromatic rings containing or linked to nitrogen often suffer protonation shifts of the same order of magnitude as aromatic N—Me groups with which they may therefore be confused. These cases arise when a good mechanism exists for transmission of the deshielding influence of protonated nitrogen to a C-methyl group. Examples are α-picoline, N-methyl-*p*-toluidine, and 2-methylindole (shifts in ppm

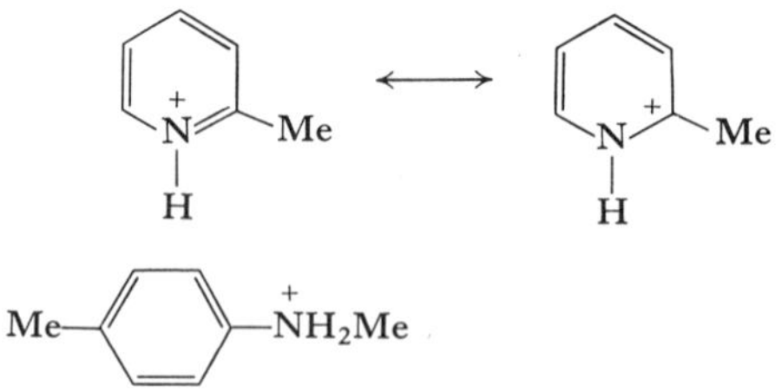

($+0.46$)

($+0.21$)

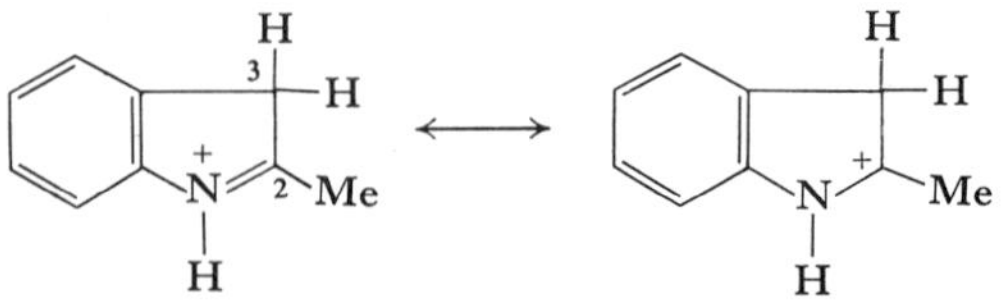

($+0.69$)

with TFA as solvent are given in parentheses alongside the formulae). The shift in the last example should be compared with the value +0·15 ppm obtained for the 3-methyl isomer (protonation is probable at C-3 in both indole derivatives) where the C—Me group is further removed from the charged nitrogen atom.

Additional information of value to the characterization of N-methyl amines is obtained from the multiplicities of N-methyl signals that arise through coupling with $\overset{+}{N}$—H protons. Thus, N-methyl signals of the protonated salt of a tertiary amine $\left(\overset{+}{\underset{H}{>}}N\text{Me} \right)$ should be a doublet while that of a secondary amine $\left(H \overset{+}{\underset{H}{>}}N\text{Me} \right)$ a triplet (the signal is a quartet in methylamine hydrochloride, Fig. 2.2). These couplings are observed and have J values of about 5–6 Hz, but are only seen under conditions of slow $\overset{+}{N}$—H proton exchange. Thus, they are observed somewhat erratically in the spectra of amine salts dissolved in CDCl$_3$ dependent on sample purity (trace acids, bases or water catalyse exchange) and anion basicity. The N—Me doublets (slow rate of exchange) or broad singlets (intermediate rate) obtained under these conditions are converted to sharp singlets without change in chemical shift when D$_2$O is added to the sample solution, a result which provides a good means of confirming the N-methyl assignment; an example is shown in Fig. 2.6. A more reliable method of observing coupling due to the $\overset{+}{N}$—H proton is to examine the base in a strong acid such as TFA (Anderson and Silverstein, 1965; Ma and Warnhoff, 1965) or concentrated sulphuric acid, the latter being necessary to achieve splitting of the N—Me resonance of aromatic amines (Thompson *et al.*, 1966). The utility of splitting pattern analysis is not, of course, confined to N—Me signals, and the multiplicities of any signal identifiable with a proton group adjacent to nitrogen (e.g. methylene in the N-benzylpiperidine, Fig. 2.8) may serve to characterize the amine class.

Catalytic debenzylation of the amino alcohol (2) provides an example of amine characterization by PMR as an aid to elucidation of a reaction pathway. On the basis of experience with the amino ketone (3), which gives the pyrrolidine derivative (4) in this reaction, the amino alcohol could lead either to the same cyclic tertiary amine or to the acyclic secondary amine (5). Among other evidence, the triplet nature (J 5 Hz) of the N—Me signal of the product as hydrochloride (Fig. 2.9) identified it as a 2° amine, not a 3° amine (Casy and Hassan, 1969).

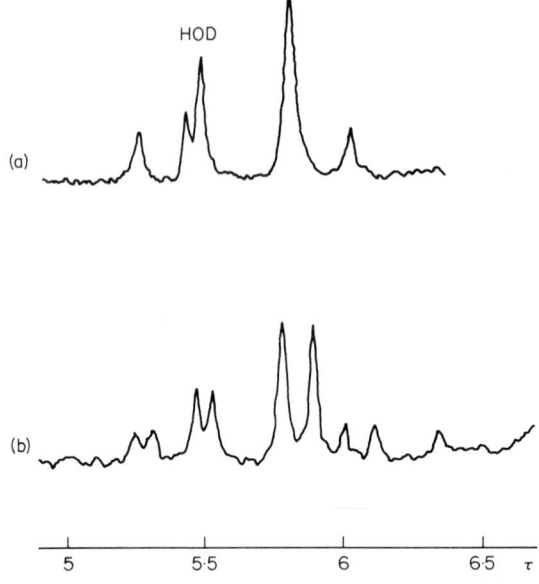

FIG. 2.8. Benzylic methylene region of the 60-MHz PMR spectrum of 1-benzyl-2-methylpiperidine hydrochloride in (a) CDCl₃—D₂O; (b) CDCl₃. In (b) each component of the PhC$\underline{H}_2$N quartet is split into a doublet by coupling with the ⁺NH proton.

Spin–spin coupling between N—H and N—Me protons may often be observed, even under neutral conditions, in cases where the N—H exchange rate is slow, as in amides and other essentially non-basic compounds in

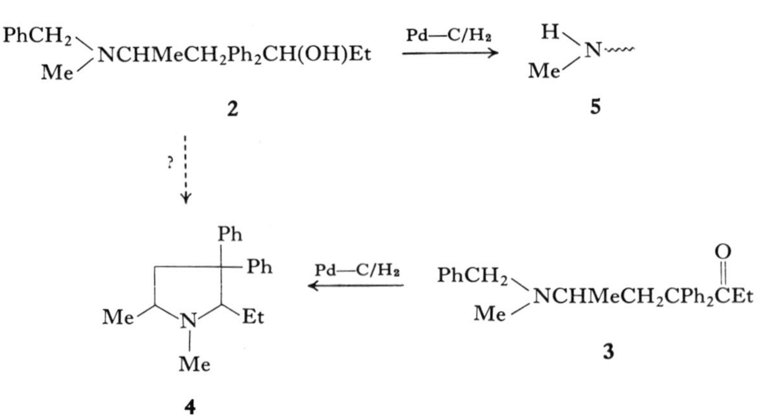

$$\begin{matrix} \text{PhCH}_2 \\ \text{Me} \end{matrix} \text{NCHMeCH}_2\text{Ph}_2\text{CH(OH)Et} \xrightarrow{\text{Pd—C/H}_2} \begin{matrix} \text{H} \\ \text{Me} \end{matrix} \text{N}\sim$$

2   **5**

?

4   **3**

Scheme I

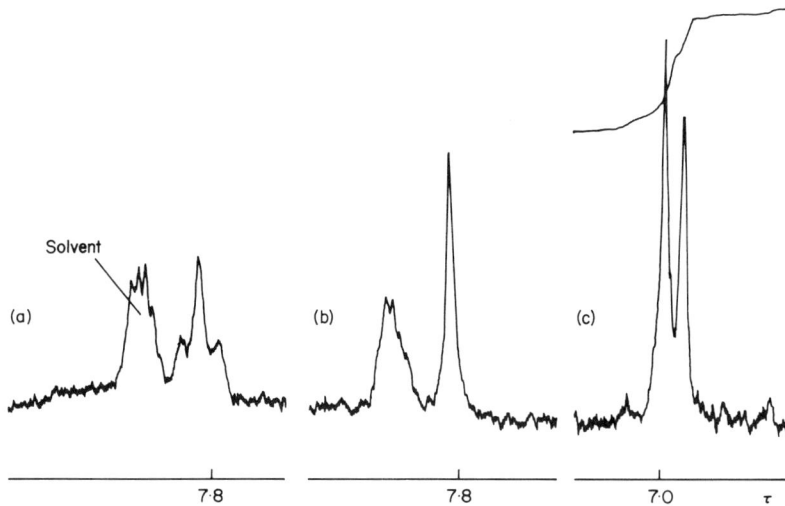

Fig. 2.9. 60-MHz PMR N-methyl signal of (a) the secondary amine (**5**) hydrochloride in DMSO-d$_6$; (b) the same compound in DMSO-d$_6$-D$_2$O; (c) the tertiary amine hydrochloride (**4**) in CDCl$_3$ (see Scheme I for formulae).

which the NHMe group is adjacent to a carbon atom double-bonded to a heteroatom (see formulae below) (Freifelder *et al.*, 1965). Doublets are

also seen in N-methylsulphonamides where the NH proton has acidic character. In neutral solvents N-methylanilines do not show N—Me doublets in spite of their weakly basic nature and nor does the even weaker base, *p*-nitro-N-methylaniline. Doublets (*J* 5·7–5·3 Hz) are seen, however, in di- and trinitro-N-methylanilines (Hedberg *et al.*, 1964).

The methylene signal of N-substituted benzylamines (**6**) also forms a doublet when R is a strongly electron-withdrawing group (e.g. *p*-BrC$_6$H$_4$

SO_2, CF_3CO, p-$NO_2C_6H_4$) but no good correlation between electro-negativity of the N-substituent and the J_{HC-NH} value exists (Rae, 1966) even though coupling constants are, in general, dependent upon electronic influences.

The dihedral angle dependence of H—N—C—H spin–spin coupling is revealed from some results upon dihydrouracils known to have favoured half-chair conformations (Rouillier *et al.*, 1966; Katritzky *et al.*, 1969). In 5-bromodihydrouracil **7** and related compounds, coupling between NH and the 6-equatorial proton proved to be four to five times greater than that involving the axial proton. The mean NH—6H dihedral angles

$J_{1/6a}$ 1 Hz

$J_{1/6e}$ 4·2 Hz

7

for equatorial and axial protons will be zero and 120° respectively, as a result of rapid inversion of the nitrogen atom. Hence a relationship between J and ϕ akin to that established by Karplus for J_{HC-CH} vicinal coupling (see p. 89) where a zero dihedral angle is associated with strong and a 120° angle with weak coupling, appears to be in operation. Another example is that of 1-benzyl-2-methylpiperidine (Fig. 2.8). The higher field doublet of the HNCH$_2$ signal, associated with the axial proton H_a (see **8**) shows stronger

8

coupling with the NH proton (ϕ 180°, J 7 Hz) than the doublet due to the equatorial proton H_e (ϕ 60°, J ~4 Hz) (Casy and Hassan, unpublished results).

AMINE SALT-BASE RATIOS AND RELATED PROBLEMS

The NMR spectrum of a mixture of two species undergoing rapid chemical exchange depends only upon the NMR parameters of the species and upon their relative amounts (Pople *et al.*, 1959). An incompletely neutralized solution of an N-methyl base represents such a system, namely

$$\underset{\overset{|}{\mathrm{H}}}{\overset{+}{>}\mathrm{N}}{-}\mathrm{Me} + \mathrm{H_2O} \;\rightleftharpoons\; >\mathrm{N}{-}\mathrm{Me} + \overset{+}{\mathrm{H}_3}\mathrm{O}$$

with the N—Me resonance having a chemical shift value δ_m intermediate between those of the two species expressed as follows:

$$\begin{aligned} \delta_m &= X_s\delta_s + (1 - X_s)\,\delta_b \\ &= \delta_b + (\delta_s - \delta_b)\,X_s \end{aligned} \tag{1}$$

where δ_b and δ_s are the chemical shifts of pure base and protonated base respectively and X_s is the mole fraction of the protonated base ($\equiv$ salt) present. Since the N—Me base/protonated base chemical shift difference may amount to about 1 ppm (i.e. 50–60 Hz at 60 MHz) a method is available for the direct determination of amine salt/base ratios provided (i) chemical shift values of the N—Me base and fully protonated base are available, and (ii) strict linear dependence of the N—Me chemical shift upon the salt/base ratio obtains.

The last requirement has been established for several bases of the simple aliphatic type (pK_a 8–10), early examples being the 1°, 2° and 3° methylamines (Grunwald *et al.*, 1957). Grunwald and co-workers made solutions of known salt/base ratios in water and measured N—Me chemical shifts relative to the Me signal of tetramethylammonium (TMA) bromide as standard. Plots of chemical shift against the protonated amine fraction were straight lines, whilst those against pH formed the S-shapes typical of titration curves (Fig. 2.10). These principles provide a means of measuring the pK_a of an organic base which is an alternative to the titration method. In the conventional procedure salt/base solutions of known composition are accurately prepared, their pH measured and the data used to calculate the dissociation constant of the protonated base from the equation (2) (Albert and Sergeant, 1962). The NMR method requires the amine or

$$pK_a = pH + \log\frac{(\text{salt})}{(\text{base})} \tag{2}$$

amine salt (accurate weights are not required) to be dissolved in a series of buffer solutions of known pH and the N—Me (or characteristic N—R) chemical shifts to be measured relative to a standard signal that is independent of pH (e.g. TMA or DSS). Buffers should be included which ensure the amine to be (i) completely free and (ii) completely protonated in order that the linearity curve may be constructed.

Practical obstacles may, of course, make the procedure difficult to apply in practice (e.g. many organic bases are sparingly soluble in the ideal solvent, water, while many alternative organic solvents will give NMR signals close to the N—R range) and no pK_a determinations following this procedure appear to have been made. There are several examples, however,

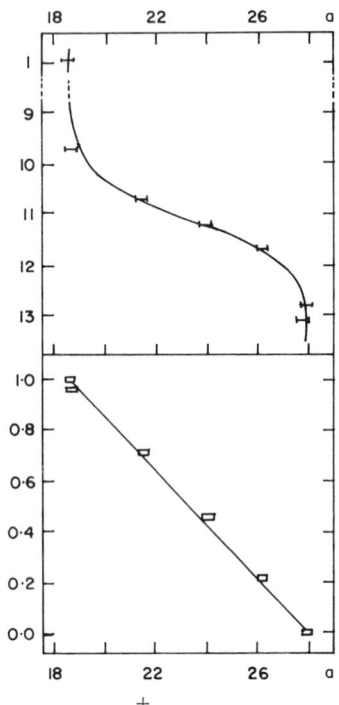

FIG. 2.10. Titration curves for MeNH$_3^+$. The chemical shift a is defined as the separation of the PMR lines resulting from the methyl protons in the MeNH$_2$— MeNH$_3^+$ equilibrium mixture and from the protons in the Me$_4$NBr standard, expressed in Hz (magnetic field 7430 oe). In the upper figure, a is shown as a function of the pH, and in the lower as a function of the concentration ratio acid to base plus acid (Grunwald *et al.*, 1957).

in which the apparent pK_a value is derived directly from the sigmoid curve obtained on plotting the chemical shift of a group near a basic centre against acid concentration. In the titration of a base $(B + H \rightleftharpoons BH^+)$ the mid-point of the curve, where $[BH^+] = [B]$, gives the pK_a value and the accuracy of the value will depend on the confidence with which the mid-point can be identified. The method has been applied in particular to very weak bases such as amides (Liler, 1969), phosphine oxides and sulphoxides

(Haake and Cook, 1968); the H_0 scale of Hammett is employed, chemical shifts being measured relative to $Me_3\overset{+}{N}H$ in mixtures of sulphuric acid and water. A typical plot is shown in Fig. 2.11 for dimethylphenylphosphine oxide (pK_a −2·2) and dimethyldithiophosphinic acid (pK_a −1·1 and −5·6). The validity of the method is supported by the fact that pK_a values for acetophenone and dimethylacetamide obtained in this way agree well with data obtained by other methods. This procedure has also been used in the biochemical field, e.g. measurement of the pK_a values of the four histidine residues of ribonuclease (see Chapter 8).

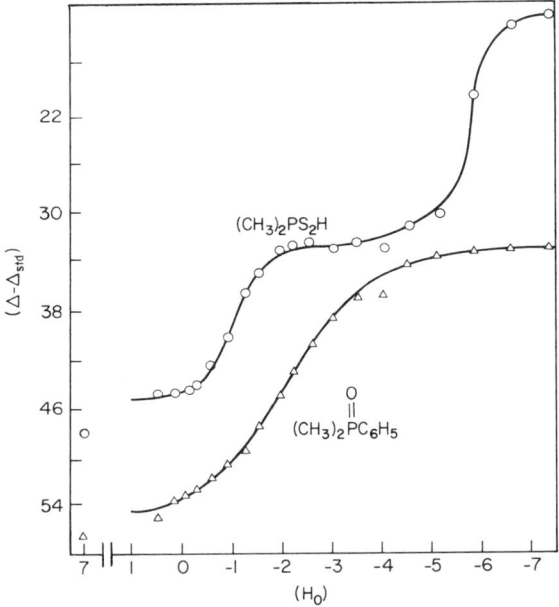

Fig. 2.11. Dependence of chemical shift of methyl groups of dimethylphenyl-phosphine oxide and dimethyldithiophosphinic acid on H_0 in aqueous sulphuric acid (spectra recorded at 60-MHz with TMS as standard) (Haake *et al.*, 1967).

Application of the same principles to the determination of salt/base ratios have also been reported. This type of mixture is encountered when an amine salt is partitioned between chloroform and an aqueous buffer and knowledge of the salt/base ratio is of importance in connection with the selection of partition systems for the separation of amines. Koch and Doyle (1967) developed an NMR assay procedure for pharmaceutical amines such as diphenylhydramine and desoxyephedrine along the lines already discussed by recording the N-methyl chemical shifts of various salt/base mixtures in chloroform with TMS as standard. Plots of δ_m versus X_s or X_b [see equation (1)], showed the required linearity. Hydrochloride salt/base ratios for diphenylhydramine measured by NMR compared closely with results obtained by a combined ultraviolet (giving total base and salt) and $AgNO_3$ titration (salt only) procedure, the former method being the more rapidly performed and requiring less material. Separate calibration curves were required for each individual salt of a particular amine because the N—Me chemical shift of the protonated base was dependent upon the anion, the signal moving downfield with increasing size of anion, e.g. 171·4 (HCl), 174·7 (HNO_3) and 178 ($HClO_4$) Hz from TMS (see also p. 60).

The related question of acid dissociation studies by NMR may be conveniently discussed in this section. Just as the chemical shifts of proton groups near protonated nitrogen suffer up-field shifts after proton loss (and, conversely, down-field shifts for the change, base to protonated base) so do the resonance positions of groups near an acid function alter when the acid releases a proton. In both cases, proton loss *increases* shielding of the adjacent proton group but the effect is usually much smaller for acids, especially carboxylic acids, as illustrated by some shielding constant data calculated for methylene groups (Table 2.2) (Sudmeier and Reilley, 1964).

TABLE 2.2

Methylene shielding constants (ppm)

CH_2CO_2H	1·1	} Diff.
$CH_2CO_2^-$	0·9	} 0·2 (12 Hz at 60 MHz)
$CH_2—\overset{+}{N}R_2H$	1·9	} Diff.
$CH_2—NR_2$	1·15	} 0·75 (45 Hz at 60 MHz)

The shielding influence of an ionized acid function is less effective in water (or D_2O) than in DMSO, the anions being strongly solvated (dispersing the negative charge) in the former but essentially free in the latter solvent. Thus, the α-methylene signal of the sodium salt of butylsulphonic acid is higher field (146 Hz, 60 MHz) in DMSO than in D_2O (172 Hz) (Biellman and Callot, 1967).

Differences in shielding between undissociated and dissociated species are pronounced in substituted anilines which behave as weak acids in a liquid ammonia-sodamide system. In the anion **9** the electronic influence of the negative charge on nitrogen is effectively felt by the aromatic protons through a resonance mechanism. Birchall and Jolly (1966) have utilized

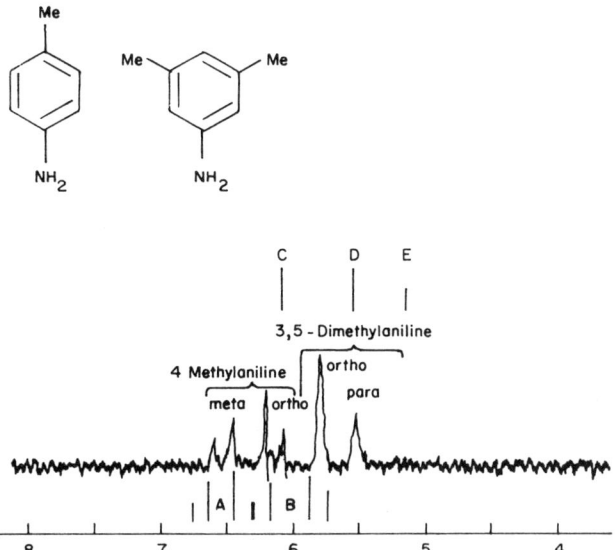

chemical shift changes consequent upon ionization to determine the relative acidities of a series of substituted anilines in the manner illustrated below. The typical AB aromatic signal of 4-methylaniline (unionized) moves upfield when the acid is fully ionized and, in addition, the *ortho* and *meta* proton resonances move apart [in Fig. 2.12, (A) is the unionized and (B) the ionized AB signal]. This behaviour is more striking in the case of 3,5-dimethyl aniline; in the unionized acid the *o*- and *p*-resonances (both

Fig. 2.12. 60-MHz PMR spectrum of a solution of 4-methylaniline, 3,5-dimethylaniline, and their anions in liquid ammonia (Birchall and Jolly, 1966).

near singlets) are almost coincident whereas in the anion they are both upfield of the acid signals and well separated [in Fig. 2.12 (C) is the unionized and (DE) is the ionized signal]. The ratio acid/anion in a mixture is obtained by measuring chemical shifts of particular aromatic protons and reading off the ratio from a graph of chemical shift versus mole proportion of acid (or anion) constructed from chemical shift data upon the free and ionized acid (linearity being assumed as usual, see p. 67). As an alternative, the separation of *o/m* (4-methylaniline) or *o/p* protons (3,5-dimethylaniline) may be used—this procedure eliminates the need for an internal standard. To determine the relative acidities of two acids HA and HB a solution of 1 mole proportion each of HA, HB and sodamide in liquid ammonia is prepared and its PMR spectrum recorded; Fig. 2.12 is a typical example. The equilibrium

$$HA + B^- \rightleftharpoons HB + A^-$$

will be set up, the concentration of B^- being greater than A^- if HB is the stronger acid, and vice versa. Knowing the ratios HA/A^- and HB/B^- (from the PMR data as explained) the pK_a difference between the two acids (ΔpK_a) is calculated from the relation

$$\Delta pK_a = \log HB/B^- - \log HA/A^-$$

Amongst the acids so ranked, those containing electron-withdrawing substituents (e.g. Cl, CN, CF_3) were stronger than aniline while acids with alkyl substituents were weaker; there was a more than 6 pK_a unit difference between the strongest (2,5-dichloro) and weakest (4-methyl-aniline) derivatives examined.

A PMR method has also been proposed for assessing the Lewis acidity of covalent metal halides (Satchell and Satchell, 1969). The two broad NH signals of m-methoxybenzamide (10) which arise as a result of restricted

	pK	δ_{cis}	δ_{trans}
$SbCl_3$	−1·76	8·01	~7·4
$ZnBr_2$	−2·94	8·6	8·00
$GaBr_3$	−3·4	9·6	8·64
BF_3	−2·93	9·3	8·15

10

rotation about the N—C bond move down field in the presence of a covalent metal halide and the extent of the displacement is generally more negative the stronger the Lewis acid (data for boron trifluoride is anomalous). The spectrum of benzamide itself fails to display distinct NH signals even when the amide is examined in FSO_3H at −85° (Birchall and Gillespie, 1963).

SITE OF PROTONATION

The problem of identifying the site of protonation in a molecule containing more than one basic centre may frequently be solved by studying the effect of proton uptake on chemical shifts and spin–spin coupling patterns of protons in the vicinity of the various possible sites. Some examples follow.

The protonation site of 2- and 4-pyridones, for long a controversial question, has now been established as oxygen rather than nitrogen by PMR studies (Katritzky and Reavill, 1963, 1965). Thus, the C-6 proton signal of 4-methyl-2-pyridone **11** in sulphuric acid is a triplet J 6·7 Hz

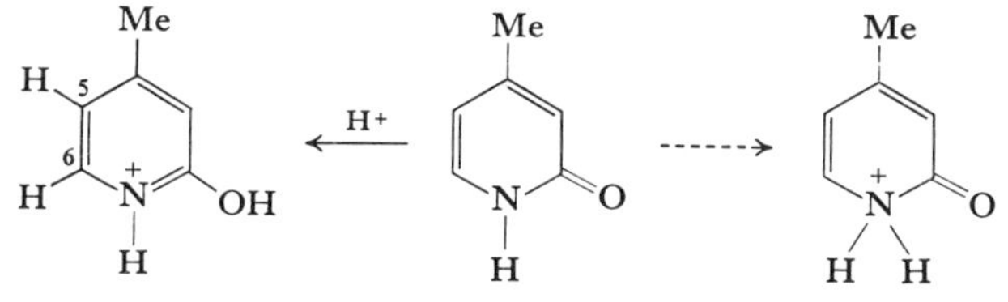

11

(doublet J 6·6 in D_2SO_4) consistent with its being equally coupled to the C-5 proton and to *one* proton on nitrogen. If O-protonation had occurred (see formulae) this signal would have been a quartet. In 5-methyl-2-pyridone

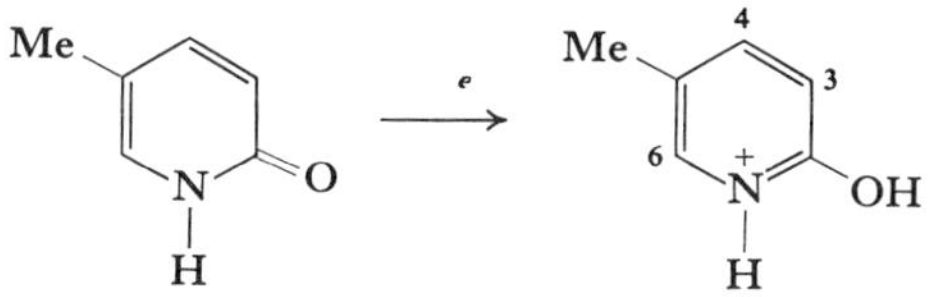

12

(**12**) the C-6 signal forms a doublet of doublets (J 6·9 and 2·2 Hz) showing vicinal coupling to NH and meta to C-4, while *meta* coupling between NH and the C-3 proton can also be distinguished. Similar results and arguments establish S-protonation in thiopyridones, and ring nitrogen

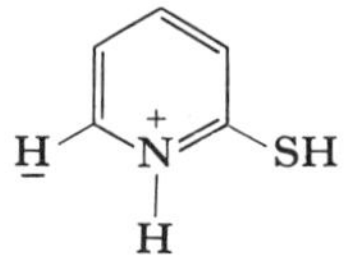

6-*H*: triplet J 5·5 Hz in H_2SO_4, doublet J 5·5 Hz in D_2SO_4

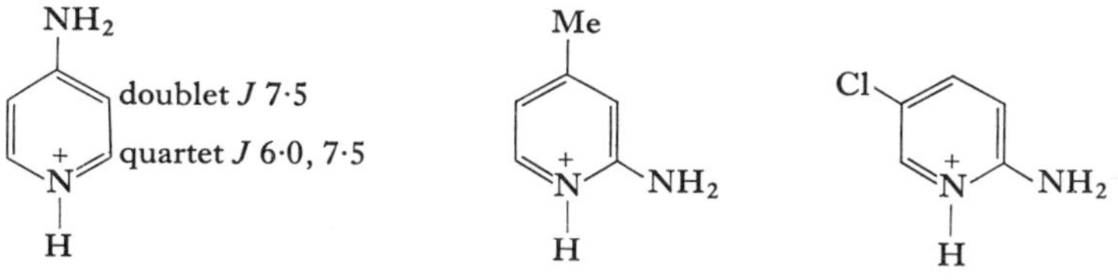

protonation in 2- and 4-aminopyridines. Of the three derivatives below examined as hydrochlorides in SO_2, all showed one proton $\overset{+}{N}H$ peaks low field of the NH_2 signal while the two stronger bases exhibited splitting of the α-CH signal by $\overset{+}{N}H$.

PMR evidence for the O-protonation of open chain amides in addition to that of cyclic amides such as the pyridones is also available (Gillespie and Birchall, 1963). The spectrum of acetamide (**13**) in fluorosulphuric

$$\underset{Me}{\overset{HO}{\diagdown}}C\overset{+}{=}\overset{}{N}\underset{\diagdown H}{\diagup H}$$

13

acid (FSO_3H) at $-92°$ displays three peaks of areas $1:2:3$ assigned to OH, NH_2 and CH_3 protons respectively (at higher temperatures the OH peak is not resolvable because of its rapid exchange rate). N-protonation would have given a 2 peak (NH_3 and CH_3) spectrum. The spectrum of dimethyl-formamide (**14**) at $-80°$ in FSO_3H shows a signal attributed to the OH

$$\underset{H}{\overset{HO}{\diagdown}}C\overset{+}{=}\overset{}{N}\underset{\diagdown Me}{\diagup Me}$$

14

proton plus two N-methyl signals showing that N—Me non-equivalence (a result of restricted rotation about the C—N bond) is maintained in the protonated amide; N-protonation would render them equivalent because rotation about the C—N bond would then be free. Indeed, N—Me-formyl proton coupling constants (J_{trans} 1·1 Hz, J_{cis} 0·7 Hz) are almost twice as large as those found in the pure liquid and are reasonably associated with a greater double bond character of the central C—N bond in the protonated form.

The problem of N versus O protonation in sulphonamides has also been studied by PMR spectroscopy. Evidence for the N-protonation of the sulphonamide (**15**) (Birchall and Gillespie, 1963) (in FSO_3H the NMe

$$p\text{-MeC}_6\text{H}_4\text{SO}_2\text{NMe}_2$$

15

signal is a doublet) is ambiguous because O-protonation could render the N—Me groups non-equivalent. Clear evidence of N-protonation is available, however, from results upon the cyclic sulphonamide (Fig. 2.13) in FSO_3H through the observation of $\overset{+}{N}H$ coupling effects upon the N—Me and adjacent methylene protons (the latter become non-equivalent when nitrogen inversion is prevented through protonation) as the temperature is lowered (see Fig. 2.13). At $0°$ the methylene signal is an octet showing geminal (*J* 15 Hz) and vicinal (*J* 5·5 Hz) coupling while the N—Me signal forms a doublet, in further proof of the presence of an NH proton. It is interesting that although NH spin–spin coupling effects are not seen at 20°, the methylene protons remain non-equivalent and this has been taken to mean that proton exchange is occurring with retention of configuration

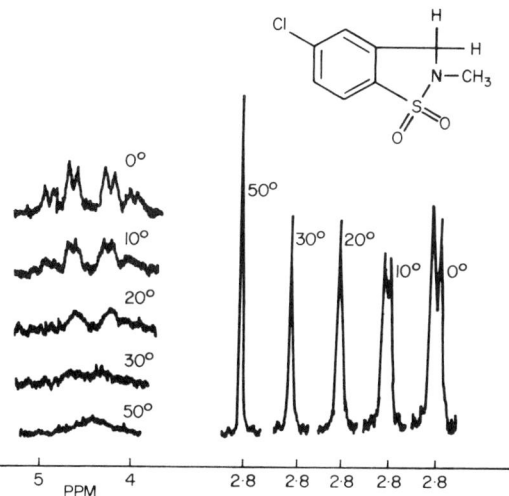

Fig. 2.13. Part of the 60-MHz PMR spectrum of N-methyl-5-chloro-1,2-benzisothiazoline 1,1-dioxide in fluorosulphonic acid at several temperatures. The spectral amplitude setting is not the same in all the spectra (Menger and Mandell, 1967). Methylene signals are to the left, and N-methyl signals to the right of the spectrum.

(Menger and Mandell, 1967). The spectrum of dibenzylmethylamine hydrochloride in water provides a similar example (Saunders and Yamada 1963).

Next, follow some examples where protonation may occur at carbon rather than nitrogen.

None of the spectra of a series of pyrrolo-(2,1-*b*)-thiazoles (**16**), as their

perchlorates in TFA, shows a broad band or triplet characteristic of an $\overset{+}{\text{NH}}$ function, hence proton uptake must occur at a carbon atom (sulphur is excluded as will be seen) (Molloy *et al.*, 1965). The thiazole ring cannot be the site because all the derivatives show a low field doublet of doublets which can only arise from the AB proton system (H_1 and H_2) present in this ring (see Fig. 2.14). All the salts show a 2 proton singlet near δ 5·4 assigned to a methylene group. In the case of the 6,7-dimethyl derivative

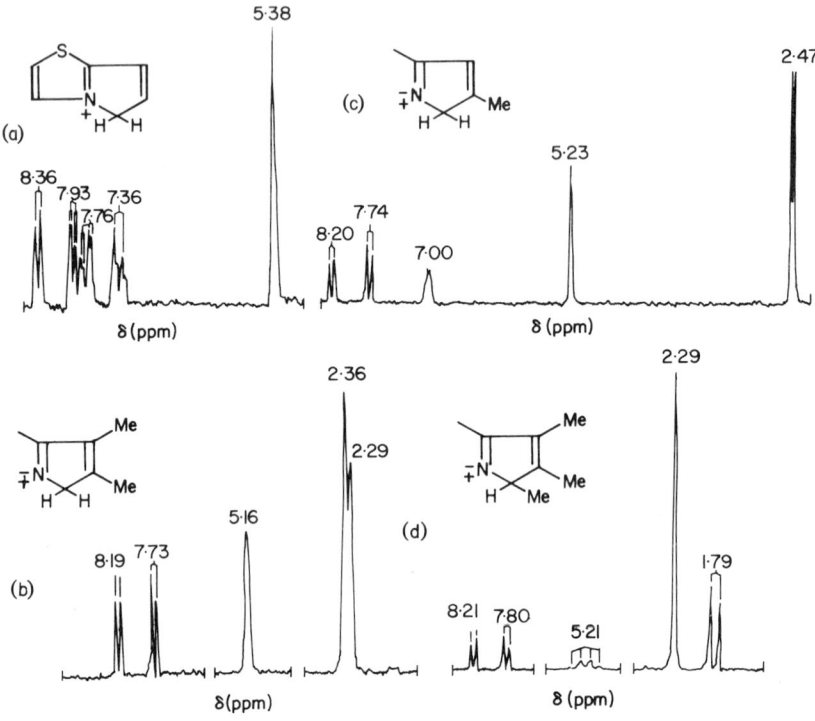

FIG. 2.14. 60-MHz PMR spectra of some pyrrolo[2,1-*b*]thiazole perchlorates in CF_3CO_2H (Molloy *et al.*, 1965).

(17) [also Fig. 2.14(b)], a methylene group can only develop at C-5 which must therefore be the protonation site. Also, since methylene signals in all the salts fall within a narrow range (δ 5·16–5·38) it is concluded that C-5 protonation occurs in all the salts. For example, C-7 is an unlikely site since its methylene signal would be expected to lie considerably upfield of that from a 5-methylene group as the latter is directly linked to positively charged nitrogen (i.e. to a powerful deshielding influence). In confirmation, the methylene signal of the 5,6-dimethyl derivative (18), which must be located at C-7, has the up-field position of δ 4·15. No methylene group may form in the case of the 5,6,7-trimethyl derivative (19) and here the

18

19

C-5 proton is preferred since its perchlorate shows a quartet (due to H—Me coupling) that falls in the same region as the methylene signals of the salts **(16)** [see Fig. 2.14(d)].

Conjugate acids of *enamines* **(20)** (related to the pyrrolothiazoles in that both classes are α/β unsaturated amines) are also conveniently studied by PMR spectroscopy. It has long been assumed that protonated enamines are C—H **(21)** rather than N—H **(22)** species and there is good evidence of

20   **21**   **22**

this being true for many enamines (mostly cyclic) in the solid state (Bláha and Červinka, 1966). However, the presence of both C and N protonated forms of some acyclic enamine derivatives has been revealed by PMR results (Elguero *et al.*, 1965). Thus, the PMR spectrum of a freshly prepared solution of the morpholino enamine of isobutyraldehyde in 6N—HCl shows (i) two C-methyl doublets (*J* about 2 Hz) at 78 and 84 Hz from the standard signal ($\overset{+}{\text{N}}\text{Me}_4\overset{-}{\text{Cl}}$, 60 MHz frequency) characteristic of

N-protonated   C-protonated

the N-protonated form, and (ii) a single doublet (*J* 7 Hz) at 114 Hz. The higher field position of the latter signal and its typical vicinal H/H *J* value shows it to arise from the C-protonated form. The relative intensities of signals (i) and (ii) showed the N-conjugate acid to preponderate. Hydrolysis of the enamine proceeded fairly rapidly as shown by the development of another secondary methyl doublet at 130 Hz (*J* 7 Hz), due to isobutyralde-hyde and/or its hydrate, and was almost complete after 2 hours. Hydrolysis was even faster at lower acid concentrations and the C-protonated form could not be detected in acids weaker than 6N.

In the next example a stable C-conjugate acid of an enamine is described which proves to be the exclusive protonated species. Treatment of the 2-pyridone (**23**) with ethyl lithium gives a product, formulated as the

enamine (**24**), which has been shown to be identical with a major metabolite of the narcotic analgesic, methadone. While the structure of the base is in doubt,† the corresponding hydriodide salt is clearly identified as the C-protonated form (**25**) from the fact that the PMR signal of the terminal methyl group of the 2-substituent (salt in $CDCl_3$) is a triplet, *J* 7·5 Hz,

typical of the system $CH_2\underline{Me}$, rather than the doublet anticipated for an ethylidene side chain. When the spectrum was recorded in $CDCl_3$—D_2O the 6-methylene signal disappeared and the 7-Me triplet became a singlet (Fig. 2.15) (Beckett *et al.*, 1968). If the methylene carbon is the site of protonation, the two protons attached to it will both be acidic and capable of exchange. Hence, their signal should disappear on deuteration and their coupling action with $CH_2\underline{Me}$ will be disrupted, as observed.

The examples so far discussed involve comparisons between predicted and observed multiplicities of PMR signals. Differential deshielding of proton groups following the protonation of a base may also provide useful information. Assignment of the three N-methyl singlets of the spectrum of

† Recent PMR studies show it to be a mixture of *cis* and *trans* **24** (Hassan and Casy, 1970).

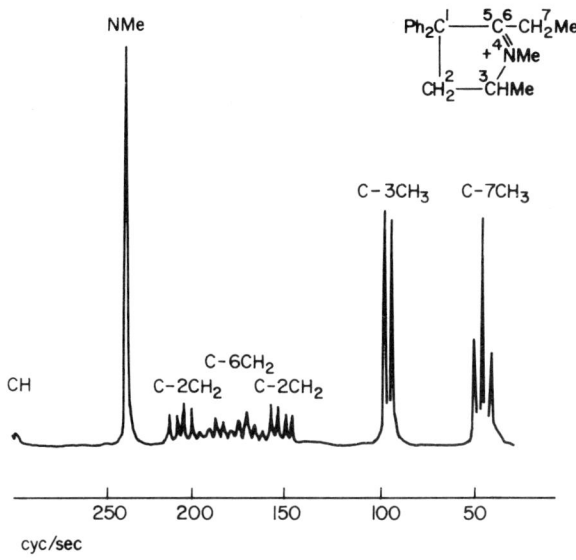

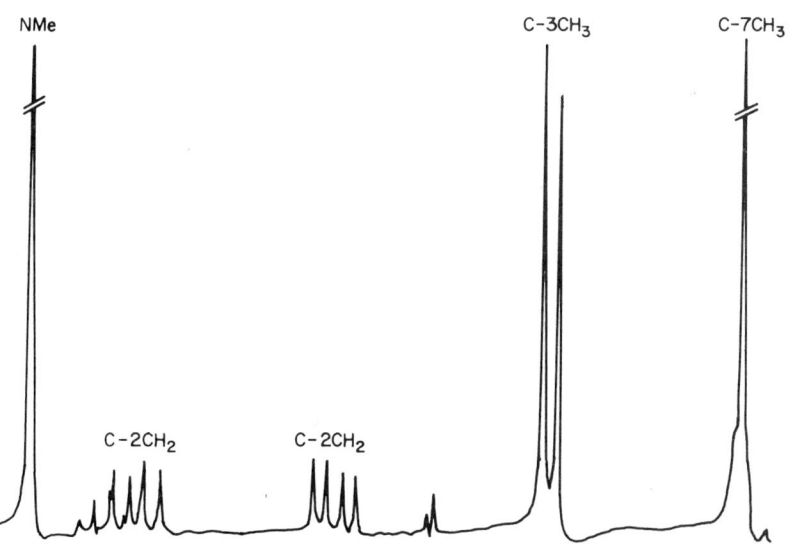

FIG. 2.15. Part of the 100-MHz PMR spectrum of the pyrroline hydriodide (**25**) in CDCl$_3$ (upper) and CDCl$_3$—D$_2$O (lower figure).

caffeine (**26**), for example, has been made by spectral comparisons in CDCl$_3$ and protonating solvents (Ottinger *et al.*, 1965). All singlets move downfield after protonation, the lowest field (C) being the most and the highest (A) the least affected. From ultraviolet studies, the initial protonation site of caffeine is known to be the unsubstituted nitrogen (N-9) of the imidazole ring (Bergmann and Dikstein, 1955); hence the N-7 methyl

	A (high)	B (mid)	C (low)
$\delta_{CDCl_3} - \delta_{TFA}$	−10·5	−12·5	−20
$\delta_{CDCl_3} - \delta_{C_6H_6}$	+4	+12	+51·5

(data in Hz at 60 MHz)

group should be the most affected (**27**) and is assigned to signal C. The least influenced signal (A) is attributed to the N-1 methyl since this group is the further removed from the imidazole ring. The magnitudes of upfield shifts induced by benzene (ASIS, see p. 267) are of the same order of ranking as the protonation effects (see data above) confirming the assignments; the imidazole rather than the pyrimidyl moiety of caffeine is more likely to be the electron acceptor in the caffeine–benzene collision complex, hence the N-7 methyl protons will be the most affected by the ring-current shielding influence of the solvent molecules.

Another example concerns the orientation of mononitrobenzoquinuclidine (Duke *et al.*, 1970). The product is known to possess a 1,2,4-trisubstituted benzene ring because its aromatic PMR signal is composed of an *o*-doublet, a *m*-doublet, and an *o–m* doublet of doublets (see p. 41). The signal of the proton at position 8 is identified by its greater downfield shift on protonation of nitrogen and since it displays no *ortho* coupling the 7-position must be substituted.

$$\Delta_5 + 0\cdot38, \qquad \Delta_6 + 0\cdot33, \qquad \Delta_8 + 0\cdot51,$$
$$(\Delta = \delta_{CDCl_3} - \delta_{CDCl_3 + TFA})$$

A more sophisticated approach, based on the same principles, has been employed to identify the preferred ionization centres of citric acid (Loewenstein and Roberts, 1960), and this example is described here in some detail since it illustrates how quantitative results may be derived from NMR data in such problems.

The methylene PMR signal of citric acid forms two superimposed AB systems due to the non-equivalence of the two hydrogens in each methylene group. Only the central doublet is well resolved and the distance between its centre and the signal of tetramethylammonium bromide (TMA) is taken as a measure of the methylene chemical shift. This value (δ) increases as the pH rises (Fig. 2.16) because the methylene groups become more shielded as the CO_2H groups ionize.

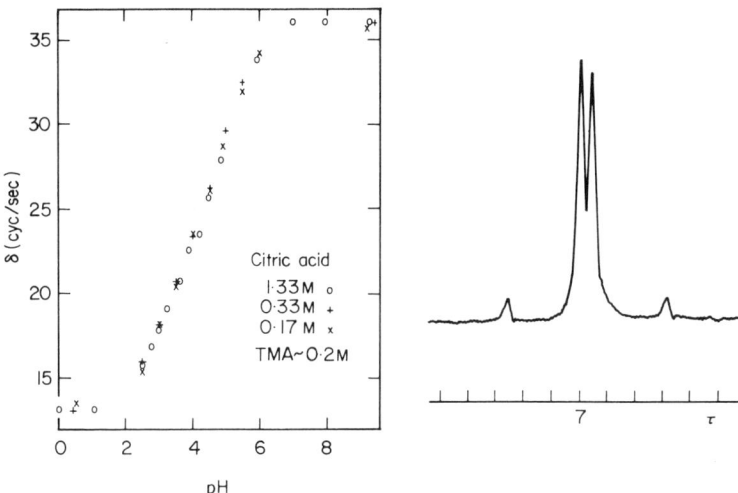

FIG. 2.16. The chemical shift between the centre of the strong methylene doublet in citric acid (see spectrum to the right of the figure) and the TMA resonance as a function of the pH in water (Loewenstein and Roberts, 1960).

It is first necessary to know the chemical shift associated with each of the three ionization processes of citric acid. At a given pH the $\underline{CH_2}$ chemical shift is given by the expression (2):

$$\delta = x_1\,\delta_1 + x_2\,\delta_2 + x_3\,\delta_3 + x_4\,\delta_4 \tag{2}$$

where x_1, x_2, x_3 and x_4 are the mole fractions of non-, mono-, di- and tri-ionized citrates respectively, and δ_1, δ_2, δ_3 and δ_4 their chemical shifts relative to TMA. δ_1 and δ_4 can be taken directly from the plot of δ/pH, since they are values of δ at very low and very high pH respectively. Values of x_1, etc. for any particular pH may be calculated from the pK_a values of citric acid ($pK_{a_1} = 3\cdot13$, $pK_{a_2} = 4\cdot76$, $pK_{a_3} = 6\cdot4$). Equation (2) may then be set up numerically for a series of points on the titration curve (pH range 3–5·5) and pairs of equations used to solve for the two unknowns δ_2 and δ_3. The value 22·5 (δ_2) and 33·5 Hz (δ_3) are obtained in this manner.

Chemical shifts associated with each ionization stage are then calculated:

Single ionization: $\delta_2 - \delta_1 = 22\cdot5 - 13\cdot3 = 9\cdot2$ Hz

Double ionization: $\delta_3 - \delta_1 = 33\cdot5 - 13\cdot3 = 20\cdot2$ Hz

Triple ionization: $\delta_4 - \delta_1 = 35\cdot9 - 13\cdot3 = 22\cdot6$ Hz

It is possible now to gain some idea of the relative concentrations of the ionized forms involved at the mono- and di-ionization stages.

The two possible *monoanion* species are as follows:

(a) $-CO_2H$ (b) $-C\bar{O}_2$

HO$-$$\bar{C}O_2$ $\rightleftharpoons$ HO$-$$CO_2H$

(a) $-CO_2H$ $-CO_2H$

mole fraction x $y(x + y = 1)$

The x species contains two methylene groups of identical environment (a), flanked by a non-ionized terminal CO_2H group. The chemical shift of this CH_2 group is obtained from data upon the di-ionized, unsymmetrical monomethyl ester (**28**) obtained by treating the symmetrical diester (**29**)

(a) $-CO_2Me$ $-CO_2Me$

HO$-$$\bar{C}O_2$ HO$-$$CO_2H$

         $-\bar{C}O_2$ $-CO_2Me$

(c)   **28**   **29**

with 2 moles of $NaOH-H_2O$. In this, the assumption is made that: (i) the magnetic influence of CO_2Me is essentially the same as CO_2H and (ii) the chemical shift of a CH_2 group is little affected by the terminal CO_2H (or $\bar{C}O_2$) located at the other end of the molecule.

The y species contains two types of CH_2 group. One is identical to the CH_2 of non-ionized citric acid and the other (b) has a similar environment to CH_2 in the fully ionized symmetrical monoester (**30**), for which data are available.

(b) $-\bar{C}O_2$

HO$-$$CO_2Me$

(b) $-\bar{C}O_2$

         30

In *di-ionized* citric acid, two species are again involved. In v the CH_2 environments are identical and correspond with (b) in the mono-ion y. They are non-equivalent in u, one corresponding with (a) in the mono anion x and the other with (c) in the di-ionized unsymmetrical ester **28** (and also with δ_{CH_2} for tri-ionized citrate).

(c) $-\bar{C}O_2$ (b) $-\bar{C}O_2$

HO$-$$\bar{C}O_2$ $\rightleftharpoons$ HO$-$$CO_2H$

(a) $-CO_2H$ (b) $-\bar{C}O_2$

Mole fraction u v $(u + v = 1)$

The so-derived chemical shift values, *relative to the methylene line of non-ionized acid taken as zero*, are as follows:

$$(a) = 7\cdot25 \text{ Hz}$$
$$(b) = 19\cdot3 \text{ Hz}$$
$$(c) = 22\cdot6 \text{ Hz}$$

As anticipated, the shielding influence of ionized CO_2H upon CH_2 is greater when the two groups are directly linked [value (b)] than when they are separated by a carbon atom [value (a)], while greatest shielding occurs when both CO_2H groups are ionized [value (c)].

The chemical shift associated with each ionization step has already been calculated. The value for mono-ionization ($\delta_2 - \delta_1 = 9\cdot2$ Hz) will be made up of contributions from species x and y as follows:

$$ax + \frac{by}{2} = \delta_2 - \delta_1 \tag{3}$$

Note that the contribution of (b) is only half that of (a), since species x contains two (a) groups while y contains only one (b) —CH_2 group.

Similarly equation (4) governs the chemical shift change associated with di-ionization.

$$bv + (a + c)u/2 = \delta_3 - \delta_1 \tag{4}$$

Substituting the values of (a), (b) and $\delta_2 - \delta_1$ in (3) and using the fact that $x + y = 1$, the equation is solved to give the result, $x = 0\cdot2$ and $y = 0\cdot8$. Thus, mono-ionized citrate ion is about 80% in the unsymmetrical form. Calculations using equation (4) (allowing for experimental error) show that the dianion must exist almost exclusively as the symmetrical form.

The protonation sites of chelating agents of the polyamine and amino-carboxylate type have also been established by PMR spectroscopy (Kula *et al.*, 1963; Sudmeier and Reilley, 1964). The PMR spectrum of ethylene-diamine tetraacetate (EDTA) tetraanion shows two sharp methylene signals, one due to the acetate groups (a) and the other to the ethylenic

$$\underset{\bar{O}_2CCH_2}{\overset{\bar{O}_2CCH_2}{\diagdown}} \overset{(b)}{NCH_2CH_2N} \underset{CH_2\bar{C}O_2}{\overset{\overset{(a)}{CH_2\bar{C}O_2}}{\diagup}}$$

methylenes (b); the area of the (b) signal is half that of (a). These signals move downfield when 2 moles of mineral acid are added to the EDTA tetraanion solution, the fact of both being affected supporting nitrogen as the protonation site because proton uptake at CO_2^- would only be expected to alter the position of peak (a), i.e. the peak due to protons immediately adjacent to the carboxylate function. A more elaborate treatment (Sudmeier and Reilley, 1964), involving the use of methylene screening constants calculated from model compounds, confirms these conclusions.

References

Albert, A. and Sergeant, E. P. (1962). "Ionization Constants of Acids and Bases". Methuen, London.

Anderson, W. R. and Silverstein, R. M. (1965). *Anal. Chem.* **37**, 1417.

Beckett, A. H., Taylor, J. F., Casy, A. F. and Hassan, M. M. A. (1968). *J. Pharm. Pharmacol.* **20**, 754.

Bergmann, F. and Dikstein, S. (1955). *J. Amer. Chem. Soc.* **77**, 691.

Biellman, J. F. and Callot, H. (1967). *Chem. Commun.* 458.

Birchall, T. and Gillespie, R. J. (1963). *Can. J. Chem.* **41**, 2642.

Birchall, T. and Jolly, W. L. (1966). *J. Amer. Chem. Soc.* **88**, 5439.

Bláha, K. and Červinka, O. (1966). *In* "Advances in Heterocyclic Chemistry" (A. R. Katritzky and A. J. Boulton, eds), Vol. 6, p. 147.

Bovey, F. A. (1967). "NMR Data Tables for Organic Compounds," Vol. 1. Interscience, New York.

Casy, A. F. and Hassan, M. M. A. (1969). *J. Med. Chem.* **12**, 337.

Duke, R. P., Jones, R. A. Y., Katritzky, A. R., Mikhlina, E. E., Yanina, A. D., Alekseeva, L. M., Turchin, K. F., Sheinker, Yu. N. and Yakhontov, L. N. (1970). *Tetrahedron Lett.* 1809.

Elguero, J., Jacquier, R. and Tarrugo, G. (1965). *Tetrahedron Lett.* 4719.

Fraenkel, G. (1963). *J. Chem. Phys.* **39**, 1614.

Fraenkel, G. and Kim, J. P. (1966). *J. Amer. Chem. Soc.* **88**, 4203.

Freifelder, M., Mattoon, R. W. and Kriese, R. (1965). *J. Org. Chem.* **69**, 3645.

Gillespie, R. J. and Birchall, T. (1963). *Can. J. Chem.* **41**, 148.

Grunwald, E., Loewenstein, A. and Meiboom, S. (1957). *J. Chem. Phys.* **27**, 641.

Haake, P. and Cook, R. D. (1968). *Tetrahedron Lett.* 427.

Haake, P., Cook, R. D. and Hurst, G. H. (1967). *J. Amer. Chem. Soc.* **89**, 2650.

Hassan, M. M. A. and Casy, A. F. (1970). *Can. J. Chem.* **48**, 3742.

Hedberg, J., Weil, J. A., Jannsonis, G. A. and Anderson, J. K. (1964). *J. Chem. Phys.* **41**, 1033.

Katritzky, A. R. and Reavill, R. E. (1963). *J. Chem. Soc.* 753.

Katritzky, A. R. and Reavill, R. E. (1965). *J. Chem. Soc.* 3825.

Katritzky, A. R., Nesbit, M. R., Kurtev, B. J., Lyapova, M. and Pojarlieff, I. G. (1969). *Tetrahedron* **25**, 3807.

Koch, S. A. and Doyle, T. D. (1967). *Anal. Chem.* **39**, 1273.

Kula, R. J., Sawyer, D. T., Chan, S. I. and Finley, C. M. (1963). *J. Amer. Chem. Soc.* **85**, 2930.

Liler, M. (1969). *J. Chem. Soc.* (B), 385.

Loewenstein, A. and Roberts, J. D. (1960). *J. Amer. Chem. Soc.* **82**, 2705.

Ma, J. C. N. and Warnhoff, E. W. (1965). *Can. J. Chem.* **43**, 1849.

Menger, F. M. and Mandell, L. (1967). *J. Amer. Chem. Soc.* **89**, 4424.

Molloy, B. B., Reid, D. H. and McKenzie, S. (1965). *J. Chem. Soc.* 4368.

Ogg, R. A. (1954). *Discuss. Faraday Soc.* **17**, 215.

Ottinger, R., Boulvin, G., Reisse, J. and Chiurdoglu, G. (1965). *Tetrahedron* **21**, 3435.

Pople, J. A., Schneider, W. G. and Bernstein, H. J. (1959). "High Resolution Nuclear Magnetic Resonance". McGraw-Hill, New York.

Rae, I. D. (1966). *Aust. J. Chem.* **19**, 1983.

Reynolds, W. F. and Priller, U. R. (1968). *Can. J. Chem.* **46**, 2787.

Roberts, J. D. (1956). *J. Amer. Chem. Soc.* **78**, 4495.

Roberts, J. D. (1959). "Nuclear Magnetic Resonance. Applications to Organic Chemistry". McGraw-Hill, New York.

Rouillier, P., Delman, J., Duplan, J. and Nofre, C. (1966). *Tetrahedron Lett.* 4189.

Satchell, D. P. N. and Satchell, R. S. (1969). *Chem. Commun.* 110.

Saunders, M. and Yamada, F. (1963). *J. Amer. Chem. Soc.* **85**, 1882.

Slomp, G. and Lindberg, J. G. (1967). *Anal. Chem.* **39**, 60.

Sudmeier, J. L. and Reilley, C. N. (1964). *Anal. Chem.* **36**, 1699

Terui, Y., Aono, K. and Tori, K. (1968). *J. Amer. Chem. Soc.* **90**, 1069.

Thompson, W. E., Warren, R. J., Zarembo, J. E. and Eisdorfer, I. B. (1966). *J. Pharm. Sci.* **55**, 110.

Varian Associates. (1962, 1963). NMR Spectra Catalogs Vols 1 and 2.

White, R. F. M. (1963). *In* "Physical Methods in Heterocyclic Chemistry" (A. R. Katritzky, ed.), Vol. II, p. 103. Academic Press, New York and London.

CHAPTER 3

The Application of PMR Spectroscopy
to Stereochemical Problems

PMR spectroscopic data is capable of providing valuable information upon both the relative dispositions of atoms or groups of atoms about the dissymmetric or rigid part of a molecule (i.e. *configuration*) and upon the favoured *conformation* of the molecule. The last term defines the specific geometry of a molecule in terms of bond distances and bond and dihedral angles, and provides a picture of the actual disposition in space of the atoms that make up the molecule. For a given configuration, an infinite number of conformations commonly result from rotation about single bonds within the molecule (Eliel, 1962; Eliel *et al.* 1965) and it is often difficult to divorce conformation from configuration when molecular geometry is studied by the NMR technique. In non-interconverting isomers, e.g. *cis–trans* alkenes, configuration is the chief problem for solution (e.g. triprolidine and its isomer, p. 223), although information upon the conformation of the molecule as a whole may be an additional aim of a PMR study. The characterization of geometrical isomers in cyclic molecules by PMR, however, invariably proceeds via identification of favoured conformers, as does also that of diastereoisomeric acyclic molecules. The study of conformation in its own right is an important application of NMR spectroscopy and progress in conformational analysis over the last decade has been closely linked to advances in magnetic resonance techniques (Eliel, 1965; Franklin and Feltkamp, 1965). The conformation of molecules in solution is of particular interest to biochemists and medicinal chemists, and it may be stated with some confidence that NMR spectroscopy provides more information about the shape of solute molecules than any other physical method (Thomas, 1968). It is important to stress that conformational interest is not restricted to dissymmetric molecules, the identity of favoured conformers and the position of conformational equilibrium in formally symmetrical molecules such as acetylcholine and histamine being of particular importance in the biological field.

Differentiation of enantiomorphs and absolute configurational assignments by PMR spectroscopy are discussed separately in Chapter 5.

Before illustrating some of the many applications of PMR spectroscopy to stereochemical problems, an account will be given of the dependence of the extent of spin–spin coupling between two protons upon their stereochemical disposition, this relationship being crucial in the use of PMR techniques in stereochemistry (Sternhell, 1969).

In 1958 Lemieux *et al.* drew attention to the fact that vicinal coupling between two axial protons was greater than between those bearing an *a/e* or *e/e* relationship. This was first evident from differences in the half-widths of resonance bands for the 1-hydrogen atom of *cis* and *trans* 4-*t*-butylcyclohexanol (**1**) and (**2**). Better resolution was achieved in later examples (initially, spectra were recorded at 40 MHz) and it proved possible to measure the *J* values themselves; values of 5–8 Hz (J_{aa}) and 3 Hz (J_{ae} and

1 *cis* W_H 7 Hz   2 *trans* W_H 22 Hz

J_{ee}) were obtained from the anomeric hydrogen signal of a series of acetylated hexose sugars, e.g. β-D-xylose tetraacetate (**3**) and α-D-glucose pentaacetate (**4**).

3   *J* 6 Hz   4   *J* 3 Hz

The stereochemical relationships between vicinal protons is best denoted in terms of a dihedral angle (ϕ), (sometimes called the angle of torsion). This is the angle between two planes, one defined by the C—C and C—H_A bonds and the other by the C—C and C—H_B bonds (Fig. 3.1) and is conveniently depicted by means of a Newman projection.

Shortly after Lemieux's reports, Karplus (1959, 1960) proposed a relationship between the dihedral angle (ϕ) and the vicinal coupling constant expressed in the form:

$$J_1 = K_1 \cos^2 \phi - c \text{ for angles of } \phi \text{ between 0 and } 90°$$
$$J_2 = K_2 \cos^2 \phi - c \text{ for angles of } \phi \text{ between 90 and } 180°$$

where K_1 and K_2 are constants depending on the nature of the C—C fragment and c is a constant which is generally very small or zero.

A plot of ϕ against J (with K_1 and K_2 assigned the values 8·5 and 9·5 respectively) is shown in Fig. 3.2. Modified versions of the original J/ϕ relationship have been reported, one due to Karplus himself (1963) and one to Williamson and Johnson (1961). The last authors used data from steroids and their equation gives a plot similar to that of Fig. 3.2 but with a higher $J_{180°}$ value (16 Hz). In these plots dihedral angles of 60 and 180° are especially important because they occur in staggered conformations

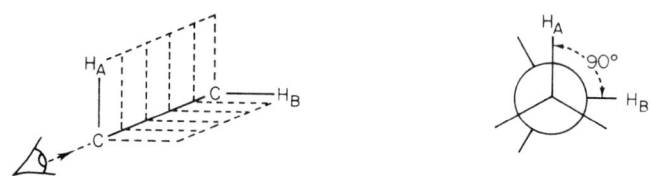

Fig. 3.1. Representation of a dihedral angle of 90° in a $H_A CCH_B$ fragment; one plane is defined by the C—C and C—H_A bonds and the other by the C—C and C—H_B bonds.

while in six-membered alicyclic compounds they are the angles obtaining in e/e and a/e ($\phi = 60°$) and a/a ($\phi = 180°$) coupling (shown alongside the plot of Fig. 3.2). The plot shows that J values for angles of ϕ near 180° are nearly four times greater than for angles near 60°, in fair agreement with the initial comparisons of J_{aa}, J_{ae} and J_{ee} values. The Karplus relationship is now over ten years old and data published during this period has generally upheld the $\cos^2\phi$ relationship. Thus J_{aa} in cyclohexane rings ($\phi = 180°$) is almost always in the range 8–14 Hz whereas a/e and e/e coupling ($\phi = 60°$) usually falls in the range 0–6 Hz (Thomas, 1968). The configuration of a pair of isomers of the type **5** and **6**, may therefore be assigned as *trans* and

H

A   B
          H

5

H

A   H
          B

6

cis respectively with confidence if J_{HH} for **5** falls in the higher and J_{HH} for **6** in the lower of the two ranges quoted above. A straightforward example is that of the configurational assignment of the appetite-depressing drug, phendimetrazine and its diastereoisomer (Dvornik and Schilling, 1965).

$J_{2,3}$ 8·8 Hz

phendimetrazine

$J_{2,3}$ 2·7 Hz

isomer

A general procedure for the fitting of experimentally obtained vicinal coupling constants to the various stereochemical options, which takes into account uncertainties in the choice of values for K_1 and K_2 in the Karplus equation, is described in a review by Sternhell (1969).

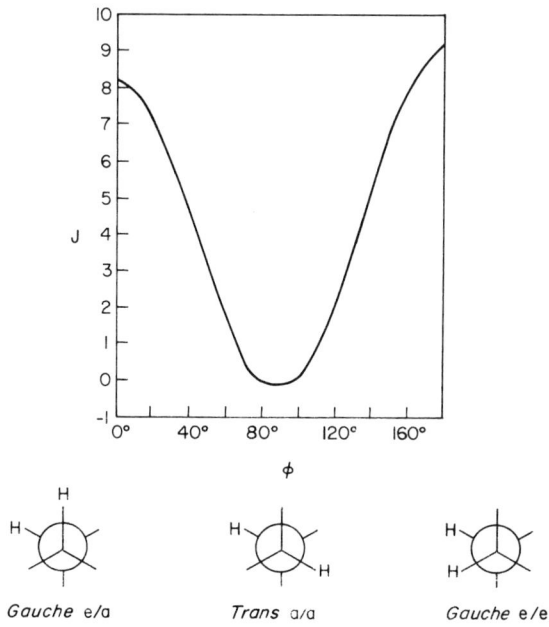

ϕ

Gauche e/a   Trans a/a   Gauche e/e

Fɪɢ. 3.2. The Karplus function describing the magnitude of vicinal proton–proton J coupling as a function of the dihedral angle ϕ in the H—C—C—H bond system (Bovey, 1969). The Newman projections denote *gauche* e/a, *trans* a/a, and *gauche* e/e coupling.

Judicious use of the $\cos^2 \phi$ relationship is necessary, however, since other factors may influence the magnitude of the coupling constant (Karplus, 1963). Among these, the influence of an electronegative substituent on the HC—CH fragment, is probably the most important; it has the effect of reducing J_{vic} values to below normal. Some substituent effects are illustrated in Table 3.1.

TABLE 3.1

(CH₃CH₂)ₙX		XCH₂CH₂Y		
X	$J_{average}$ Hz $\pm$ 0·1	X	Y	$J_{average}$ Hz $\pm$ 0·1
H	8·0	Cl	Cl	6·83
Me	7·26	Br	OH	6·00
OH	6·97	MeO	OH	5·33
F	6·9	MeO	MeO	5·3
N	6·9	HO	HO	5·27

From Abraham and Pachler (1963).

The value for ethane (X = H in Table 3.1) was obtained from the ^{13}C satellite signal (a similar case is described on p. 220). There is a linear relationship between the magnitude of J_{vic} and the total electronegativity of the substituent groups (Fig. 3.3) and it is to be noted that carbon itself has an electronegative effect in this respect. The influence of a substituent

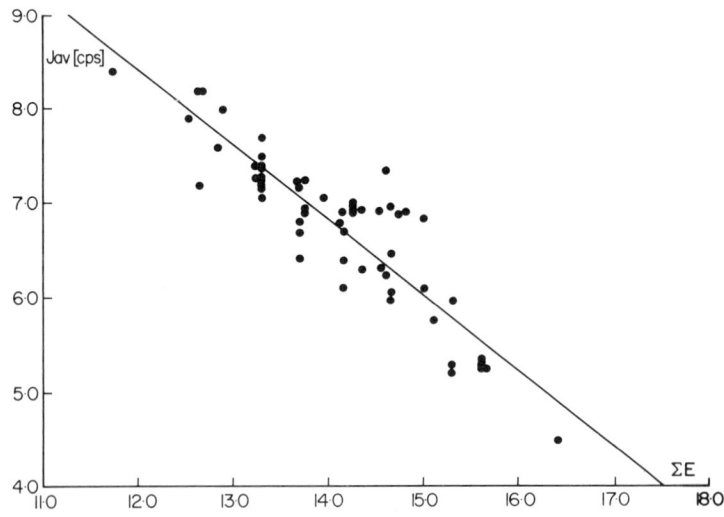

FIG. 3.3. The vicinal proton–proton coupling constant (J) versus the sum of the electronegativities ($\sum E$) of the substituents (after Abraham and Pachler, 1964). The data were obtained from compounds of type CH₃CH₂X, XCH₂CHXY, XCH₂CH₂Y and cyclic compounds.

X upon J_{vic} is greatest when X is *trans* coplanar to one of the coupling protons. This is seen in some results quoted by Booth (1965). In the steroid **8** both coupling protons are *trans* coplanar with electronegative substituents (H_a/X, H_e/carbon), while in **7** only one proton is so orientated

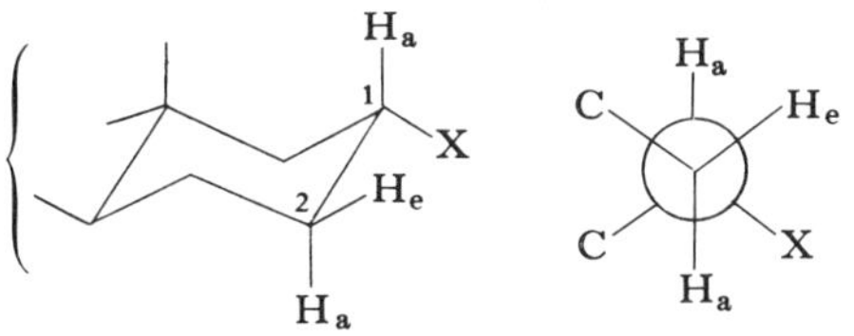

7

$J_{ae} \simeq 5\cdot5$ Hz (± 1 Hz) for e-X = OH, OAc

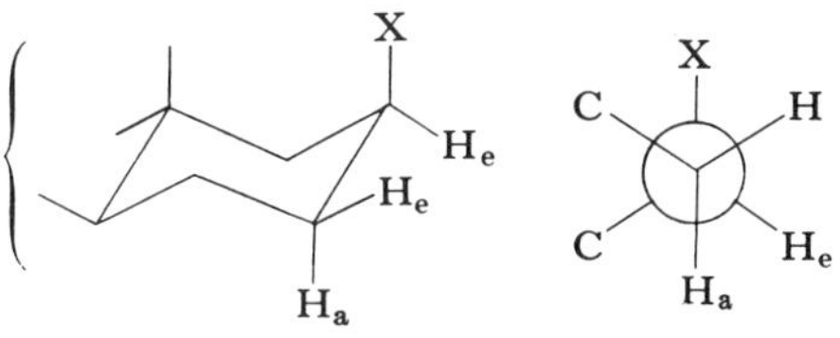

8

$J_{ae} \simeq 2\cdot5$ Hz (± 1 Hz) for a-X = OH, OAc

(H_e/carbon) and this fact is held responsible for the lower J_{ae} values found in the former case. The general finding that J_{ae} values are greater than J_{ee} in cyclohexyl derivatives may be explained on the same basis. Thus two *trans* carbon substituents are involved in J_{ee} of **10** but only one in J_{ae} of **9**.

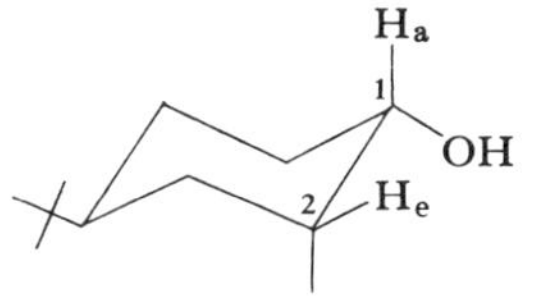

9

1-H_a *trans* coplanar with H
2-H_e *trans* coplanar with carbon

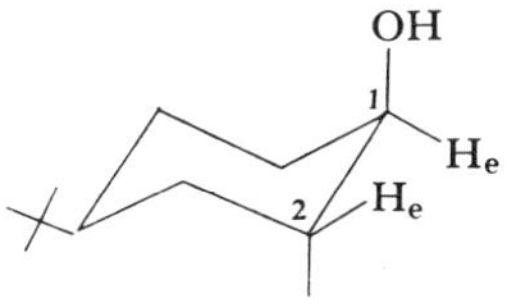

10

1-H_e *trans* coplanar with carbon
2-H_e *trans* coplanar with carbon

It is clear, therefore, that J_{ae}–J_{ee} differences have a potential value in stereo chemical assignments.

Long-range coupling over four bonds ($^4J_{HH}$) is also sensitive to stereochemical influences and appears to be most effective when the system can assume a W-configuration with the coupling protons approximately

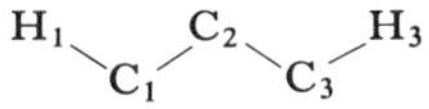

coplanar and linked by a zig-zag path (Sternhell, 1964, 1969); *meta* coupling in aromatic systems is a special case of the W fragment. The magnitude of $^4J_{HH}$ coupling is small and is often seen only in the broadening of a proton signal. Protons of the axial methyl group of a cyclohexane derivative (**11**)

11

may adopt, in turn, a W configuration with axial protons four bonds removed, and there are many reports of line broadening in this type of signal, e.g. that of angular methyl groups of steroids (Bhacca *et al.*, 1965; Shoppee *et al.*, 1966). In some cases clear splitting of the *t*-methyl group is seen as in the spectrum of 4,4-d$_2$-5α-androstan-2-one (Fig. 3.4) in which

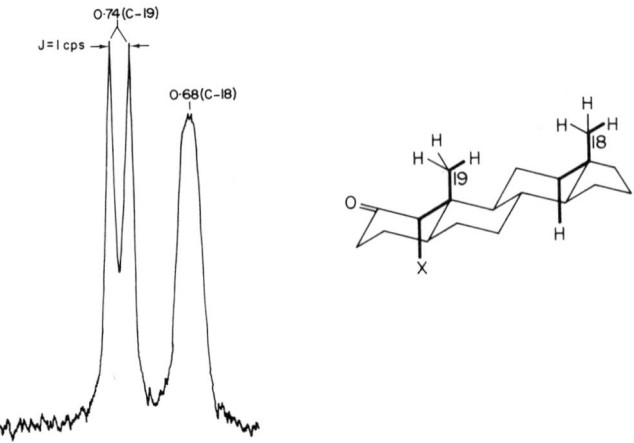

Fig. 3.4. Expanded 60-MHz PMR spectrum of the 0·6–0·8 ppm region of 4,4-d$_2$-5α-androstan-2-one (X = H). The spectrum of the non-deuterated compound has the same C-18 and C-19 features (Bhacca *et al.*, 1965).

C-19 appears as a doublet (*J* 1 Hz) and C-18 as a broad singlet (Bhacca *et al.*, 1965). Evidence for coupling between the C-19 methyl group and the 1-α (axial proton) of 5α-androstan-2-one (see formula in Fig. 3.4, X = H) is provided by the sharpending of the C-19 methyl signal when the 1α-H is replaced by deuterium [width at half height (*W*$_H$) reduced from 2·6 to 1·8 Hz]. Shoppee *et al.* (1966) have also detected $^4J_{HH}$ coupling by recording the difference between the *W*$_H$ of the signal investigated and *W*$_H$ of the TMS standard. Thus a difference of 0·6–0·9 Hz was found for the *cis* *t*-butylcyclohexane derivatives (**12**), compared with 0·2–0·3 Hz for the

cis-**12**

X = OH, OCOMe

trans-**12**

trans (*t*-Bu/Me) isomers in which the coupling protons cannot be linked by the favoured zig-zag path. An example of the use of $^4J_{HH}$ stereochemical requirements in a configurational assignment is given in the section on narcotic analgesics (p. 190).

MAGNETIC ANISOTROPIC EFFECTS

Another aspect of NMR spectroscopy that should be mentioned before specific stereochemical applications are discussed is the shielding effects of magnetically anisotropic groups.

The fact that the shielding influence of certain structural features (in particular those with π-electron systems), is dependent on the spatial relationship of the shielded proton, and the function in question (Bovey, 1969) is clearly of potential stereochemical value. The shielding effects of several of these so-called magnetically anisotropic groups has been treated in a semi-quantitative fashion and some plots of shielding contributions are available that make it possible to assess the shielding suffered by protons located in various positions about a functional group. Johnson and Boveys' calculation of the anisotropic shielding constants of benzene is the best known example (1958) and their data have been much utilized in the analysis of PMR spectra.

The shielding zones about a benzene ring are illustrated diagrammatically in Fig. 3.5. In general terms, protons close to the aromatic plane are de-shielded while those above the ring are shielded by the phenyl group. A quantitative assessment of shielding due to benzene, developed from the free electron model of Pauling, is shown in Fig. 3.6 (Johnson and Bovey, 1958). In this plot the proton position is identified by two coordinates, namely z, the axis normal to the plane of the aromatic ring at its centre, and ρ, the axis connecting the centre of the ring and the intersection of the z axis with the ring plane, as illustrated. In Fig. 3.6, z and ρ values are measured in ring radii, the radius of the benzene hexagon being taken as the (aromatic) C—C distance, 1.39 Å. The plot represents one quadrant of a plane passing normally through the centre of the benzene ring and shows

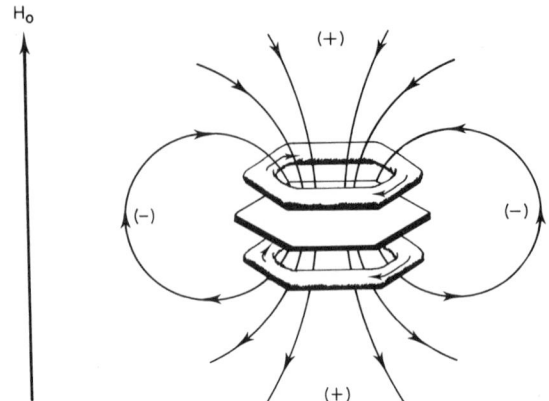

FIG. 3.5. Electron density, ring currents, and magnetic lines of force about a benzene ring. The symbol (−) denotes enhancement and (+) diminution of the applied field H_0; protons in the former are subjected to paramagnetic shielding, and in the latter regions to diamagnetic shielding (Bovey, 1969).

the contours of equal shielding (isoshielding) over a region extending outward 5·0 ring radii (6·59 Å) in the plane of the carbon ring and 3·5 ring radii (4·87 Å) along the symmetry axis; shielding values are in parts per million. Negative isoshielding values (σ_g) correspond to an increase in the apparent field (proton signal is then moved down-field) and positive to a decrease (signal moved upfield). For a given value of ρ of 1 or larger,

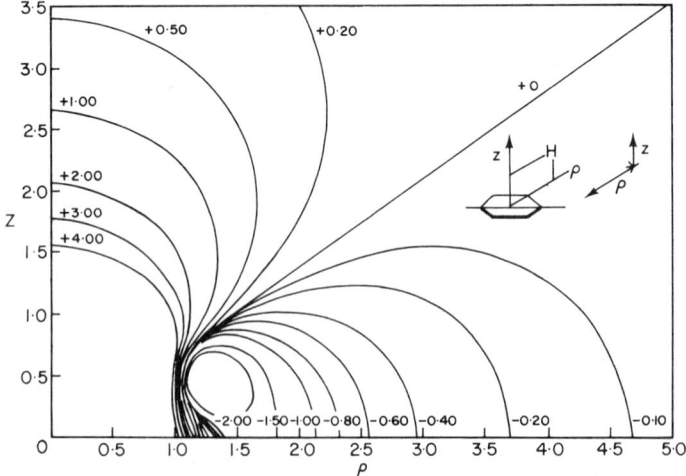

FIG. 3.6. The shielding zone about the benzene ring. The "isoshielding lines" were calculated from the model represented in Fig. 3.5 (Bovey, 1969).

highest proton deshielding results when $z = 0$ (i.e. when the proton is directly in the aromatic plane), while highest shielding occurs just above the ring when $\rho = 0$; both effects decrease rapidly as values of z and p increase. The line for $\sigma_g = 0$ represents the cross-section of the nodal surface separating the shielding and deshielding regions.

From a practical point of view z and ρ values may be measured with reasonable accuracy by the use of scale models of the Dreiding type and an appropriately graduated ruler. ApSimon (1968) has described a measuring device that can be attached to the benzene ring of a Dreiding model in such a way that the ring centre corresponds with the pivot point of the

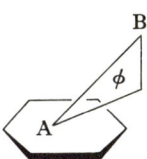

measuring arm. The device provides the A–B distance and the angle ϕ, and thus allows z and ρ to be calculated. The corresponding aromatic shielding value may then be found either from the plot (Fig. 3.6) or, more conveniently, from a set of tables included as an appendix to Bovey's text on NMR spectroscopy (1969).

Studies of the anisotropic effects of the following groups have also been made:

- (i) the C—C and C—H bonds, especially when present in alicyclic system (ApSimon *et al.*, 1967a);
- (ii) the C—C double bond (ApSimon *et al.*, 1967b);
- (iii) the carbonyl (C=O) group (ApSimon *et al.*, 1966; Karabatsos *et al.*, 1967).

All these studies are of potential stereochemical value and examples of their application are given throughout this book.

An account of studies upon various classes of stereoisomers will now be given.

ALKENES

The configurations of a pair of *cis-trans* disubstituted alkenes (**13** and **14**)

13

14

may readily be assigned on the basis of the greater magnitude of *J trans* (12–18 Hz) compared with *J cis* (7–10 Hz). This difference is of the order expected from the $\cos^2\phi$ relationship, since dihedral angles of 0° and 180° obtain in *cis* and *trans* isomers respectively (see J/ϕ plot, Fig. 3.2). As an example, coupling constant data show that the decarboxylation of isomeric α-fluorocinnamic acids proceeds with retention of configuration (Elkik and Francesch, 1968). In the β-fluorostyrene (**15**) derived from the *cis* (Ph/CO$_2$H) acid, $J_{H_AH_B}$ is greater than $J_{H_AH_B}$ in the isomeric compound

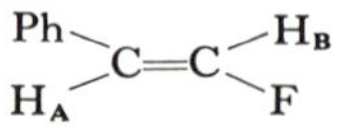

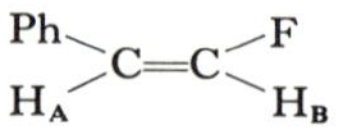

	15	**16**
$J_{H_AH_B}$	11·4 Hz	5·4 Hz
J_{H_AF}	21·3 Hz	43 Hz

(**16**) showing H$_A$ and H$_B$ to be *trans* in the former. In confirmation, J_{H_AF}, which arises from a *cis* interaction in the *trans* H$_A$H$_B$ compound (**15**), is smaller than J_{H_AF} in the *cis* H$_A$/H$_B$ (*trans* H$_A$/F) isomer (**16**).

A related case of interest is that of HNCH coupling constant differences seen in the spectra of *cis* and *trans* 3-aminoacrylic esters (**17**) (Bottomley

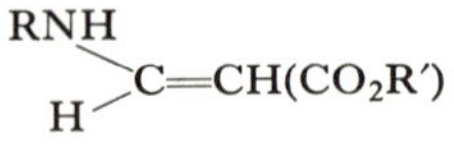

$$\text{RNH} \diagdown \text{C}=\text{CH(CO}_2\text{R}') \diagup \text{H}$$

(**17**)

et al., 1967). A relatively large coupling exists between the NH proton and that of the adjacent methine proton, and values are generally higher and less variable (12–15 Hz) for the *cis* esters than for the *trans* isomers (7·5–15 Hz). Conjugation of the nitrogen lone pair with the acrylic ester moiety will induce a partial double bond character in the —N̈—C= bond so that the amino and methine protons will tend to take up either a *syn* or an *anti* configuration with respect to each other. In the *cis* compounds intra-

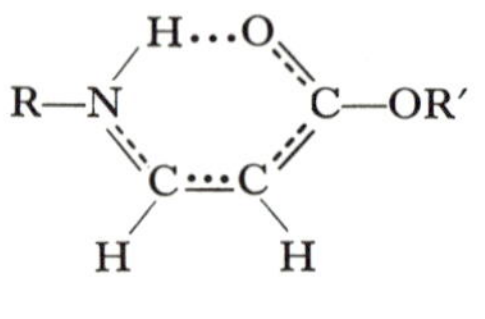

molecular hydrogen bonding will favour the *anti* configuration, with the result that strong HNCH coupling is observed. This configuration is less probable in *trans* esters (no intramolecular bonding) but becomes more favoured with increasing size of R, as reflected in the following J_{HNCH} values for *trans* esters: R = Me, 7·5 Hz; R = n-Pr, 8 Hz; R = iso-Pr, 9 Hz; R = cyclohexyl, 10 Hz; R = t-Bu, 13·2 Hz.

In tri-substituted ethylenes configurational assignments must, in general, be based on the chemical shift of the single vinylic proton. Tobey (1969) has described a procedure applicable to a wide variety of di- and tri-substituted ethylenes which warrants discussion in some detail. His method is based on the independence and additivity of vinyl substituent shielding effects, which had previously been derived for 36 functional groups by the statistical analysis of over 4000 chemical shifts (Pascual *et al.*, 1966; Matter *et al.*, 1969). The resonance position of the vinyl proton in molecules as below, where X, Y and Z are symmetrical substituents generally smaller in size than bromine, can be calculated from equation (1).

$$X_{cis} \diagdown C = C \diagup{}^{H} Y_{trans} \diagup \diagdown Z_{gem}$$

$$\sigma_{ppm} = -5·32 + \sigma_{cis\text{-}X} + \sigma_{trans\text{-}Y} + \sigma_{gem\text{-}Z} \qquad \dots (1)$$

In this equation −5·32 ppm is the resonance position of $CH_2{=}CH_2$ in CCl_4 and $\sigma_{cis\text{-}X}$, $\sigma_{trans\text{-}Y}$ and $\sigma_{gem\text{-}Z}$ are the shielding constants X, Y and Z from the *cis*, *trans* and *gem* substituent locations (TMS at 0·0 ppm).† The three vinyl shielding values of a substituent X may be obtained by comparing the totally analysed ABC spectrum of a mono-substituted ethylene $CH_2{=}CHX$ with that of ethylene. Thus, taking vinyl bromide as an example, it is seen that introduction of bromine lowers the resonance

$$\begin{array}{c} -5·75 \\ Br \diagdown C = C \diagup{}^{H} \\ H \diagup \diagdown H \\ -6·36 \qquad -5·83 \end{array}$$

position of the *gem* proton −1·04 ppm [−6·36−(−5·32)] relative to ethylene. Therefore, $\sigma_{gem\text{-}Br} = -1·04$ ppm. Similarly $\sigma_{cis\text{-}Br} = -0·43$ and $\sigma_{trans\text{-}Br} = -0·51$ ppm. Sigma values are best obtained, however, by averaging the apparent shielding effects of a given function in a series of compounds bearing substituents of varying size and electronic properties so that allowance for steric and electronic interactions between substituents may be made. Thus data on **18** and **19** give σ_{trans} (−0·54 ppm), **18** and **20** σ_{cis}

† By giving chemical shifts in ppm a negative sign, positive shielding constants correspond with upfield, and negative with downfield shifts.

$$Br_2C=CBr(H) \quad -7.13$$

18

$$(H)(Br)C=C(Br)(H) \quad -6.59$$

19

$$(Br)(H)C=C(Br)(H) \quad -6.98$$

20

$$Br_2C=C(H)(H) \quad -6.26 \quad \text{(all ppm)}$$

21

(−0·15 ppm), and **18** and **21** σ_{gem} (−0·87 ppm). Average values from data on thirteen compounds are: $\sigma_{cis\text{-}Br}$ −0·33 ± 0·09, $\sigma_{trans\text{-}Br}$ −0·53 ± 0·04 and σ_{gem} −1·02 ± 0·11 ppm. For all three positions bromine moves the proton resonance downfield, the effect being greatest from the *gem* position, while $\sigma_{trans\text{-}Br}$ is significantly greater than $\sigma_{cis\text{-}Br}$; an average uncertainty of ±0·1 ppm (6 Hz at 60 MHz) is assumed for each value. Sigma values for the substituents Cl, Me and CN are derived from model compounds in the same way (*cis* and *trans* Me shields but *gem* Me deshields). Insertion of these values into Equation (1) closely reproduced the 50 PMR positions of 39 model compounds; in these cases uncertainties in the values tended to compensate rather than propagate when used additively. For differentiation purposes isomers should differ by more than 0·2 ppm in their vinylic chemical shifts owing to the σ uncertainty of ±0·1 ppm. An example is the assignment of the single dichloroacrylonitrile (vinylic resonance −7·23 ppm) obtained upon ammonolysis of tetrachlorocyclopropene; calculated values for the three possible isomers are as follows:

$$\underset{-5.84}{Cl_2C=C(H)(CN)} \qquad \underset{-6.99}{(H)(Cl)C=C(Cl)(CN)} \qquad \underset{-7.16}{(Cl)(H)C=C(Cl)(CN)}$$

(all ppm ± 0·18)

The β,β-dichloro structure can be eliminated and the ammonolysis product assigned a *cis* in preference to a *trans* configuration.

Application of this procedure to phenyl substituents is complicated by the fact that aromatic shielding depends not only on the location of the substituent relative to the vinylic proton but also on the time-averaged orientation of the aromatic group relative to the ethylenic double bond plane (see p. 93). However, the deshielding influence of phenyl does not decrease seriously until the angle between the two planes exceeds 20° as is readily confirmed by use of Bovey's shielding Tables (1969). Chemical shift data on a series of models shows that a phenyl group causes a very large (−1·42 ppm) down-field shift in the resonance position of a *gem* vinylic proton and a moderate (−0·38 ppm) shift for a *cis* proton, while *trans* protons are hardly affected. Shielding effects in polyphenylated ethylenes such as **22** and **24** (Scheme I) are virtually indistinguishable from those in which a single phenyl group exists, as in styrene (**23**); this result shows that *cis* and *gem* Ph–Ph interactions do not seriously disturb the coplanar orientation of

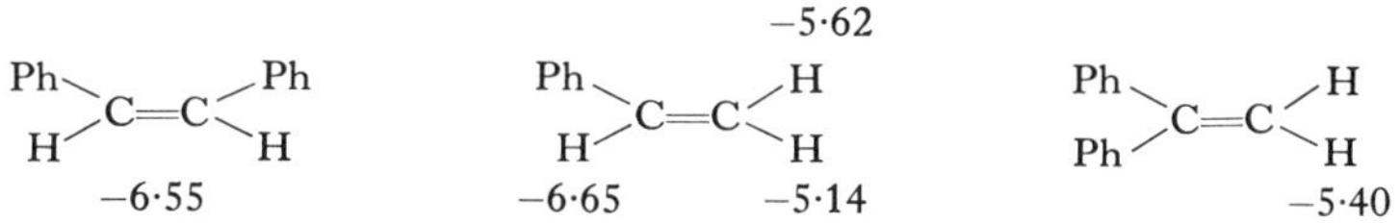

(in ppm)

Scheme I

phenyl and double-bond planes. The *trans* sigma value = +0·1 ppm from **22** and **23** and +0·22 ppm from **23** and **24**, while σ_{cis} = −0·26 ppm from **23** and **24**; values calculated from **23** and ethylene are σ_{trans} +0·18 ppm and σ_{cis} −0·3 ppm.

Cis-Me/Ph interactions also appear to have only a small influence upon the coplanarity of the styrene unit; thus the up-field shift of the vinylic

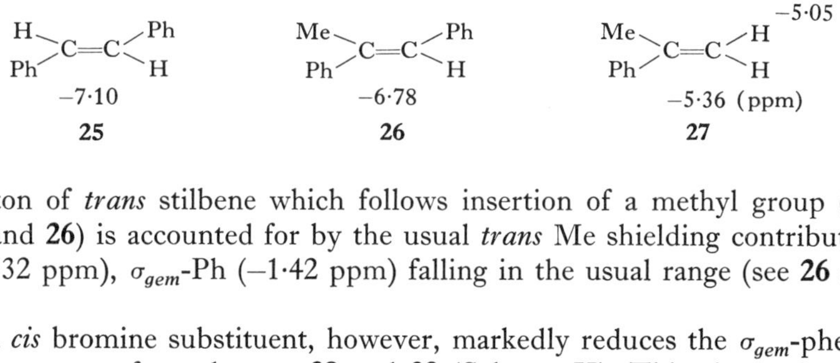

proton of *trans* stilbene which follows insertion of a methyl group (see **25** and **26**) is accounted for by the usual *trans* Me shielding contribution (+0·32 ppm), σ_{gem}-Ph (−1·42 ppm) falling in the usual range (see **26** and **27**).

A *cis* bromine substituent, however, markedly reduces the σ_{gem}-phenyl value as seen from data on **28** and **29** (Scheme II). This change is con-

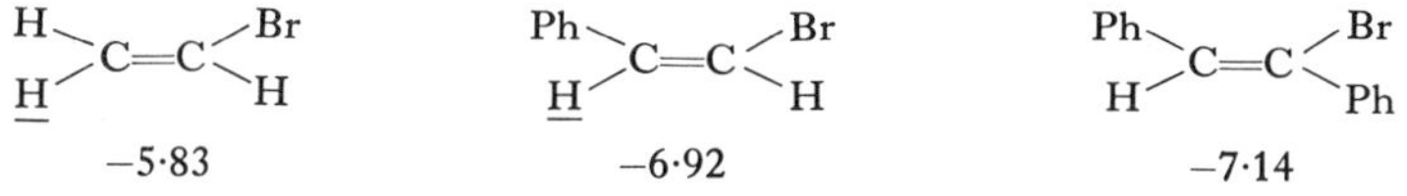

$\sigma_{gem\text{-}Ph}$ = −1·09 (usual value is −1·43 ppm)

Scheme II

sidered a result of the phenyl group being twisted more than 20° out of the ethylenic plane whereby the vinylic proton falls outside the aromatic deshielding zone. Phenyl *gem* to Br is not forced appreciably out of co-planarity, the value $\sigma_{cis\text{-}Ph} = -0.22$ ppm from results on **29** and **30** being in the normal range. An assignment involving σ values for Br, Me and Ph is shown below. In these isomers the four-bond $J_{Me/H}$ couplings differ

	δH obsd. ppm	δH calcd. ppm
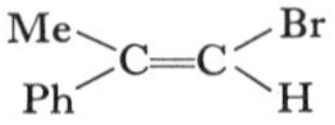 Me, Ph C=C Br, H	−6·30	−6·33 ± 0·17
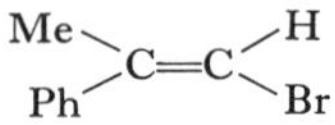 Me, Ph C=C H, Br	−6·03	−5·92 ± 0·19

only by 0·25 Hz and thus provide no indication of geometry (Davis and Roberts, 1962).

TABLE 3.2

Vinylic chemical shifts of some α-substituted stilbenes[a] in CDCl₃

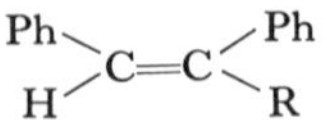

$$Ph\underset{H}{\overset{}{>}}C=C\underset{R}{\overset{Ph}{<}} \qquad Ph\underset{H}{\overset{}{>}}C=C\underset{Ph}{\overset{R}{<}}$$

cis 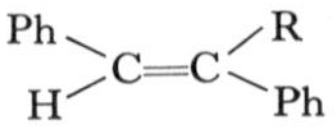   *trans*

R	Chemical shift (ppm)		Difference (ppm) (*trans–cis*)
	trans-vinylic	*cis*-vinylic	
H	7·1	6·67	0·43
Me	6·82	6·5	0·32
Et	6·63	6·37	0·26
*iso*Pr	6·38	6·33	0·05

[a] Casy *et al.* (1968).

PMR data upon a series of *cis* and *trans* α-substituted stilbenes (Table 3.2) provide a further example of the interpretation of aromatic screening influences. *Trans* stilbene (R = H, see Table 3.2) is a completely flat molecule in its preferred conformation and the vinylic protons are subject to maximum deshielding by the conjugated aromatic–double-bond system. The higher field vinylic signal of the *cis* isomer may be attributed to: (i) reduced π-electron deshielding as a result of a less-favoured planar molecule, and/or (ii) absence of a $\sigma_{cis\text{-}Ph}$ contribution. The results of Tobey

(1969) suggest that only the second factor is significant. In the *trans* series, an α-methyl group moves the vinylic resonance up field, due to the $\sigma_{trans\text{-}Me}$ shielding contribution rather than a change in shape of the molecule. The more pronounced upfield shifts seen in the *trans* α-ethyl and α-isopropyl derivatives, however, are almost certainly due to a progressive divergence of aromatic and ethylenic planes. When R is isopropyl, screening influences upon vinylic protons must be similar in the two isomers, because they have almost identical chemical shifts. Configurational assignments to a series of aminoalkyl stilbenes and styrenes with antihistaminic properties have been made by application of these principles (see p. 234).

Styrene isomers of the type **31** and **32** are differentiated by chemical shift

trans RCH₂/R′CH₂   *cis*

31   **32**

positions of the vinylic proton and the two methylene groups (R$\underline{C}$H₂ and R′$\underline{C}$H₂)—all lower field in the more planar *cis* Ar/H isomer—and also by separation of the A_2B_2 doublets of the aromatic signal (Barbieux *et al.*, 1964). This is greater in the *cis* than in the *trans* isomer (25–30 Hz versus 13–16 Hz) because the deshielding influence of the double bond upon the A_2 aromatic protons (*ortho* to vinyl group, see **31**) is greater in the more planar molecule with the result that the A_2 signal moves downfield from the B_2 doublet which has a similar chemical shift in both isomers. This interpretation is supported by the A/B chemical shift difference of 14 Hz found for the distinctly non-planar analogue, stilboestrol.

stilboestrol

Anisotropic effects of aromatic groups often provide evidence for, or corroborate, configurational assignments even when the proton group affected is formally well removed from the aromatic substituent. An example in the case of alkene isomers is provided by work upon adducts formed from amines and propargyl sulphones (Pink *et al.*, 1965). The ester methyl signal ($CO_2CH_2\underline{Me}$) of the *cis-trans* mixture (**33**) is a pair of triplets

PhSO$_2$\C=C(Me)NHCH$_2$CO$_2$Et

33

separated by 2·1 Hz (at 100 MHz) and the higher field set is attributed to the *trans* H/Me isomer for the following reason. The population of conformers in which the terminal methyl group of the flexible side chain lies above the benzene ring (i.e. within the aromatic shielding zone) is considered likely to be greater in the *trans* H/Me isomer because approach of the side chain of the *cis* isomer towards the aryl nucleus should be restrained by hydrogen bonding between the NH and sulphonyl groups. A similar effect is seen in the isomers (**34**), where the adduct mixture shows two

PhSO$_2$\C=C(Me)NH(CH$_2$)$_3$OMe

34

OMe signals; only one signal is found for the related pair in which Ph is replaced by *t*-Bu, a result supporting the assumption that chemical shift differences arise as a result of differential aromatic influences.

Configurational assignments to the oxazolinone (**35**) and its isomer

35

H\Ph/C=C(NHCOPh)CO$_2$Me

36

illustrate the application of carbonyl deshielding effects to stereochemical problems (Brocklehurst *et al.*, 1968). The isomeric vinylic chemical shifts differ (2·73 for the stable and τ 2·32–2·61 for the labile form), that at lower field being attributed to **35** (β-H *cis* to C=O). In confirmation, vinylic chemical shifts of the derived α-benzamidocinnamates (**36**) (obtained by a reaction which does not alter configuration) differ in the same sense.

APPLICATION OF THE NUCLEAR OVERHAUSER EFFECT (NOE)

The integrated intensity of the resonance band due to a particular proton or proton group depends, among other factors, upon the spin–lattice relaxation time T_1 of the proton(s) in question. A proton X may influence

T_1 of a second proton Y by an intramolecular dipole–dipole interaction, and this contribution is particularly important if the two protons are close together. It can be shown that complete saturation of the X nucleus will result in a 50% enhancement of the integrated intensity of the band due to the Y nucleus, and this is known as a Nuclear Overhauser Effect (NOE). For its observation, the spectrum has to be recorded by a frequency sweep method rather than the normal magnetic field sweep procedure.

Anet and Bourn (1965) first drew attention to the stereochemical potential of the NOE phenomenon which results from allowing a distinction to be made between closely placed and remote proton groups and gave the case of β,β-dimethylacrylic acid as an example. The spectrum of this acid

$$\underset{Me}{\overset{Me}{\diagdown}}C=C\underset{CO_2H}{\overset{H(1)}{\diagup}}$$

showed two well separated methyl doublets ($J = 1\cdot3$ Hz for both), that at higher field being assigned to methyl *cis* to the vinylic proton. The band of H(1) was a near-septet through equal coupling to each methyl group. Irradiation of either of the two methyl bands caused the H(1) band to change to a 1:3:3:1 quartet, but, more significantly, the integrated intensity of the quartet was higher when the high-field methyl group (*cis* to H) was irradiated (relative intensity 117) than when the low-field group (*trans* to H) was irradiated (relative intensity 96) or than when the irradiation frequency was offset from the methyl groups (relative intensity 100). The decoupling which also occurs does not affect integrated intensities when the spectrum is observed by means of a frequency sweep.

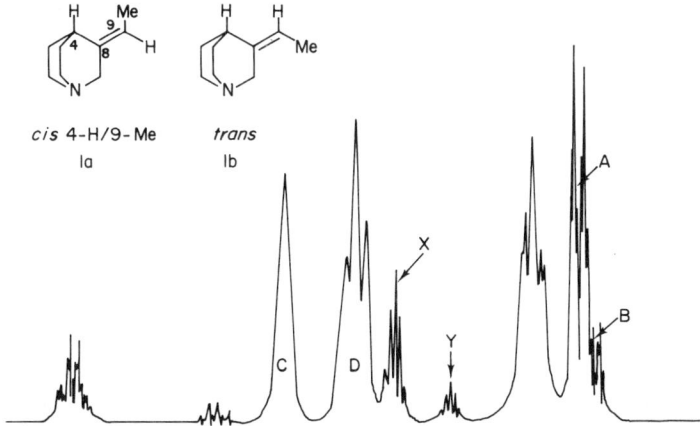

cis 4-H/9-Me   *trans*
    Ia   Ib

Fig. 3.7.   100 MHz spectrum of 3-ethylidene-1-azabicyclo(2,2,2) octanes (70:30 mixture of 1a and 1b) in CDCl$_3$ plus 10% CF$_3$CO$_2$H (Nouls *et al.*, 1967).

Another example concerns signal assignments to the *cis* and *trans* quinuclidine derivatives shown in Fig. 3.7 (Nouls *et al.*, 1967). The 100-MHz spectrum of a mixture of both isomers (major 70:minor 30) in $CDCl_3$—CF_3CO_2H shows signals of 4-H as a quintet (it is coupled to four protons), labelled X for the major and Y for the minor isomer, and signals of 9-Me as a doublet of triplets, labelled A (major) and B (minor). When the B group is irradiated at the centre of the multiplet the integrated intensity of the Y group (measured relative to the sum of the signal strengths of the C and D groups, see Fig. 3.7) does not change; irradiation of the A group, however, enhances the intensity of the X signal by about 30% and leaves the Y signal unaltered. Thus, 4-H must be close to 9-Me in the major isomer which then has the *cis* configuration. The four-proton signal of the *cis* isomer has the lower field position and must therefore be deshielded by the 9-methyl group. This result correlates with the deshielding of an axial proton by a β-axial methyl group in cyclohexane derivatives (Booth, 1966),

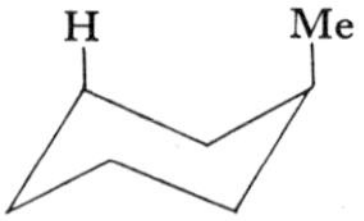

H deshielded by 0·18 ppm

relative H—Me dispositions being similar in the two cases. A review upon the applications of NOE to organic chemistry is available (Moreau, 1969) and examples of pharmacological interest are given in Chapter 6.

ACYCLIC DIASTEREOISOMERS

Diastereoisomers of the type shown below (the *erythro* isomer being defined as the one in which two pairs of like groups may be eclipsed)†

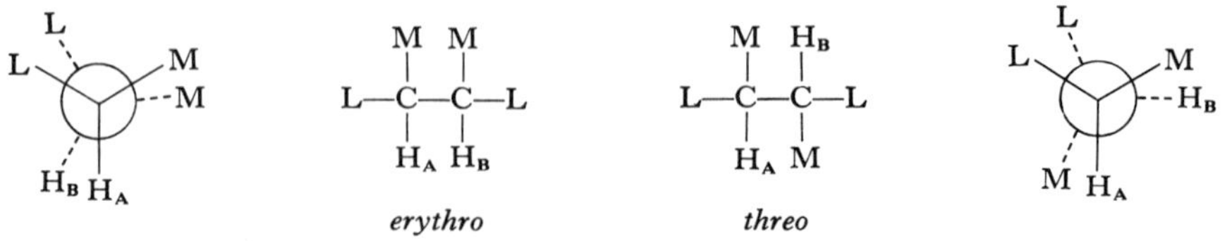

erythro   *threo*

† The following procedure is recommended as an aid to deciding whether a particular Newman projection (fully staggered) represents an *erythro* or a *threo* isomer: Draw (or mentally visualize) the equivalent conformation in which two identical substituents are eclipsed. If two other identical (or similar) substituents

frequently differ in the value of their J_{AB} coupling constants (A refers to H_A and B to H_B in these and related formulae) and this difference may provide evidence of configuration on the following basis. When L and M differ markedly in size the two L groups will be *trans* in the favoured conformation of both isomers, with the result that H_A and H_B have a preferred *trans* orientation in the *erythro* and a gauche orientation in the

erythro

threo

threo isomer. Thus a J_{AB} value of about 10–12 Hz is expected for the *erythro* isomer and 1–3 Hz for the *threo* form, consistent with dihedral angles of 180° and 60° respectively. In practice, contributions from other conformers usually lower the *erythro* and raise the *threo* range (also electro-negative substituents will lower *J gauche* values), but a J_{AB} coupling constant difference between isomers of greater than 3–4 Hz often enables con-figurational assignments to be made. The use of such data is, however, by no means straightforward as the following examples illustrate (Kingsbury, 1968; Kingsbury and Thornton, 1966; Kingsbury and Best, 1967). A clear example is the *erythro–threo* pair (Table 3.3, Nos 6 and 7) which involve vicinal isopropyl groups. Confirmation of the favoured nature of **37** and **38** for *erythro* and *threo* isomers respectively is provided by differences in the J_{AC} values (C is the methine proton of the isopropyl group). In the *trans* H_A/H_B conformation **37**, a *gauche* H_A/H_C orientation is more probable than a *trans* arrangement, since Ph/Me 1-3 interactions would be generated in the latter. Hence J_{AC} should be small ($\phi = 60°$), as is in fact observed

---

also become eclipsed, the isomer is an *erythro* form, if not, a *threo*. In the example below, the staggered projection is an *erythro* form because two sets of identical

substituents (H/H and M/M) are eclipsed in the non-staggered form. Note also that Fischer projections denote eclipsed conformers.

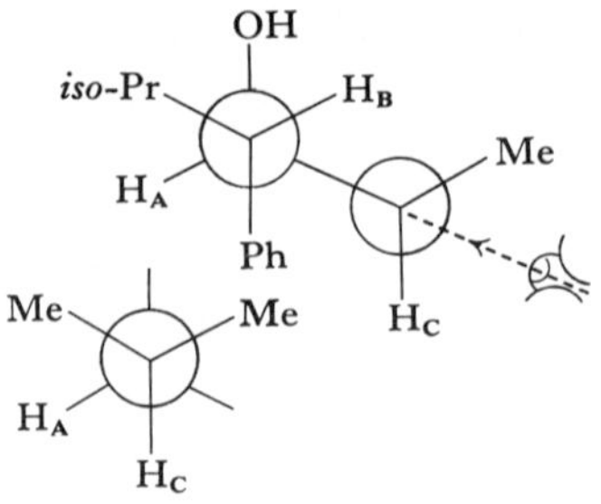

erythro

37

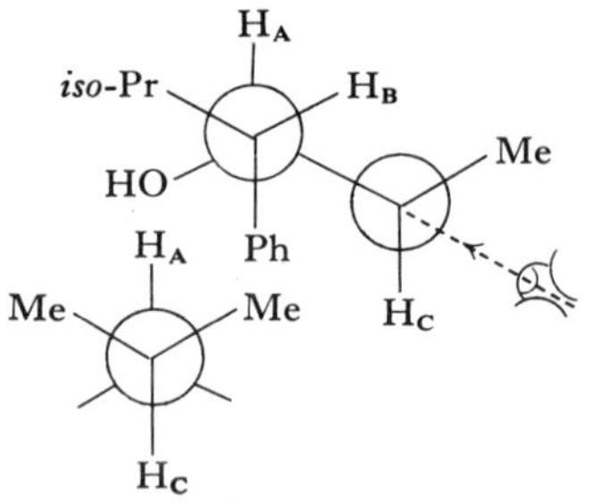

threo

38

($J = 1 \cdot 6$ Hz). In the *threo* conformer (**38**) (one of the two possible *gauche* H_A/H_B forms is shown), H_A and H_C are *trans* and a larger J_{AC} value is anticipated. The value obtained (7·6 Hz) is distinctly larger than the *erythro* value but its size shows that conformational preferences are less in the *threo* than in the *erythro* isomer.

TABLE 3.3

J_{AB} coupling constant data for some isomeric 2-phenylethanols in CCl₄

$$\begin{array}{cc} OH & Ph \\ | & | \\ RCH_A & -CH_BR' \end{array}$$

No.	Isomer	R	R′	J_{AB} (Hz)
1	*erythro*	Me	Me	6·7
2	*threo*	Me	Me	6·3
3	*erythro*	Me	Et	6·7
4	*threo*	Me	Et	6·0
5	*erythro*	Et	Et	7·3
6	*erythro*	*iso*Pr	*iso*Pr	10·3
7	*threo*	*iso*Pr	*iso*Pr	3·6

From Kingsbury and Thornton (1966).

When only one group in the molecule is large compared with the others or when no outstandingly large group is present, significant populations of *trans* and *gauche* H_A/H_B conformers obtain and averaged J_{vic} values near 6–7 Hz result (Table 3.3, Nos 1–5). In such cases isomeric ΔJ_{vic} values are too small to allow firm configurational assignments to be made; the *erythro J* value is, however, generally somewhat greater than the *threo* value.

When both L and M are large *trans* H_A/H_B conformers are favoured in both isomers because the total number of gauche interactions between bulky groups is a minimum in these forms; in these cases both diastereoisomers will have large J_{AB} values, see example below (Kingsbury and Best, 1967).

$$
\begin{array}{cc}
\text{Me} & \text{Ph} \\
| & | \\
\text{Ph—CH}_A\text{—CH}_B \\
| \\
\text{X}
\end{array}
$$

X = Br J_{AB} *erythro* 9·7, *threo* 9·5 Hz
X = I J_{AB} *erythro* 10·5, *threo* 10·1 Hz

In the 1,2-dihalogen derivative (**39**) (R = Me or *iso*Pr), significant $\varDelta J_{AB}$ differences are seen between isomers [e.g. R = *iso*Pr, X = Br; J_{AB}

$$
\begin{array}{cc}
\text{X} & \text{X} \\
| & | \\
\text{R—CH}_A\text{CH}_B\text{—Me}
\end{array}
$$
39

10·6 Hz (*erythro*), 3·5 Hz (*threo*)]; these isomers fit the case where L and M differ markedly in size (see p. 105). When R in **39** is *t*-butyl, however, there is a drastic fall in the J_{AB} value in both isomers [e.g. R = *t*Bu, X = Br; J_{AB} 2·02 Hz (*erythro*), 1·6 Hz (*threo*)] showing gauche H/H conformers to be favoured (Kingsbury and Best, 1967). In this case it is probable that the *trans* conformer (**40**) is destabilized by the 1,3 interactions between

trans H/H   gauche H/H
40   **41**

bromide and one of the *t*-butyl methyl groups that occur in this conformation and which are reduced in gauche conformers (e.g. **41**) and in the isopropyl analogue (**42**).

trans H/H

42

(all *erythro* isomers)

In the 1,2-diphenylcyanides (**43**) pronounced differences between the J_{AB} values of isomers are only seen when R is *iso*Pr or *t*-Bu (e.g. R = *t*-Bu,

$$\begin{array}{cc} R & CN \\ | & | \end{array}$$
PhCH$_A$CH$_B$Ph   R = Me, Et, *iso*-Pr, *t*-Bu

43

J_{AB} 10·2 and 3·7 Hz) (Kingsbury, 1968). The isomers with the larger J_{AB} values must have favoured *trans* H$_A$/H$_B$ conformers. Of the two arrange-

threo	*erythro*	*erythro*
44	**45**	**46**

ments **44** and **45**, the former is the more probable because this places the bulkiest groups (*t*-Bu/Ph *not* Ph/Ph) *trans*; hence the isomer with the higher H$_{AB}$ value must have the *threo* configuration in terms of the definition already given (p. 104). The favoured conformation of the *erythro* isomer **46** must also have a *trans t*-Bu/Ph arrangement. The *gauche* Ph/Ph interactions of the cyanides (**43**, R = Me and Et) do not appear to seriously exceed those of Me (and Et)/Ph, since J_{AB} value for such isomers [e.g. **43**, R = Me J_{AB} 7·2 Hz (*erythro*), 6·8 Hz (*threo*)] indicate neither *trans* nor *gauche* isomers to be particularly favoured.

The H$_A$/H$_B$ coupling constants of the isomeric amino-alcohols [**47**(a)] differ by almost 5 Hz and their *J* values (4·2 and 9·3 Hz respectively in formic acid) show both isomers to exhibit conformational preferences.

$$\begin{array}{cc} H_2N & OH \\ | & | \end{array}$$
ArCH$_A$CH$_B$Ar   Ar = (a) 3,4-methylenedioxyphenyl
                              (b) Ph

47

The related diphenyl derivatives [**47**(b)], of established stereochemistry, show similar J values 4·2 Hz (*erythro*) and 9·4 Hz (*threo*) and the configurations of the compounds **47**(a) are assigned accordingly (Lyle and Durand, 1967; Huffman and Elliot, 1965). Favoured conformers consistent with the observed J_{AB} values are the anti H_A/H_B form (**48**) for *threo* and the

48-*threo*   **49**-*erythro*

gauche form (**49**) for *erythro*. Similar J_{AB} differences are found for the diastereoisomeric series (**50**) in which NR_2 is a tertiary amino group, e.g. **50**(a), $NR_2 = 1$-piperidino, J_{AB} 10·4 (*threo*), 4·8 Hz (*erythro*) bases in

PhCHCHPh with NR$_2$ and OR substituents

	(a) R = H
PhCHCHPh	(b) R = Me
	(c) R = COMe

50

CDCl$_3$, preferred conformers again being **48** and **49** respectively (Munk *et al.*, 1968). In the *erythro* case, hydrogen bonding of the type $OH \cdots NR_2$ for bases and $HO \cdots H\overset{+}{N}R_2$ for conjugate acids (**47** a and b were examined in an acid solvent) appears to control conformer preference.† In the *threo* examples, hydrogen bonding may also operate in the conformation **51** and the evident preference for **48** seems to be due to its division of the four

51   *threo (gauche H/H)*   **52**   *erythro (trans H/H)*

† The validity of invoking hydrogen bonding interactions to account for conformational preferences may be challenged when hydrogen-bonding solvents rather than aprotic solvents are employed, because of the possibility of intermolecular bonding of solute to solvent in the former case. Some justification for using such arguments when solvents such as formic acid are used is provided, however, by evidence that intra- are stronger than inter-molecular hydrogen bonds (Jardetzky, 1963).

bulkiest groups into two pairs separated from one another by hydrogen atoms (cf. p. 107). The space occupied by the NR_2 and solvated $^+NH_3$ function is judged to be large; Mateos and Cram (1959) indicate the following order of steric requirements for the $RCH(Ph)CH(OH)R$ system: $Ph > OR$ ($R = $ acyl or Me) $> H$ and $R_2N > Ph > H$. The *anti-erythro* conformer **52** is more favoured at the expenses of the gauche form **49** when hydrogen bonding is eliminated as in esters and ethers, for example, **50**(b), $NR_2 = 1$-piperidino, J_{AB} 9·2 (*threo*), 7·7 Hz (*erythro*). The *gauche-erythro* form (**49**) is most favoured in quaternary salts, e.g. the methiodide of **50a**, $NR_2 = 1$-piperidino, J_{AB} 10·7 (*threo*) 2·9 Hz (*erythro*), since $^+NR_2Me/Ph$ steric interactions in the *anti* form **52** will be very severe. *Erythro* J_{AB} values for all 1-pyrrolidino derivatives examined were smaller than those of corresponding dimethylamino, 1-piperidino and 4-morpholino derivatives [e.g. *erythro*, **50**(c), $R = 1$-pyrrolidino, J_{AB} 4·4 Hz, *erythro*, **50**(c), $R = 1$-piperidino, J_{AB} 8·7 Hz; the lower value shows that **52** is particularly unfavoured for the former base] and this result indicates that the steric requirements of the pyrrolidine group are unusually high.

Schmid (1968) has studied the PMR characteristics of some diastereo-isomeric 1,2-disubstituted-1-aryl propanes (**53**). In cases where X and Y

$$PhCH_A(X)CH_B(Y)Me \text{ isomers}$$

(a) *erythro-trans* H/H   (b) *threo-gauche* H/H   (c) *threo-trans* Me/Ph

53

were Cl or Br, isomeric J_{AB} values (8–10 Hz for *erythro*, near 5 Hz for *threo*) supported the *trans* (H_A/H_B) and *gauche* forms (a) and (b) respectively as preferred conformations. In these, the two highly electronegative halogen atoms are *trans* to each other. Further evidence derived from methyl chemical shifts—these were consistently higher field for *threo* derivatives and suggest that methyl is close to (and shielded by) phenyl in these, and removed from phenyl in the *erythro* isomers. A high population of the *threo* conformer (c) (also *gauche* H_A/H_B but *trans* Me/Ph) would be in-consistent with an isomeric chemical shift difference because the methyl environments in *erythro* (a) and *threo* (c) are similar. Other isomeric pairs, e.g. those in which $X = OH$, $Y = Br$ showed little evidence of conforma-tional preference as judged by the small differences between their J_{AB} values and methyl chemical shifts.

Ambiguity in stereochemical assignments based on the J_{AB} values of diastereoisomers may sometimes be overcome by converting the acyclic isomers to cyclic forms (of greater conformational specificity) especially when this can be done without involving bonds attached to the asymmetric centres. Thus the diastereoisomeric amino-alcohols **54** and **55** (J_{AB} 7 and 9·1 Hz respectively) yield the cyclic isochromans **56** and **57** when their

56	**54**	**55**	**57**
J 3 Hz (H/H)	*erythro*	*threo*	J 9 Hz (H/H)

58

threo trans H/H

methiodides are heated; the latter are clearly assigned *cis* and *trans* configurations respectively from their J_{AB} values (Randall *et al.*, 1965). The precursor amino-alcohols may be assigned the same configurations as their derived isochromans, because the cyclization step is unlikely to alter configuration. Looking now at the J_{AB} values of the amino-alcohols themselves it is seen that while the *erythro* value (7 Hz) indicates a lack of conformational preference amongst *trans* and *gauche* H_A/H_B conformers, the value of 9·1 Hz for the *threo* isomer is consistent with the preferred *trans* H_A/H_B conformer (**58**). There is evidence of strong intramolecular hydrogen bonding between OH and NMe$_2$ in the *threo* isomer, hence **58** is probably stabilized by an OH—N bond.

Examples of configuration problems in diastereoisomers of pharmacological interest, solved by their conversion to cyclic analogues, are given elsewhere (p. 211).

PMR spectroscopy provides an elegant means of differentiating between *meso* and *racemic* (dl) diastereoisomers. Consider first the molecule **59** in which methylene (C-2) is adjacent to an asymmetric carbon centre (C-3); the methylene protons form an AB system typical of CH$_2$ in an asymmetric environment, a phenomenon which is well known [see van Gorkom and

$$\overset{\text{H}}{\underset{\text{Me}}{\overset{1}{\text{Me}}\overset{2}{\text{CH}_2}\!-\!\overset{3}{\text{C}^*}\!-\!\text{Ph}}}$$

59

Hall (1968) for a recent review]. If C-1 methyl in **59** is now replaced by a similar asymmetric unit, the methylene protons become equivalent *provided the configurations of the C-1 and C-3 centres are identical,* as in the Fischer projections (**60**), and will behave as an A_2 group in their coupling with vicinal protons. This configurational arrangement **60**(a) and (b) gives molecules which lack a plane of symmetry and are therefore optically active; the PMR spectra of the enantiomorphs [**60**(a) and (b)] are indistinguishable and identical with that of a racemic mixture when optically inactive solvents are used (see Chapter 5). When the C-1 and C-3 configurations differ, as

$$\begin{array}{ccc} \text{Me} & \text{H} & \text{Ph} \\ |1 & |2 & |3 \\ \text{Ph---C---C---C---Me} & & \text{(a)} \\ | & | & | \\ \text{H} & \text{H} & \text{H} \end{array}$$

-   } optically active enantiomorphs

$$\begin{array}{ccc} \text{H} & \text{H} & \text{H} \\ | & | & | \\ \text{Ph---C---C---C---Me} & & \text{(b)} \\ | & | & | \\ \text{Me} & \text{H} & \text{Ph} \end{array}$$

60

$$\begin{array}{ccc} \text{Me} & \text{H} & \text{Me} \\ | & | & | \\ \text{Ph---C--C--C---Ph} \\ | & | & | \\ \text{H} & \text{H} & \text{H} \end{array}$$

61

in **61**, the methylene protons have an asymmetric environment and behave as an AB group. The arrangement **61** has a plane of symmetry and is therefore the optically inactive *meso* diastereoisomer. Thus *meso* isomers of the type *RCH$_2$R* may be differentiated from (+), (−) and (±) forms by the greater complexity of the *meso* methylene PMR signal. This is illustrated from data upon isomers of 2,4-diphenylpentane, the example used above to demonstrate general principles. In the spectrum of racemic material the methylene signal is a triplet, the A_2 group being split by the methine protons at C-2 and C-4 (the identity or otherwise of proton groups adjacent to the central methylene group is discussed below), while in the *meso* spectrum the methylene signal is clearly far more complex (Bovey, 1968; Fig. 3.8).

Another example is that of the isomeric 2,4-pentanediols (Fukuroi *et al.*, 1968). In the spectrum of the *meso* isomers each line of the methylene AB quartet is split into three by the flanking equivalent methine protons, giving a well-resolved 12-line signal [Fig. 3.9(a)]. A complex methine signal is also

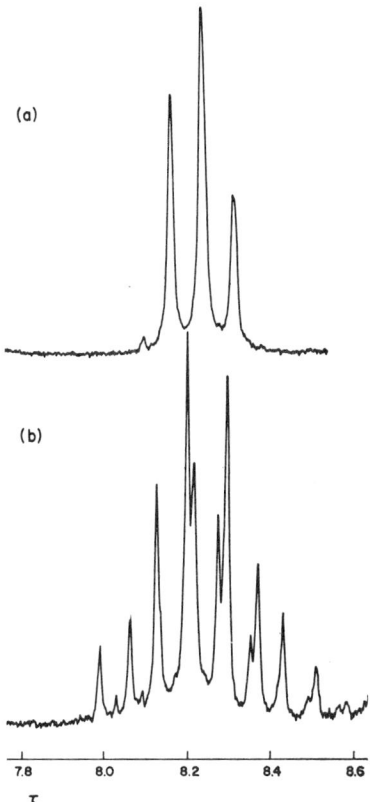

(a)

(b)

| | | | | |
| 7.8 | 8.0 | 8.2 | 8.4 | 8.6 |

τ

FIG. 3.8. The 100-MHz methylene PMR signal of (a) racemic and (b) *meso*-2,4-diphenylpentane, 10% (v/v) in chlorobenzene at 35° (Bovey, 1968).

obtained (not shown). In the spectrum of the racemic diol, the methylene signal is a triplet which proves, on higher resolution, to comprise a pair of overlapping doublets [Fig. 3.9(b)]. These arise as a result of small chemical shift differences between the C-1 and C-3 methine protons, now to be discussed. In *meso* isomers of type **61**, the C-1 and C-3 methine protons (and also the C-1 and C-3 methyl groups) have identical environments and therefore do not give rise to separate PMR signals. In racemates of type

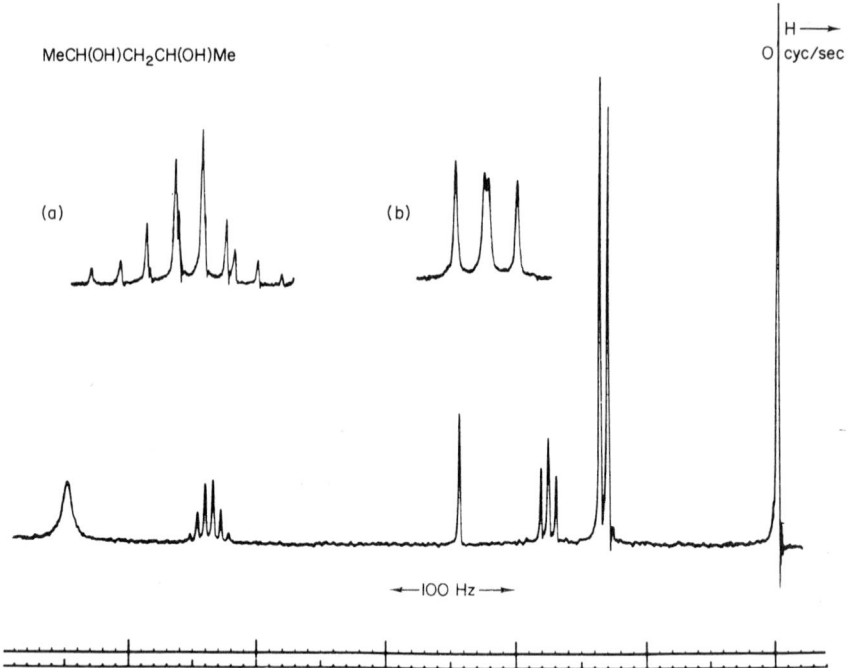

FIG. 3.9. 100-MHz PMR spectra of racemic 2,4-pentanediol in pyridine. The expanded methylene signals of the *meso* and racemic isomers are shown in inserts (a) and (b) respectively. In (a), the solvent is D_2O (Fukuroi *et al.*, 1968).

60, however, the C-1 methine and methyl environments are *not* the same as those of the corresponding groups at C-3 and hence each may produce a unique signal. Environmental differences between *meso* and racemic isomers may thus be summarized:

$$
\begin{array}{ccc}
X & H & X \\
| & | & | \\
C\!-\!C\!-\!C \\
| & | & | \\
Z & H & Z
\end{array}
\quad
\begin{array}{l}
\textit{meso} \text{ methylene non-equivalent} \\
\quad\text{X, Z equivalent}
\end{array}
$$

$$
\begin{array}{ccc}
X & H & Z \\
| & | & | \\
C\!-\!C\!-\!C \\
| & | & | \\
Z & H & X
\end{array}
\quad
\begin{array}{l}
\textit{racemic} \text{ methylene equivalent} \\
\quad\text{X, Z non-equivalent}
\end{array}
$$

The extents to which the signals of non-equivalent groups differ varies greatly and depends to a large degree on conformational preferences; resort to a high frequency spectrometer or to a search for an appropriate solvent may be necessary to detect spectral differences, remarks which apply to AB systems in general (van Gorkum and Hall, 1968). A further example where spectral differences are seen is that of the isomeric dimethyl 3-acetoxy-2,4-dimethylglutarates [**62**(a), (b) and (c)] (Enterman, 1969).

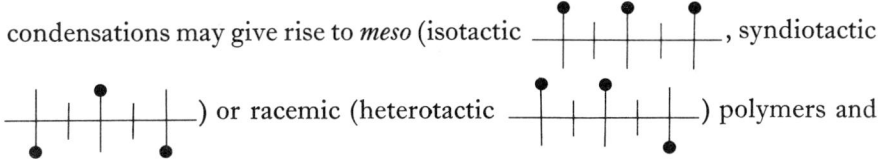

(R = CO_2Me; t = triplet; d = doublet; dd = doublet of doublets)

Two of the three possible optically inactive isomers have identical spectra. In these, the C-2 methine signal (low field because of the acylation shift) forms a sharp triplet, and this fact shows it to be adjacent to equivalent C-1 and C-3 methine protons. Hence these spectra must arise from *meso* forms; this is confirmed by the presence of a single methyl doublet near τ 9 [the two *meso* forms **62**(a) and (b) cannot be differentiated by means of this data]. By elimination, the unique spectrum must be due to the racemic mixture and this is borne out by its display of doublet of doublet signals for both the C-2 methine and C-1(3) methyl groups.

The differentiation of meso and racemic diastereoisomers has been largely developed to aid elucidation of the configurations of polyvinyl compounds and other polymers (Bovey, 1968, 1969). Methyl acrylate condensations may give rise to *meso* (isotactic ⎯⎤⎤⎤⎤⎤⎯, syndiotactic ⎯⎤⎤⎤⎤⎤⎯) or racemic (heterotactic ⎯⎤⎤⎤⎤⎤⎯) polymers and each type has distinctive PMR characteristics. Chemical shift differences between methylene protons of isotactic polymers, for example, are greater than those of the syndiotactic type. This is illustrated in Fig. 3.10 which shows the spectrum of a predominantly isotactic (CH_2 resonance is an AB quartet, $J \sim 15$ Hz) and a predominantly syndiotactic polymer (CH_2 resonance is a broad singlet).

The identification of preferred conformers in acyclic molecules is often of importance in its own right and may be studied in cases where isomer

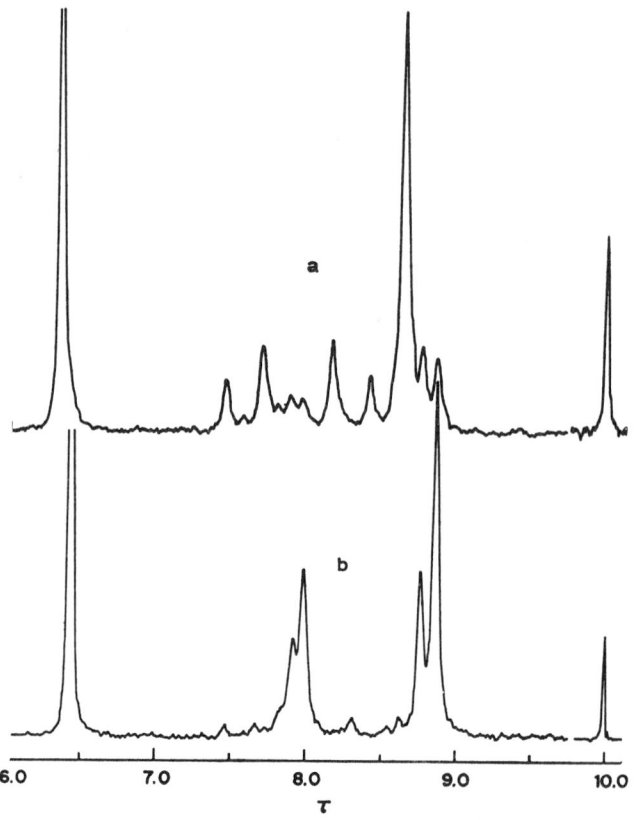

Fig. 3.10. 60-MHz PMR spectra of 15% (w/v) solutions in chlorobenzene of polymethyl methacrylate prepared with (a) an anionic initiator (PhMgBr) and (b) a free-radical initiator. The former sample is a predominantly isotactic polymer and the latter a syndiotactic polymer (Bovey, 1968).

configurations are already established by other means. Examples of biological interest, where molecular conformation may be related to drug-receptor uptake etc., are given elsewhere; the case discussed below is of particular interest because it illustrates the use of deuterated analogues.

The more stable conformer of the 1,2-dibromoethane, $BrCH_2CHBrR$, is the *gauche* bromine form (**63**) when R is *t*-butyl, and the *trans* form (**64**)

Br

H⟍ ⟋Br

H⟋ ⟍H

R

63 *gauche Br/Br*

Br

R⟍ ⟋H

H⟋ ⟍H

Br

64 *trans Br/Br*

when R is phenyl or carbethoxy (Buza and Snyder, 1966; Jablonski and Snyder, 1968); these findings have been made by methods in which the three coupled protons are individually identified, as outlined below for the compound, where R is CO_2Et. Analysis of the spectrum of ethyl 2,3-dibromopropionate, $BrCH_2CHBrCO_2Et$ (40% in benzene) shows that one of the non-equivalent methylene protons has a chemical shift of 5·18 Hz (relative to the highest field line of the CH_3CH_2 quartet) while the other has the value 35·14 Hz; the former is strongly (J 11·16 Hz) and the latter weakly (J 4·4 Hz) coupled to the α-methine proton. To use this data for conformational assignments, it is necessary to allocate chemical shift values to specific methylene protons. This is done as follows. The reaction sequence:

$$HC{\equiv}CCO_2Et \xrightarrow[\text{Lindler Catalyst}]{D_2} \begin{array}{c} Br \\ D{\cdots}C{=}C{\cdots}D \\ H{\blacktriangleright}(\ +\ ){\blacktriangleleft}CO_2Et \\ {\searrow}Br \end{array} \longrightarrow \begin{array}{c} Br\ \ D \\ D{\cdots}\diagup{\cdots}CO_2Et \\ H\ \ Br \end{array} + 13\% \textit{ erythro}$$

threo (major product)

EtO₂C⟍ ⟋Dα

H⟋ ⟍D

Br

65 *threo (trans Br/Br)*

Br

D⟍ ⟋Br

H⟋ ⟍D

CO_2Et

66 *threo (gauche Br/Br)*

provides a deuterated ester ($BrCHDCDBrCO_2Et$) mixture in which the *threo* diastereoisomer preponderates as a result of the steric course of reaction. The chemical shift of the sole β-proton of this isomer is 7·0 Hz and thus corresponds with the methylene proton (ν 5·18 Hz) of the undeuterated ester that has the larger J_{vic} value. The relationship of the β-hydrogen to the α-bromine and deuterium atoms is known (because the compound has a *threo* configuration) and this may be represented by either of the conformers **65** and **66**, analogous to probable *trans* and *gauche* Br/Br conformers of the normal ester respectively. Replacement of α-D by H in **66** (*gauche* Br/Br) leads to a weakly coupled ($\phi = 60°$) and in **65** (*trans* Br/Br) to a strongly coupled ($\phi = 180°$) vicinal proton pair; the latter is consistent with the observed J_{vic} value, hence the preferred conformation of both the normal and deuterated ester must be the *trans* conformer. The β-proton of *erythro* $BrCHDBrCO_2Et$ (v 35·7 Hz) corresponds with the weaker coupled methylene proton of the minor product of the deuteration sequence, a result which leads to the same conclusion of conformational preference. Stereospecific dideuterated derivatives have also been used for the conformational analysis of 3-phenylpropanol (Snyder, 1969).

CYCLIC DERIVATIVES†

6-Membered Saturated Rings

The room temperature PMR spectrum of cyclohexane is a sharp singlet since each proton experiences the same average environment in the rapidly

interconverting system **67**. On cooling, however, the signal broadens and splits into two distinct bands at −70° (Jensen *et al.*, 1962), axial and equatorial environments being separately detected when the chair–chair interconversion rate is NMR slow. The higher field band is due to the axial protons and the lower to equatorial protons and the signals are broad because each proton is coupled to its geminal (non-equivalent) neighbour and to the two vicinal protons of different conformation, e.g. H_e in **68** is coupled with the axial protons of C_1, C_2 and C_3. Two sharp signals of 28 Hz separation are seen in the 60-MHz spectrum of cyclohexane-d_{11} **(69)** at −100°, after the H—D coupling has been removed by double irradiation (Anet and Bourn, 1967).

† See Booth (1969) for a recent review of the applications of PMR spectroscopy to the conformational analysis of cyclic derivatives.

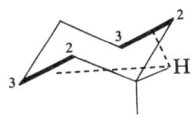

transverse   longitudinal screening component

Axial-equatorial chemical shift differences are due to long-range shielding effects associated with the anisotropies of carbon–carbon single bonds bearing a 2,3 relationship to the protons in question (see above); the net influence is deshielding (cf. C_6H_{12} τ 8·5; RCH_2R τ 8·75 for *n*-hexane), equatorial protons being the most affected (Jackman, 1959; Apsimon *et al.*, 1967a). A good example of temperature effects upon the PMR signals of axial and equatorial protons is provided by spectra of piperidine-3,3,5,5-d_4 (Fig. 3.11). At room temperature the ring protons of this derivative give deuterium broadened singlets (α- at lower field and twice the intensity of γ-) all vicinal proton coupling being eliminated. As the temperature is lowered the signals broaden and both give AB spectra at $-85°$. In methanol-d_4, geminal proton chemical shift differences are: α, 26·1 (J 11·9) and γ, 24·8 (J 13·1) (Hz, 60-MHz frequency).

Garbisch and Griffith (1968) have recorded the spectrum of the octa-

$$\begin{array}{c} D_2 \\ D_2C-C \diagdown \quad \begin{matrix} H \\ H \end{matrix} \\ | \\ D_2C-C \diagup \quad \begin{matrix} H \\ H \end{matrix} \\ D_2 \end{array}$$

70

deuterated cyclohexane (**70**) in CS_2 (an AA'BB' spectrum at $-103°$) and obtained the following parameters: J_{aa} 13·12; J_{ee} 2·96; J_{ae} 3·65 Hz.

The ring proton signals of cyclic derivatives are more complex when substituents are present and when rings are heterocyclic, and are particularly difficult to resolve in compounds which show pronounced conformational preferences, e.g. *t*-butyl derivatives. In these cases, axial and equatorial environments are not averaged and complex coupling patterns arise. A single proton geminal to an electronegative substituent may, however, often be moved sufficiently down field to be clear of the main proton band and data of stereochemical value often accrue from the resolution of such signals (see later).

Fused polycyclic systems such as steroids typically show complex ring proton resonances because ring fusions (commonly *trans*) immobilize the molecule. Chair–chair interconversions are still possible, however, in *cis*

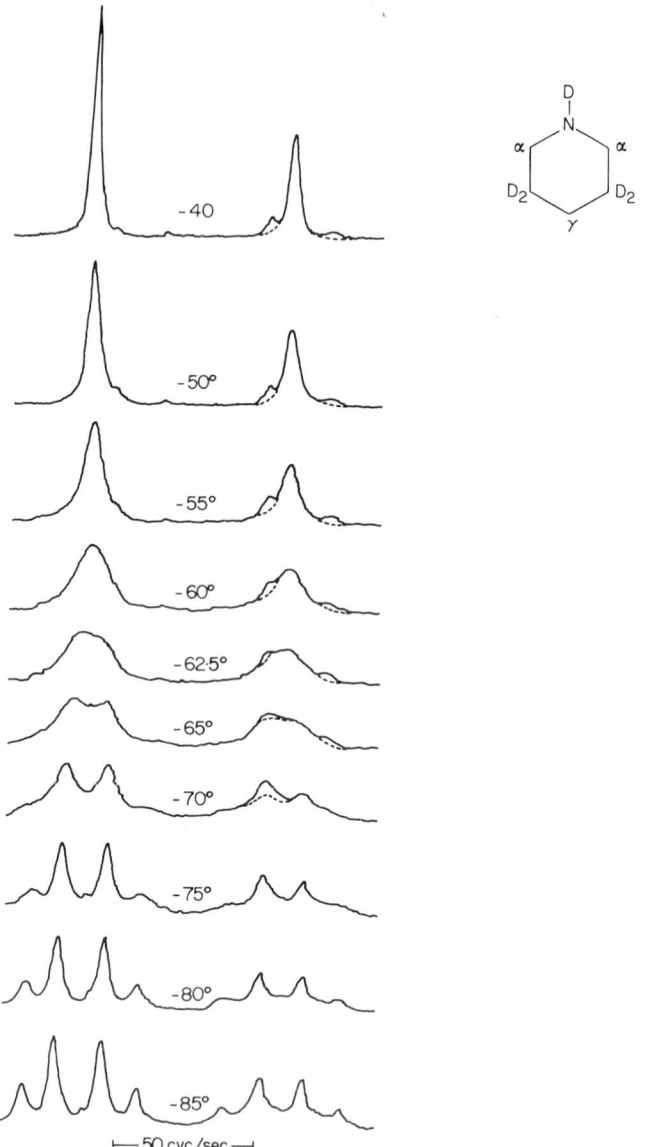

FIG. 3.11. 60-MHz PMR spectrum of piperidine 3,3,5,5-d₄ as a function of temperature in methanol-d₄; α-protons fall to the left of each spectrum (Lambert *et al.*, 1967).

fused systems and the relative complexity of the PMR spectra of *cis* and *trans* fused isomers may serve as a means of differentiation, e.g. the spectrum of *trans* decalin, a rigid molecule, forms a poorly resolved broad band, whereas that of *cis* decalin (mobile) is a singlet. The fortuitous chemical

trans decalin

cis decalin

equivalence of the 1, 2 and 3 proton groups of *cis* decalin contributes to spectral simplicity in addition to the averaging of *a* and *e* proton environments (Musher and Richards, 1958).

The elucidation of the configurations of 6-membered asymmetric cyclohexane derivatives rests heavily upon the application of the $\cos^2\phi/J_{\text{vic}}$ relationship of Karplus. Examples are given throughout this book, particular attention being given to 6-membered rings containing nitrogen in Chapter 4.

Stereochemical assignments to cyclic compounds with fewer than six ring members may also be made on the basis of the Karplus relationship but the interpretation of coupling constant data cannot necessarily be made so readily. In cyclopropane derivatives it is observed that *cis* couplings are larger than *trans* as is consistent with a *cis* dihedral angle (near zero) corre-

sponding with a larger J_{vic} value than the *trans* angle (near 120°) in the eclipsed system of a rigid 3-membered ring conformation (see above). However, if the *cis* and *trans* angles rise as a result of ring deformation (to relieve *cis* interactions), the *cis* and *trans* J values approach one another, and in fact the vicinal coupling constant ranges overlap (J_{cis} 6·6–12·5 Hz and J_{trans} 3·9–8·6 Hz) (Sternhell, 1969). Oxygen in a 3-membered ring reduces all couplings significantly (epoxides are discussed in Chapter 7, p. 254). In 4- and 5-membered rings which approach planarity, J_{cis} is generally significantly larger than J_{trans}, as expected from the Karplus relationship (e.g. **71** and **72**), but in most cases the flexibility of these ring systems demands that each problem be analysed separately (Sternhell,

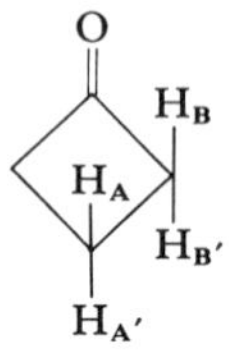

71

$J_{AB} = 10·0, J_{AB'} = 6·4$

(Sutcliffe and Walker, 1967)

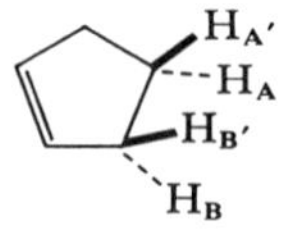

72

$J_{AB} = 7·4, J_{AB'} = 4·6$

(Abraham, 1967)

1969). Isomeric camphane-2,3-diols provide models for coupling constants in 5-membered rings constrained to an envelope conformation (see **73**) (Anet, 1961).

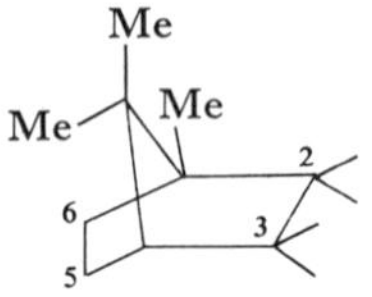

73

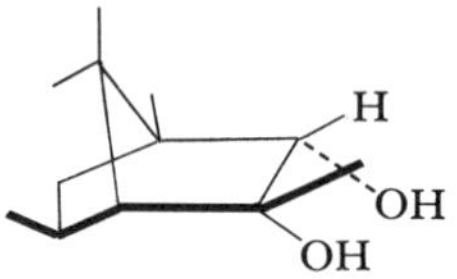

74

2,3-OH substituent	2H/3H geometry	$J_{2,3}$ (Hz)
2-*exo*, 3-*exo*	*cis*	7·7
2-*endo*, 3-*endo*	*cis*	8·0
2-*exo*, 3-*endo*	*trans*	2·3
2-*endo*, 3-*exo*	*trans*	2·2

Endo-substituents lie on the same, and *exo*-substituents on the opposite side of the 5,6-bimethylene bridge.

Long-range coupling between the 2,6 and 3,5 protons was observed in the 2-*endo*, 3-*endo* diol but not in the 2-*exo*, 3-*exo* form, the W-pathway usually required for coupling across four bonds only being available in the former derivative (see **74** and p. 91).

An example of the problems that may be involved in interpreting 5-membered ring coupling constant data is the case of *trans* hydroxy-L-proline and its *cis* (allo) isomer in which substituents are identical, but corresponding couplings vary widely (Abraham and McLauchlan, 1962). Comparing 2,3

hydroxy proline

J_{2a-3a} (*cis*) 7·66

J_{2a-3b} (*trans*) 10·44

allo isomer

10·48

3·84

couplings (above), it is seen that the *cis* value is the greater in alloproline (as normal) but the smaller of the two values in hydroxyproline itself. Similar results are also seen in 3,4 and 4,5 couplings and the results suggest that the conformation of the allo isomer differs from hydroxyproline, possibly because a strong intramolecular hydrogen bond between OH and CO_2^- can operate in the former compound. Another complication is the fact that electronegative ring members may increase, rather than decrease, the magnitude of vicinal coupling constants when in certain situations. Thus replacement of methylene α- to the double bond of the cyclopentene (**72**) by oxygen raises J_{AB} to 10·7 and $J_{AB'}$ to 8·3 Hz (Abraham, 1967). It has been suggested that the lone pair electrons of oxygen (and other heteroatoms) contribute a positive increment to J_{vic} when in certain orientations with respect to the HCCH system and that this increment is approximately +2·3 Hz for each perfect eclipsing (Anteunis, 1966). This effect is not seen in 3-membered heterocyclic compounds such as epoxides.

The analysis of some spectral data upon isomeric 3,3-diphenyltetrahydrofuran derivatives, which further illustrates configurational problems in 5-membered cyclic derivatives, is given elsewhere (Chapter 6, p. 211).

CONFORMATIONAL FREE ENERGY VALUES

Estimation of the conformer populations of substituted 6-membered alicyclic molecules may be made if the free energy difference between likely conformers (usually only the two chair forms are considered) is known.

Take the case of a mono-substituted cyclohexane $C_6H_{11}X$ in which the equatorial —X and axial —X conformers will be in equilibrium. Chair–

e-X   *a*-X

chair interconversion readily occurs because the energy requirements of this change is low, but at any given instant in time the population of the more stable (here the *e*-X conformer) will exceed that of the axial conformer. This equilibrium is expressed by the equation

$$K = \frac{[E]}{[A]} \tag{1}$$

where K is the conformational equilibrium constant, $[E]$ is the concentration of the equatorial conformer and $[A]$ the concentration of the axial conformer. The greater the free energy difference between the *e*-X and *a*-X forms, the smaller $[A]$ and the greater the K value.

The free energy difference (ΔG_x^0) and K are related by the expression

$$-\Delta G_x^0 = RT \log_e K \tag{2}$$

where R is the gas constant and T the absolute temperature. The subscript x denotes the substituent in question. The value $-\Delta G_x^0$ is called the conformational free energy difference, or simply the conformational energy. If K or $-\Delta G_x^0$ is known, the percentage of the more stable conformer may be calculated. Thus, a K value of unity ($-\Delta G_x^0 = 0$) is indicative of an absence of conformational preference while $K = 1\cdot5$ corresponds with 60 and $K = 4\cdot0$ with 80% of the more stable conformer (see table in Eliel, 1962).

NMR methods are available for the measurement of K (and hence $-\Delta G_x^0$) values. The ideal procedure is illustrated using the conformational equilibrium shown above. The chemical shift of the methine proton geminal to X will generally differ in the two conformers (lower field when it is equatorial), a weighted average being obtained when the chair–chair interconversion rate is NMR-fast. On cooling, signals due to the two conformers become apparent (usually temperatures below −80° are required) and a measure of their relative intensities provide the ratio $[E]/[A]$

required for the evaluation of K. This technique has been developed chiefly by Jensen and his colleagues (1969); it requires more demanding instrumentation and technical skill than other methods (to be described). Although it does not provide data that is valid at room temperatures, it does allow the accurate comparison of relative steric bulk if determinations are all taken under the same conditions and at the same temperature. An extrapolation procedure, giving ΔG_x^0 values at room temperature is

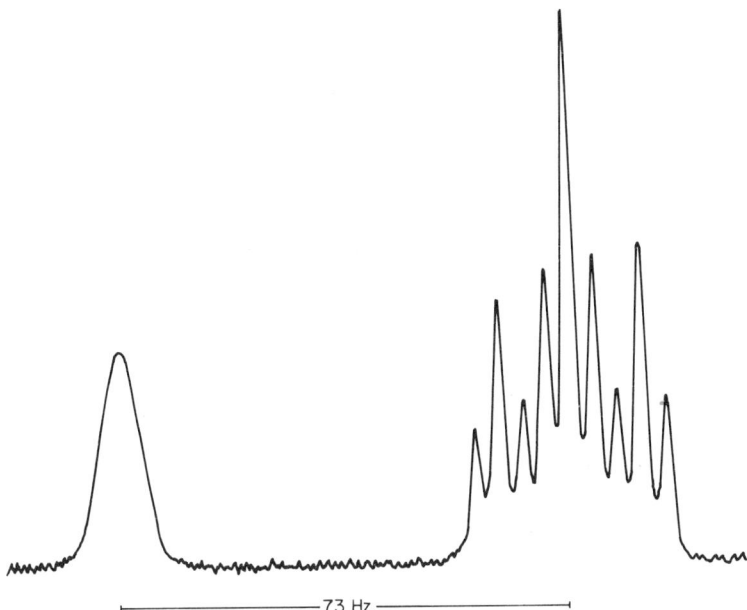

FIG. 3.12. 100-MHz PMR signal of the methine proton of iodocyclohexane at 80° in carbon disulphide (Jensen *et al.*, 1969).

mentioned later. Over twenty low temperature ΔG_x^0 values are listed in a recent paper (Jensen *et al.*, 1969); a typical example of spectral appearance is given in Fig. 3.12 the higher field signal showing the more extensive coupling pattern (triplet of triplets, 2 *a/a*, 2 *a/e* couplings) characteristic of an axial proton. The values obtained reveal some interesting variations in the steric requirements of substituents in cyclohexane derivatives. For example a steric bulk sequence F < I < Br < Cl is found which identifies chlorine, not iodine, as the "largest" halogen [this result is explained on the basis of bond length and polarizabilities (Berlin and Jensen, 1960)], while substituents with cylindrical symmetry such as nitrile and ethinyl have lower conformational preferences than groups such as carbomethoxy and nitro.

Substituent	$-\Delta G^0$ (−78 to −90°)	Stabler isomer (%)
F	0·276	67·5
I	0·468	77·5
Br	0·476	77·7
Cl	0·528	80·2
CN	0·24	65·3
C≡CH	0·41	74·7
CO$_2$Me	1·31	96·9
NO$_2$	1·05	94·1

A less accurate but more readily applicable method has been introduced by Eliel (1959) and is described below. The observed chemical shift, δ, of the methine proton of the cyclohexyl derivative $C_6H_{11}X$ (see p. 124) may be calculated from the equation

$$\delta = N_e \delta_e + N_a \delta_a \tag{3}$$

where δ_e and δ_a are the chemical shifts (at normal temperatures) of the methine proton of the purely equatorial and axial conformers respectively, and N_e and N_a are the respective mole fractions of these conformers (Gutowsky and Saika, 1953; Eliel, 1959).

Multiply (3) by $[C]$, the total concentration.

$$\delta[C] = N_e[C]\delta_e + N_a[C]\delta_a \tag{4}$$

Since $[C] = [E] + [A]$, $N_e[C] = [E]$, and $N_a[C] = [A]$, (4) becomes

$$\delta[E] + \delta[A] = \delta_e[E] + \delta_a[A] \tag{5}$$

divide (5) by $[A]$

$$\delta[E]/[A] + \delta = [E]/[A]\delta_e + \delta_a \tag{6}$$

or

$$\delta K + \delta = \delta_e K + \delta_a$$

whence

$$K = \frac{\delta_a - \delta}{\delta - \delta_e} \tag{7}$$

To use this equation δ_a and δ_e must be evaluated; this can not be done accurately and approximate methods must be used. In the most usual one, shifts are measured in a 4-*t*-butyl substituted analogue, the *cis* isomer giving

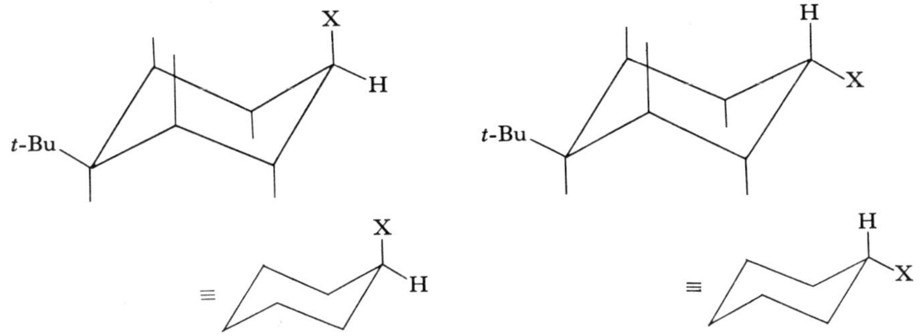

cis

trans

δ_a and the *trans*, δ_e. It must be assumed that the *t*-butyl group of these essentially "frozen" models, does not affect the 4-H methine environment, i.e. that the methine chemical shifts of pairs with and without the *t*-butyl group are identical (see later for criticism). Data upon bromocyclohexane (Eliel and Gianni, 1962) is given as an example:

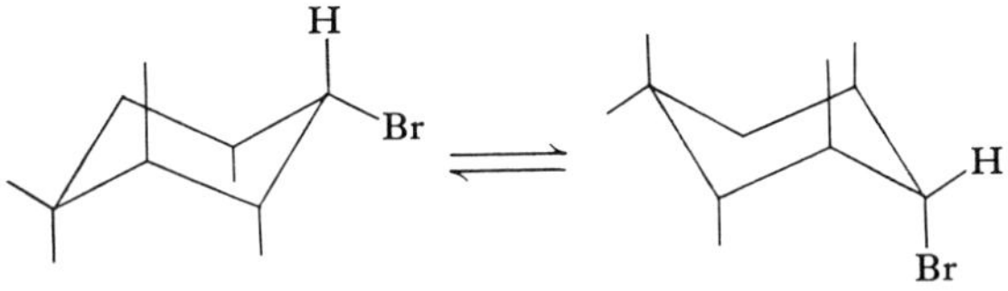

δ (methine) in Hz from TMS (40 MHz spectra) no solvent

δ 166·5

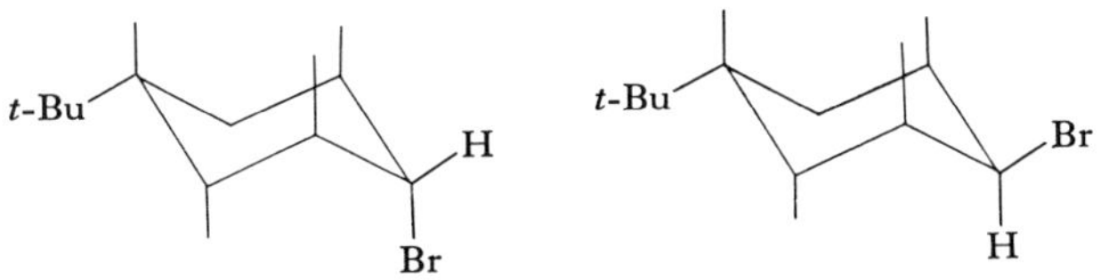

cis

δ 185

trans

δ 152·5

(N.B. higher field)

whence

$$K = \frac{(\delta_a - \delta)}{(\delta - \delta_e)} = \frac{185 - 166 \cdot 5}{166 \cdot 5 - 152 \cdot 5} = \frac{18 \cdot 5}{14} = \underline{1 \cdot 3}$$

Thus the e-Br:a-Br conformer ratio in cyclohexyl bromide is about $3:2$. The K value for *cis*-4-methylcyclohexyl bromide, calculated as above and assuming the e-Br conformer to be the more stable, leads to the improbable

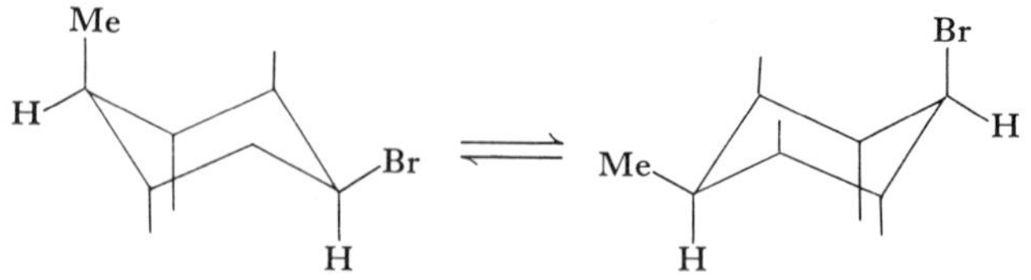

δ (H—C—Br) 179·5 (same conditions)

value of 0·2. This means that the a-Br conformer is, in fact, the more stable form and its population is about 83% ($K \sim 5$). Eliel's method becomes less accurate when δ approaches δ_e [since the quotient $\delta - \delta_e$ is involved in equation (7)] and is not suitable for large X-substituents which have a pronounced preference for the equatorial position. A large number of $-\Delta G_x^0$ values have been measured by this method and data are available in table form (Eliel *et al.*, 1965). Recent examples include the formyl (Buchanan *et al.*, 1967) and trifluoromethyl group (Della, 1967) and the halogens, fluorine, chlorine and bromine (Eliel and Martin, 1968b).

The assumption made in Eliel's method (1959) that a *t*-butyl group has no effect upon the chemical shift of a remote proton has been challenged by a number of workers. Jensen and Beck (1968) measured the separate a and e methine resonances of a series of cyclohexane derivatives at low temperatures ($< -80°$) and found values which differed from those of the methine protons of *t*-butyl analogues examined at the same temperature. Differences were seen for both a- and e-protons and ranged from 2·6 Hz for the axial proton of acetoxy derivatives to 8·5 Hz for the axial resonance of cyano derivatives. Data for bromo derivatives are shown below.

x	Temp. (°C)	Cyclohexyl		4-*t*-Butylcyclohexyl	
		ν_e	ν_a	ν_e	ν_a
Br	+24	406·2		452·9	380·4
	−47 to −49	397·6		454·2	380·9
	−84	459·0	387·2	455·3	381·7
	−91 to −92	459·7	387·4	455·7	381·9
	−102 to −100	460·0	387·8	456·0	382·1

Conditions: 0·5 M in CS_2. ν in Hz from TMS, 100 MHz.

Jensen and Beck estimated a and e methine shifts at room temperature by extrapolation and used these values to calculate $-\Delta G_x^0$. Values obtained showed better agreement with those calculated from peak area measurements (at about $-83°$) than did values derived from data upon 4-t-butyl derivatives, e.g. $-\Delta G_{Br}^0$, 0·49 (peak area), 0·55 (temperature corrected chemical shifts) and 0·35 Kcal/mol (t-butyl data). To get these differences into perspective, however, it must be noted that this range of values represents only a small population difference; thus a ΔG^0 value (at 25°) of 0·5 kcal/mol corresponds with about 70 and 0·35 kcal/mol with just under 65% of the more stable conformer population. Eliel and Martin (1968a), in a general examination of the validity of the NMR method of establishing conformational equilibria, compared $-\Delta G_x^0$ values ($x = $ OH, F and Cl) calculated from data upon 3- and 4-t-butyl derivatives. They found that the values based upon the 4-t-butyl derivatives agreed much better with literature values (using methods such as electron diffraction and low temperature NMR) than those in which the chemical shifts of 3-t-butyl isomers had been employed. These authors concluded that whereas 3-t-butyl isomers were not suitable models, the use of 4-substituted analogues as originally proposed (Eliel, 1959) was still justified for determining conformational equilibria in cyclohexanes. The question remains a controversial one.

Just as the chemical shift of the 1-methine proton in the mobile system $I \rightleftharpoons II$ is a time-average value between the purely axial and equatorial

I
mole fraction (N)

II
(1 − N)

orientations, so too are the J_{values} weighted time-averaged values between the values operative in the two chair conformations. If the 1-proton (H_x) and the adjacent methylene protons (H_A and H_B) represent an A_2B_2X system:

$$\text{Average } J_{AX} = N J_{AX\,(I)} + (1 - N) J_{AX\,(II)}$$
$$= N J_{ea} + (1 - N) J_{ae} \qquad (8)$$

$$\text{Average } J_{BX} = N J_{BX\,(I)} + (1 - N) J_{BX\,(II)}$$
$$= N J_{aa} + (1 - N) J_{ee} \qquad (9)$$

In principle, J_{AX} and J_{BX} may be obtained from the H_x signal of the cyclohexane derivative $C_6H_{11}X$, and J_{ea}, J_{ae}, J_{aa} and J_{ee} from the spectra of conformationally rigid models (e.g. *t*-butylcyclohexyl derivatives), whence N, K and $-\Delta G_x^0$ may be calculated. The various J values are only obtained, however, if the methine proton signals are resolvable and show first order splitting. Deviations from these conditions usually occur, due especially to long-range coupling of the α-proton to γ-protons. Sometimes the long-range couplings may be removed by deuteration (Anet, 1962; Anet and Henrichs, 1969). Thus Anet found that although the α-proton of cyclohexanol was a non-resolved multiplet, that of the 3,3,4,4,5,5-hexadeutero analogue was the anticipated triplet of triplets (α-H is coupled to two pairs of identical protons) (see Fig. 3.13). This nonet corresponded with J_{AX}

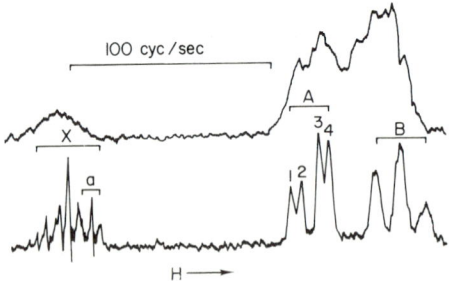

Fig. 3.13. 60-MHz PMR spectra of cyclohexanol (top) and 3,3,4,4,5,5-hexadeuteriocyclohexanol (bottom) in D_2O (Anet, 1962).

and J_{BX} values of 3·93 and 9·77 Hz respectively in solvent CCl_4. The methine signals of the rigid analogues **75** and **76** were also sharply resolved and allowed the model J_{values} to be measured (see below). Use of equation

<table>
<tr><td align="center">

75

J_{ae} 3·00 Hz

J_{ee} 2·72 Hz
</td><td align="center">

75

J_{aa} 11·07 Hz

J_{ea} 4·31 Hz in CCl_4
</td></tr>
</table>

(8) gives the result, $K = 2 \cdot 6$, while equation (9) leads to the value $5 \cdot 25$, namely

$$9 \cdot 77 = 11 \cdot 07 N + 2 \cdot 72 (1 - N)$$

$$N = 0 \cdot 84$$

$$K = \frac{N}{1 - N} = \frac{0 \cdot 84}{0 \cdot 16} = 5 \cdot 25$$

The latter calculation is probably the more reliable because it involves use of the J_{values} (aa and ee) of greatest separation.

In most applications of the average coupling constant method, however, proton signals are unresolvable and band width measurements are used instead; coupling constant sums, rather than individual values are required in this case. Thus, in the system $I \rightleftharpoons II$, if W_I and W_{II} are the separations of

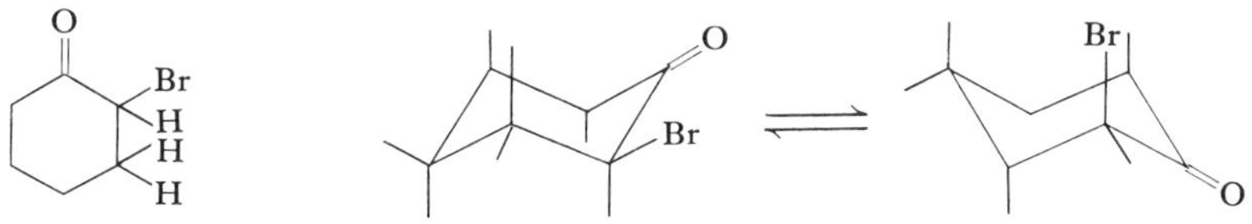

I
(*N*)

II
$(1 - N)$

the terminal lines in the methine signals of I and II respectively while W_{obs} is the measured value:

$$W_I = J_{\text{aa}} + J_{\text{ae}}$$
$$W_{II} = J_{\text{ee}} + J_{\text{ea}}$$

and

$$W_{\text{obs}} = W_I N + W_{II}(1 - N)$$

whence

$$N = \frac{W_{\text{obs}} - W_{II}}{W_I - W_{II}}$$

Again, W_I and W_{II} are obtained from the spectra of rigid analogues. The data of Garbisch (1964) are shown (Fig. 3.14). Note that assignments of the terminal peaks must take into account long-range effects so the full signal width is not measured. Feltkamp has made much use of band width measurements in calculating K values (Franklin and Feltkamp, 1965).

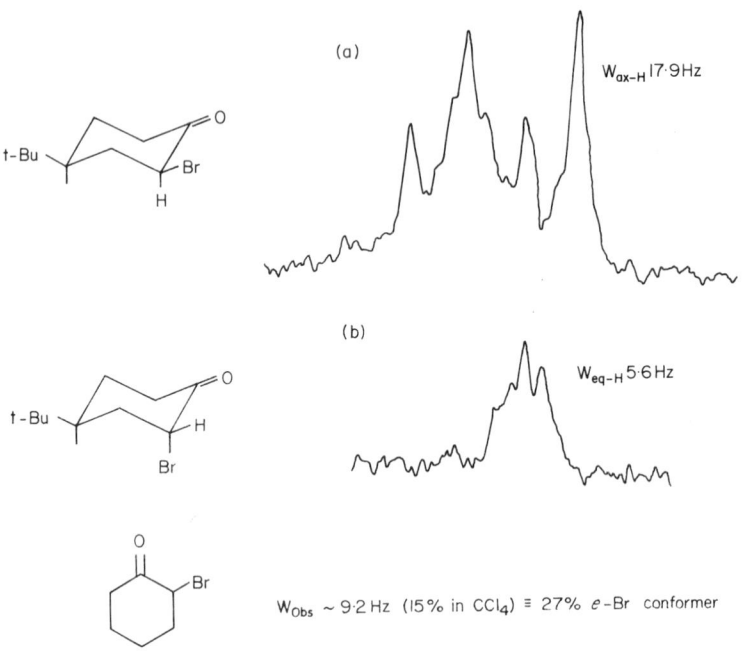

Fig. 3.14. 60-MHz PMR signals of methine (II—C—Br) protons in carbon tetrachloride of (a) *cis* and (b) *trans* 2-bromo-4-*t*-butylcyclohexanone. Coupling parameters (terminal widths) of these signals and the corresponding one of 2-bromocyclohexanone are also given. Data due to Garbisch (1964), spectra recorded at University of Alberta at sweep width 250 Hz.

Booth has also discussed this method and advocates the use of band widths measured at fractions of the band height rather than the complete width (1964).

References

Abraham, R. J. (1967). *In* "Nuclear Magnetic Resonance for Organic Chemists" (D. W. Mathieson, ed.). Academic Press, New York.
Abraham, R. J. and McLauchlan, K. A. (1962). *Mol. Phys.* **5**, 195.
Abraham, R. J. and Pachler, K. G. R. (1963). *Mol. Phys.* **7**, 165.
Anet, F. A. L. (1961). *Can. J. Chem.* **39**, 789.
Anet, F. A. L. (1962). *J. Amer. Chem. Soc.* **84**, 1053.
Anet, F. A. L. and Bourn, A. J. R. (1965). *J. Amer. Chem. Soc.* **87**, 5250.

Anet, F. A. L. and Bourn, A. J. R. (1967). *J. Amer. Chem. Soc.* **89**, 760.
Anet, F. A. L. and Henrichs, P. M. (1969). *Tetrahedron Lett.* 741.
Anteunis, M. (1966). *Bull. Soc. chim. Belges.* **75**, 413.
ApSimon, J. W. (1968). *Can. J. Chem.* **46**, 808.
ApSimon, J. W., Craig, W. G., Demarko, P. V., Mathieson, D. W., Nasser, A. K. G., Saunders, L. and Whalley, W. B. (1966). *Chem. Commun.* 754.
ApSimon, J. W., Craig, W. G., Demarco, P. V., Mathieson, D. W., Saunders, L. and Whalley, W. B. (1967a). *Tetrahedron* **23**, 2339.
ApSimon, J. W., Craig, W. G., Demarco, P. V., Mathieson, D. W., Saunders, L. and Whalley, W. B. (1967b). *Tetrahedron* **23**, 2357.
Barbieux, M., Defay, N., Pecher, J. and Martin, R. H. (1964). *Bull. Soc. Chim. Belg.* **73**, 716.
Barfield, M. and Chakrabarti, B. (1969). *Chem. Revs.* **69**, 757.
Berlin, A. J. and Jensen, F. R. (1960). *Chem. Ind.* 998.
Bhacca, N. S., Gurst, J. E. and Williams, D. H. (1965). *J. Amer. Chem. Soc.* **87**, 302.
Booth, H. (1964). *Tetrahedron* **20**, 2211.
Booth, H. (1965). *Tetrahedron Lett.* 411.
Booth, H. (1966). *Tetrahedron* **22**, 615.
Booth, H. (1969). *In* "Progress in Nuclear Magnetic Resonance Spectroscopy" (J. W. Emsley, J. Feeney and L. H. Sutcliffe, eds), Vol. 5. Pergamon Press, Oxford.
Bottomley, W., Phillips, J. N. and Wilson, J. G. (1967). *Tetrahedron Lett.* 2957.
Bovey, F. A. (1968). *Accounts Chem. Res.* **1**, 175.
Bovey, F. A. (1969). "Nuclear Magnetic Resonance Spectroscopy". Academic Press, New York.
Brocklehurst, K., Price, H. S. and Williamson, K. (1968). *Chem. Commun.* 884.
Buchanan, G. W., Stothers, J. B. and Wu, S-T. (1967). *Can. J. Chem.* **45**, 2955.
Buza, M. and Snyder, E. I. (1966). *J. Amer. Chem. Soc.* **88**, 1165.
Casy, A. F., Parulkar, A. P. and Pocha, P. (1968). *Tetrahedron* **24**, 3031.
Davis, D. R. and Roberts, J. D. (1962). *J. Amer. Chem. Soc.* **84**, 2252.
Della, E. W. (1967). *J. Amer. Chem. Soc.* **89**, 5221.
Dvornik, D. and Schilling, G. (1965). *J. Med. Chem.* **8**, 466.
Eliel, E. L. (1959). *Chem. Ind.* 568.
Eliel, E. L. (1962). "Stereochemistry of Carbon Compounds". McGraw-Hill, New York.
Eliel, E. L. (1965). *Angew. Chem. Int. Ed. Engl.* **4**, 761.
Eliel, E. L., Allinger, N. L., Angyal, S. J. and Morrison, G. A. (1965). "Conformational Analysis". Wiley, New York.
Eliel, E. L. and Gianni, M. H. (1962). *Tetrahedron Lett.* 97.
Eliel, E. L. and Martin, R. J. L. (1968a). *J. Amer. Chem. Soc.* **90**, 682.
Eliel, E. L. and Martin, R. J. L. (1968b). *J. Amer. Chem. Soc.* **90**, 689.
Elkik, E. and Francesch, C. (1968). *Bull. Soc. chim. France* 1371.
Enterman, W. (1969). Ph.D. Thesis, University of Alberta.
Franklin, N. C. and Feltkamp, H. (1965). *Angew. Chem. Int. Ed. Engl.* **4**, 774.
Fukuroi, T., Fujiwara, Y., Fujiwara, S. and Fujii, K. (1968). *Anal. Chem.* **40**, 879.
Garbisch, E. W. (1964). *J. Amer. Chem. Soc.* **86**, 1780.
Garbisch, E. W. and Griffith, M. G. (1968). *J. Amer. Chem. Soc.* **90**, 6543.
Gutowsky, H. S. and Saika, A. (1953). *J. Chem. Phys.* **21**, 1688.
Huffman, J. W. and Elliot, R. P. (1965). *J. Org. Chem.* **30**, 365.
Jablonski, R. J. and Snyder, E. I. (1968). *J. Amer. Chem. Soc.* **90**, 2316.

Jackman, L. M. (1959). "Applications of Nuclear Magnetic Resonance Spectroscopy to Organic Chemistry"; Jackman, L. M. and Sternhell, S., 2nd ed., 1969. Pergamon, Oxford.

Jardetzky, O. (1963). *J. Biol. Chem.* **238**, 2498.

Jensen, F. R. and Beck, B. H. (1968). *J. Amer. Chem. Soc.* **90**, 3251.

Jensen, F. R., Bushweller, C. H. and Beck, B. H. (1969). *J. Amer. Chem. Soc.* **91**, 344.

Jensen, F. R., Noyce, D. S., Sederholm, C. H. and Berlin, A. J. (1962). *J. Amer. Chem. Soc.* **84**, 386.

Johnson, C. E. and Bovey, F. A. (1958). *J. Chem. Phys.* **29**, 1012.

Karabatsos, G. J., Sonnichsen, G. C., Hsi, N. and Fenoglio, D. J. (1967). *J. Amer. Chem. Soc.* **89**, 5067.

Karplus, M. (1959). *J. Chem. Phys.* **30**, 11.

Karplus, M. (1960). *J. Chem. Phys.* **33**, 1842.

Karplus, M. (1963). *J. Amer. Chem. Soc.* **85**, 2870.

Kingsbury, C. A. (1968). *J. Org. Chem.* **33**, 1128.

Kingsbury, C. A. and Best, D. C. (1967). *J. Org. Chem.* **32**, 6.

Kingsbury, C. A. and Thornton, W. B. (1966). *J. Org. Chem.* **31**, 1000.

Lambert, J. B., Keske, R. G., Carhart, R. E. and Jovanovich, A. P. (1967). *J. Amer. Chem. Soc.* **89**, 3761.

Lemieux, R. V., Kullnig, R. K., Bernstein, H. J. and Schneider, W. G. (1958). *J. Amer. Chem. Soc.* **80**, 6098.

Lyle, G. G. and Durand, M. L. (1967). *J. Org. Chem.* **32**, 3295.

Mateos, J. L. and Cram, D. J. (1959). *J. Amer. Chem. Soc.* **81**, 2756.

Matter, U. E., Pascual, C., Pretsch, E., Pross, A., Simon, W. and Sternhell, S. (1969). *Tetrahedron* **25**, 691.

Moreau, G. (1969). *Compt. Rend.* 1770.

Munk, M. E., Meilahn, M. K. and Franklin, P. (1968). *J. Org. Chem.* **33**, 3480.

Musher, J. and Richards, R. E. (1958). *Proc. Chem. Soc.* 230.

Nouls, J. C., Van Binst, G. and Martin, R. H. (1967). *Tetrahedron Lett.* 4065.

Pascual, C., Meier, J. and Simon, W. (1966). *Helv. Chim. Acta* **49**, 164.

Pink, R. C., Spratt, R. and Stirling, C. J. M. (1965). *J. Chem. Soc.* 5714.

Randall, J. C., Vaulx, R. L., Hobbs, M. E. and Hauser, C. R. (1965). *J. Org. Chem.* **30**, 2035.

Schmid, G. H. (1968). *Can. J. Chem.* **46**, 3415.

Shoppee, C. W., Johnson, F. P., Lock, R. E., Shannon, J. S. and Sternhell, S. (1966). *Tetrahedron Suppl.* 8, Part II, 421.

Snyder, E. I. (1969). *J. Amer. Chem. Soc.* **91**, 2579.

Sternhell, S. (1964). *Rev. Pure Appl. Chem.* **14**, 15.

Sternhell, S. (1969). *Quart. Rev.* **23**, 236.

Sutcliffe, L. H. and Walker, S. M. (1967). *J. Phys. Chem.* **71**, 1555.

Thomas, W. A. (1968). *In* "Annual Review of NMR Spectroscopy" (Mooney, E. F., ed.), Vol. 1, p. 44. Academic Press, London.

Tobey, S. W. (1969). *J. Org. Chem.* **34**, 1281.

van Gorkom, M. and Hall, G. E. (1968). *Quart. Rev.* **22**, 14.

Williamson, K. L. and Johnson, W. S. (1961). *J. Amer. Chem. Soc.* **83**, 4623.

CHAPTER 4

Studies of Alicyclic Derivatives
Containing Nitrogen

In this chapter attention is focused upon some stereochemical questions
relating to 6-membered nitrogen-containing rings that may be investigated
by PMR spectroscopy. These fall into four groups, namely:

1. Configurational and conformational problems.
2. Stereochemistry of N-protonation.
3. Stereochemistry of quaternization.
4. Lone-pair dimensions and stereochemistry.

CONFIGURATION

Classical methods of differentiation between *cis* and *trans* isomers of

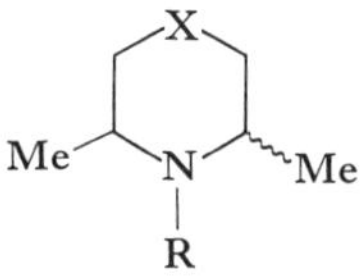

1

the type **1** rest upon the potential resolution of the *trans* isomer and the
lack of resolvability of the *cis* form which has a plane of symmetry. Examples
of PMR analyses which have superseded this approach are given below.
Commercial 2,6-dimethylmorpholine is a *cis–trans* mixture. The *cis*

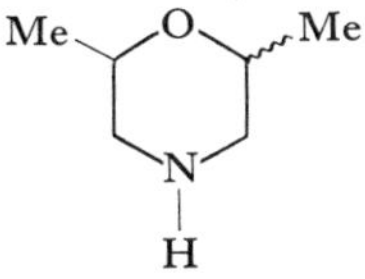

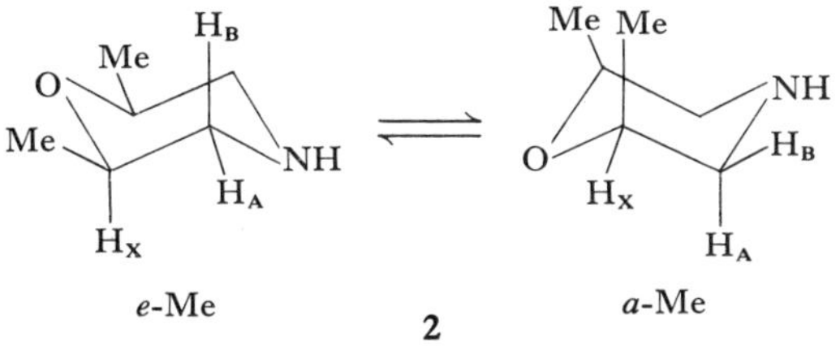

e-Me   **2**   *a*-Me

135

isomer (**2**) is effectively a rigid molecule, because the population of the diaxial methyl conformer (**2**, *a*-Me) will be very low. Hence analysis of its methylene (H_A and H_B) and methine (H_x) PMR signal should yield typical $J_{a/a}$ and $J_{a/e}$ vicinal coupling constant values. Signals obtained for the major isomer in benzene (Fig. 4.1) are near first order and do, in fact, yield J_{vic} values of the expected order (2 and 10·3 Hz) (Booth and Gidley, 1965). The analysis was based on the signal due to the equivalent methine protons at C-2 and C-6, which form a symmetrical multiplet of 11 lines. As the X part of an $ABXK_3$ system, the resonance of the 2- (and 6-) proton should

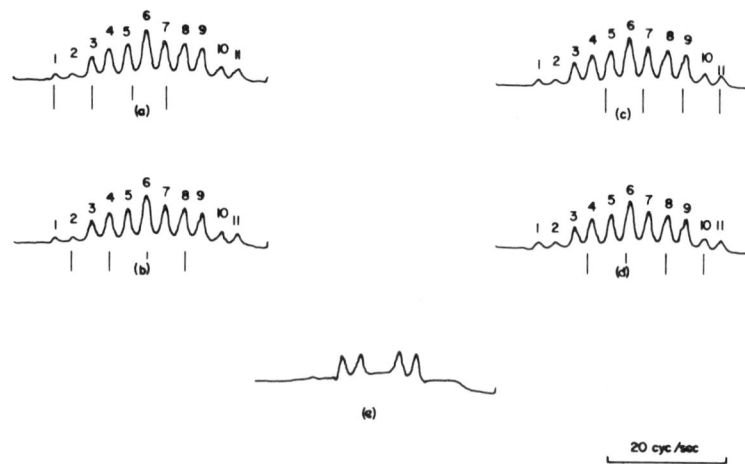

Fig. 4.1. The 60-MHz PMR signal of the 2,6-methine protons of 2,6-dimethyl-morpholine in benzene. Multiplets a–d demonstrate identification of the four 1,3,3,1-HCMe quartets and e shows the multiplet appearance recorded with simultaneous irradiation at a point 144 Hz to high field (centre of the methyl doublet) (adapted from Booth and Gidley, 1965).

Note. To identify the four 1,3,3,1-quartets (J_{HCMe} ~6·5 Hz) *either* (a) starting from the extreme LH peak mark off four lengths = 6·5 Hz; these fall near peaks 1, 3, 5 and 7 (see a). Mark off a second set starting from peak 2 (see peaks 2, 4, 6 and 8 of b). The starting position of the third set is ambiguous, therefore start at the extreme RH peak and mark lengths of 6·5 Hz in the opposite direction. The third set falls near peaks 11, 9, 7 and 5 (c) and the fourth near peaks 10, 8, 6 and 4. Note that peaks 4, 5, 6, 7 and 8 each include two closely placed lines. The centre of each set gives the X quartet of the ABX system from which J_{AX} and J_{BX} may be derived (strictly these lines only allow knowledge of the quantity $|J_{AX} + J_{BX}|$, but first-order treatment is a reasonable approximation for the purpose in hand) *or* (b) having ascertained that peaks 1 and 2 do not form part of the same 1,3,3,1-quartet (their separation is less than 6·5 Hz), mark a distance 9·75 Hz ($3 \times 6·5/2$) from peak 1 and then from peak 2. This gives the positions of two of the ABX lines; the same measurement inwards from peaks 10 and 11 give the other two lines.

consist of four lines (the X part of an ABX system), each of which is split into a 1,3,3,1-quartet because of first-order coupling with the three protons of the methyl group. A maximum of 16 lines is thus possible for the X proton but the coincidence or near-coincidence of some lines allows only 11 lines to be observed. The position of the four 1,3,3,1-quartets is readily fixed, since the separation between each line of the quartets equals the separation of the C-2 methyl doublet (see note to Fig. 4.1). Removal of the H—C—CH$_3$ coupling by double irradiation caused the methine multiplet to collapse to a quartet, the lines of which corresponded with the positions obtained in the first treatment [Fig. 4.1(e)]. (An ABX analysis in which the AB signal is utilized to derive J_{AB} and J_{AX} is described for β-methylacetylcholine, p. 221).

Turning now to the *trans* isomer, it is seen that this constitutes a conformationally mobile system since the two chair conformers [3(a) and 3(b)] being equivalent, have equal energies. Thus averaged J_{vic} values are anticipated, calculated as follows:

$$J_{AX} = 1/2(J_{e_3a_2} + J_{a_3e_2})$$

and

$$J_{BX} = 1/2(J_{a_3a_2} + J_{e_3e_2})$$

with H$_{BX}$ the larger because it contains a $J_{a/a}$ contribution. The same analysis of the spectrum of the minor isomer yields the values 3·36 and 5·8 Hz for J_{AX} and J_{BX} respectively. These are clearly average values and establish that the minor isomer has the *trans* configuration. Evaluation of $J_{a_3e_2}$ and $J_{e_3e_2}$ using the $J_{e_3a_2}$ and $J_{a_3a_2}$ values of the *cis* isomer (assumed the same in the two isomers) give the results 4·75 and 1·29 Hz respectively. The fact that protons involved in the $J_{e_3e_2}$ coupling are both *trans* coplanar with strongly electronegative atoms (N and O, see **4**) explains why this coupling is the smallest of all the four (cf. electronegative influences on J values, p. 91).

The same principles apply to establishing the configurations of *cis*- and *trans*-2,6-dimethylpiperidine, but spectral analysis is not so simple, because only the 2,6 ring protons may be resolved. These are taken as the X portion of an ABC_3X system (Booth and Little, 1968). In the *cis* isomer ("frozen"), the expected width of the X signal is $J_{AX} + J_{BX} + 3\,J_{CX}$. Taking $J_{AX} = 3$ and $J_{BX} = 10$ Hz, and $J_{CX} = 6$ Hz (the latter value from the methyl doublet),

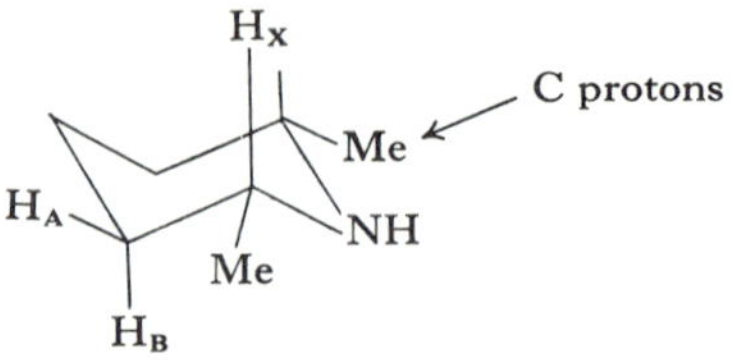

cis 2,6-dimethylpiperidine

the value 31 Hz is obtained; this corresponds with the measured value obtained for the major isomer. The width of the 2,6 proton signal of the minor isomer is 29·2 Hz while J_{CX} is 6·5 Hz. Hence $J_{AX} + J_{BX} + 3 \times 6·5$ ($3\,J_{CX}$) = 29·2 Hz, whence $J_{AX} + J_{BX} = 9·7$ Hz, a value expected if averaging of coupling constants occurs as must be the case for the mobile *trans* isomer (cf. the J_{AX} and J_{BX} values obtained for *trans*-2,6-dimethylmorpholine).

Other configurational arguments are available for N-substituted 2,6-dimethylpiperidines. In the N-benzyl derivatives (**5** and **6**), the methylene protons of the benzyl group have a symmetrical environment in the *cis* isomer (**5**) and their resonance signal is a singlet (Hill and Chan, 1965). The corresponding protons of the *trans* isomer (**6**), however, differ in their environments and give rise to an AB quartet (*J* 14 Hz) typical of a non-equivalent methylene group. The preferred conformation of the *trans*

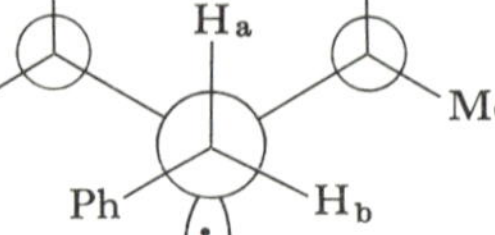

5   *cis*

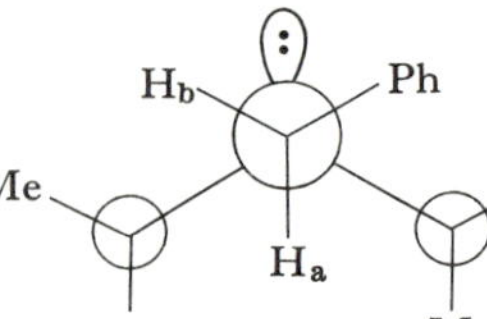

6   *trans*

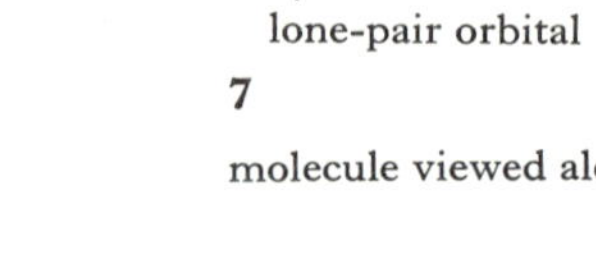

lone-pair orbital

7

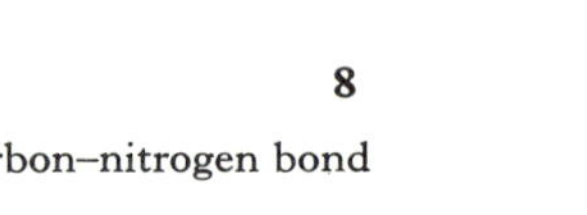

8

molecule viewed along benzylic carbon–nitrogen bond

isomer is (7) (steric interactions preclude high populations of conformers in which Ph occupies the H_a or H_b positions in 7), and the differing orientation of H_a and H_b in relation to the lone-pair orbital of nitrogen is probably the chief reason for their large chemical shift difference. Note that H_a is still *trans* and H_b still *gauche* to this orbital in the inverted chair conformation (8) as a result of rapid nitrogen inversion maintaining an equatorial benzyl group, hence chair–chair interconversion of conformers equal in energy does not, in this case, bring about averaging of the $PhCH_2$ environments. If a single axial methyl group is α- to nitrogen in a piperidine

9   10

derivative, the conformers 9 and 10 will have similar populations (steric interactions between Me and Ph are low in both conformers) with the result that the H_a and H_b environments are averaged. The observation of

11   (*cis* Me—Me)   12   13

$PhC\underline{H}_2N$ singlets rather than quartets in the asymmetric cyclic bases 11 and 12 therefore suggests that the α-methyl substituents are axial in their preferred conformations, e.g. 13 (Lyle and Thomas, 1969; Hassan and Casy, 1970). Isomeric 1-benzyl-3,5-dimethylcyclic bases do not satisfy the conformational requirements for observable $PhC\underline{H}_2N$ non-equivalence (equatorial methyl α- to nitrogen) and cannot be differentiated by this method. Asymmetric N-methyl piperidines are discussed later.

14

Only one 2,4,6-trimethylpiperidine is available in a pure condition (Booth and Little, 1968). Its spectrum displays an unusually high field 1:3:3:1 quartet ($J \sim 11$ Hz). This is the signal anticipated for the 3,5 diaxial protons of the all-*cis* ("frozen") isomer (**14**), since each should be coupled in equal degree to two vicinal (axial) protons and one geminal proton when all methyl substituents are equatorial. The high-field position of this signal is consistent with the 3,5 protons being screened by the flanking equatorial methyl groups (Booth, 1966).

To conclude this section the PMR spectral analysis of 1,2,5-trimethyl-4-piperidone (**15**) is considered in some detail, since it illustrates several general principles of value to the configurational assignment of cyclic bases (Hassan

and Casy, 1970). This piperidone is made by condensing the *bis* α-β unsaturated ketone (**16**) with methylamine, and is the intermediate for the synthesis of promedol (**17**), a narcotic analgesic used clinically in the U.S.S.R. (Prostakov and Mikheeva, 1962). The product derived from **16** is a *cis–trans* mixture, as is evident from the presence of at least three doublets in the C-methyl resonance region of its PMR spectrum [Fig. 4.2(a)]. One component must preponderate, because two of the doublets (those centred at δ 0·95 and 1·15) are intense while the third (at δ 0·99) is weak; the strong signals are attributed to the major isomer and the weak signal to the minor isomer. The second doublet of the minor form is not well resolved but is probably responsible for the shoulder apparent at the base of the highest field doublet. These assignments are confirmed by examining the spectra of the base derived from a recrystallized hydrochloride of **15** and that of the base enriched in the minor isomer by treatment with alumina. The secondary methyl signal at δ 0·99 is of reduced intensity in the spectrum of the former and of enhanced intensity in that of the latter sample, in comparison with intensities of the δ 0·95 and 1·15 signals.

Spectral analysis of signals due to the major isomer is consistent with its *trans* 2,5-dimethyl configuration (**18**). This analysis was aided by study of the fully deuterated analogue (**19**) obtained by treating the ketone in carbon tetrachloride with Na—D_2O; this treatment did not disturb the isomeric composition as seen by the spectral identity of the mixture (**15**)

18   **19**

before and after treatment with CCl_4—Na—H_2O. Specific points in the analysis are as follows:

(i) the quartet near δ 3 and the triplet at δ 2 (lower field component obscured) collapse to doublets with a typical geminal coupling constant ($J \sim 10$ Hz in the normal and 11·5 Hz in the deuterated ketone) after deuteration [Fig. 4.2(b) and (c)], and are thus due to the C-6 methylene protons. The higher field signal is assigned to the axial member because it displays the larger vicinal coupling constant (5·5 for the lower and 11 Hz for the higher field signal), the near equality of J_{gem} and J_{vic} in this case leading to the observed triplet (a quartet was apparent at higher resolution). The higher field chemical shift of the axial C-6 proton must be due to its being flanked by two equatorial methyl groups (see p. 140); in addition, the axial proton will be shielded by the lone pair on nitrogen, as discussed later (Hamlow *et al.*, 1964). After D/H exchange the higher field doublet shows fine structure not seen in the lower field signal; this difference is understandable in terms of configuration **18** since an axial deuterium atom at C-5 would be expected to couple more strongly with the axial than the equatorial C-6 proton (Brignell *et al.*, 1968).

(ii) The doublet at δ 0·95 [Fig. 4.2(a)] collapses to a singlet after deuteration and is therefore assigned to 5-methyl adjacent to the carbonyl group. Its equatorial conformation follows from ring proton assignments already made (namely, the 6-methylene group has been shown to be adjacent to an axial methine proton) and from the fact that its chemical shift is close to that of methyl in the analogues (**20**) of known preferred conformation.

$R = Me, 0·97$
$R = CH_2CH=CH_2, 0·98$
$R = CH_2Ph, 0·94$

(all δ values in solvent $CDCl_3$)

20

(a)

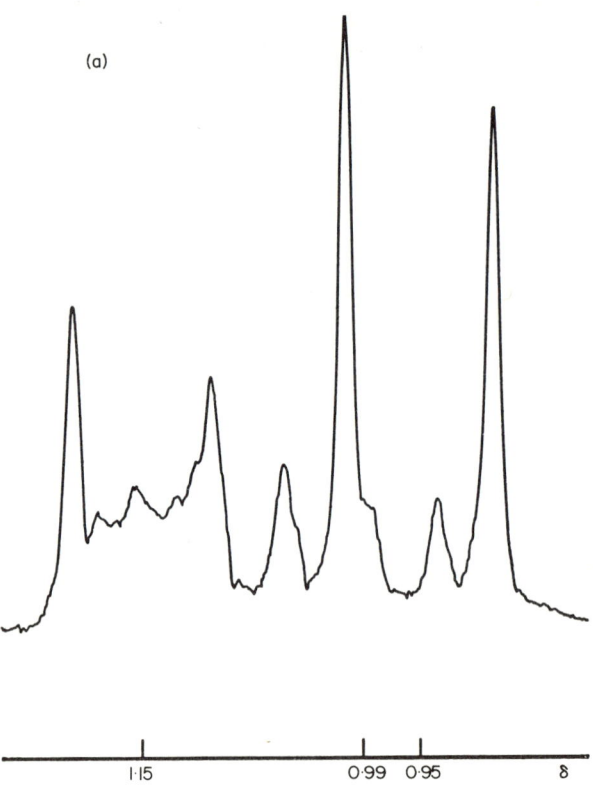

1·15   0·99 0·95   δ

(b)

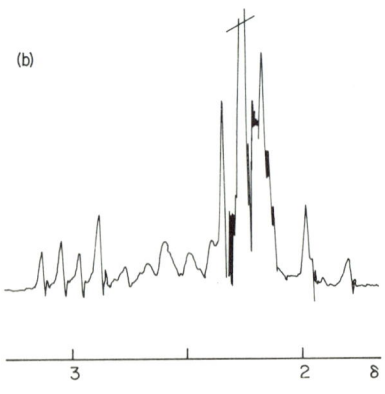

3   2   δ

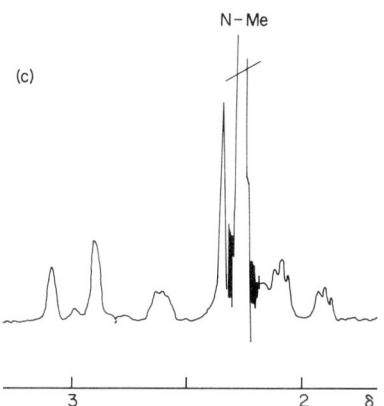

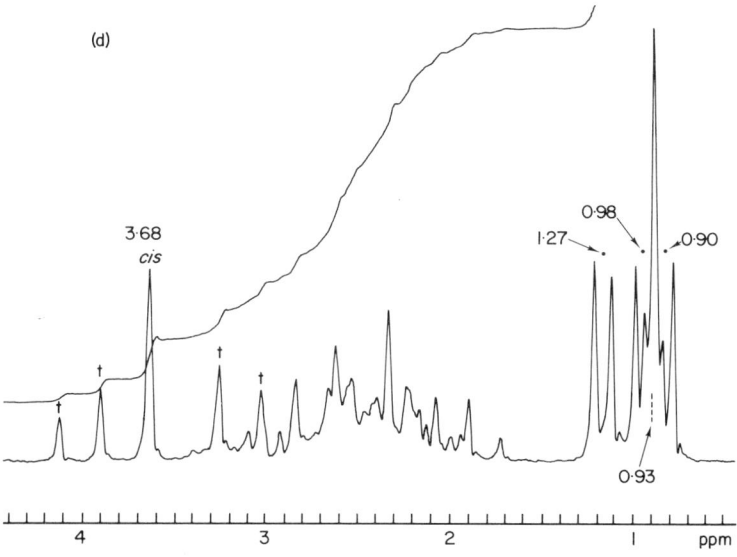

FIG. 4.2. 60-MHz PMR spectra of 2,5-dimethyl-4-piperidone isomers. 1,2,5-Trimethyl-4-piperidone mixture: (a) secondary methyl region (CDCl$_3$ 100-Hz sweep width); (b) ring proton region (CCl$_4$); (c) ring proton region after treatment with Na—D$_2$O (CCl$_4$); (d) 1-benzyl-2,5-dimethyl-4-piperidone mixture (in CDCl$_3$). PhCH$_2$N signals marked *t* (quartet, major isomer) and *cis* (singlet, minor isomer) (Hassan and Casy, 1970).

(iii) The doublet at δ 1·15 remains after D/H exchange and is assigned to the C-2 methyl group. Its equatorial orientation follows from its chemical shift identity with the analogues (**21**), and the fact [revealed by its deformed appearance, Fig. 4.2(a)] that it is virtually coupled to the C-3 protons (Musher and Corey, 1962). Virtual

21

R = Me, 1·07
R = CH$_2$Ph, 1·13

22

(all δ values in solvent CDCl$_3$)

coupling in the system (**22**) is most likely when a large α-β vicinal coupling is involved, as provided by the C-2 and C-3 axial protons in configuration **18** (see also p. 204). After deuterium exchange, this coupling is removed and the 2-methyl signal appears, significantly, as a clear doublet.

The lack of a pure sample of the *cis* isomer of **15** prevented analysis of its PMR characteristics, but evidence about the conformation of the *cis* ketones was obtained by study of the corresponding N-benzyl derivatives (**23**) obtained by an exchange reaction between **15** methiodide and benzyl-amine. The PMR spectrum of the base derived from a recrystallized

23

hydrochloride of the mixture showed it to be the pure *trans* isomer. Thus, the C-methyl resonance region showed two clear doublets, one at δ 1·27 (2-Me) and the other at δ 0·9 (5-Me), while a quartet near δ 3 (*J* 5·5 and 11 Hz) due to the equatorial C-6 proton and a triplet at δ 2 (*J* 11 Hz, axial C-6 proton) were also present. The lower field position of the 2-methyl signal, compared with the corresponding signal of *trans* **15** must be due to the deshielding influence of the benzyl substituent because 2-methyl chemical shifts in the analogues (**21**) differed in the same way. The methylene signal of the benzyl group forms an AB quartet (δ 3·24 and 4·09, *J* 13·5 Hz) almost identical with that seen in the spectrum of 1-benzyl-2-methyl-piperidine (p. 148) and is typical of methylene in an asymmetric environment. Two additional secondary methyl doublets (at δ 0·93 and 0·98) are clearly apparent in the spectrum of the *cis–trans* mixture (**23**) [Fig. 4.2(d)]; the higher field signal is assigned to the *cis* 5-methyl group (it collapsed to a singlet after D/H exchange), and the lower field doublet (unchanged after exchange) to the 2-methyl group. It was anticipated that the benzyl methylene signal of the *cis* isomer would be a quartet (as in the *trans* spectrum) since the environment of this group is asymmetric in either isomer, and its appearance as a singlet at δ 3·68 was surprising [Fig. 4.2(d)]. There is evidence, however, that chemical shift differences between methylene protons in 1-benzyl-2-methylpiperidines only arise when the methyl substituent is equatorial (p. 139). Hence the singlet nature of the *cis* N—CH$_2$Ph signal indicates that the 2-methyl group is axial in the preferred *cis* conformation (see **13**). The clear separation of *cis* and *trans* NCH$_2$Ph signals in the mixture (**23**) allow their separate integration and from these measurements the *trans*:*cis* ratio is 16:13.

Other examples of pharmacological interest are given in Chapter 6.

PROTONATED N-EPIMERS

Isomers of this nature arise in cyclic compounds containing a *t*-nitrogen ring member as a result of two modes (axial and equatorial) of proton uptake

at the basic centre (see above). Such epimers may be detected in appropriate cases by duplication of the N—R PMR signal or the signal of a ring carbon substituent since the environments of N and/or C substituents may differ in the two isomers. Detection of protonated epimers generally requires a slow proton exchange rate. If the rate is faster than the NMR experiment averaged, not duplicate, signals are usually observed since isomers of the type shown will be rapidly interconverting. In some cases, however, distinct epimeric signals are observed even when proton exchange is fast, e.g. 1,2-dimethylpiperidine hydrochloride in water (Fig. 4.4) (see also p. 75, Chapter 2), and in these examples proton exchange must occur *with retention of configuration at the nitrogen centre*. Some of the general principles involved are illustrated by the original example of pseudotropine (Closs, 1959). The PMR spectrum of this base in water at pH 1 shows two doublets in the N-methyl resonance region (Fig. 4.3). The higher field doublet (doublets arise as a result of HNCH coupling) is much more intense than that at lower field and the intensity ratio is a measure of the relative stabilities of the

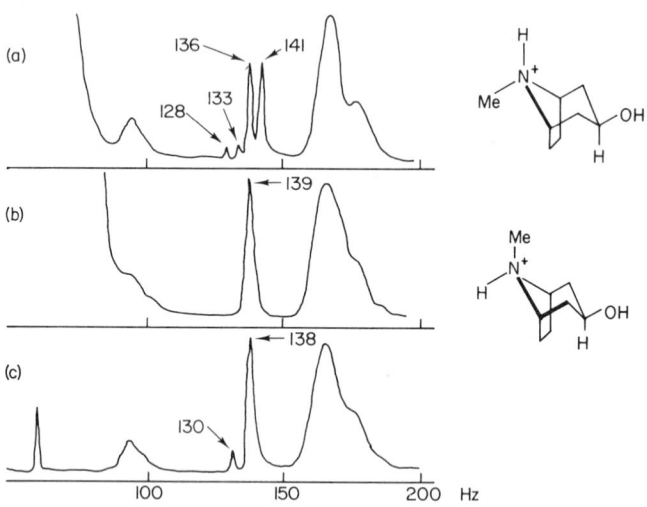

FIG. 4.3. 40-MHz PMR spectra of pseudotropine: (a) in water at pH 1; (b) in water at pH 6; (c) in D_2O at pH 1. Chemical shifts are in Hz up field from the aromatic signal of toluene (Closs, 1959).

two epimers shown in the Figure. The proton exchange rate is much faster at pH 6 and the spectrum recorded at this pH shows a single N—Me band.

In N-methyl epimeric salts, assignment of N—Me signals to specific isomers may be made on the basis of the more intense signal arising from the more stable epimer when the latter can be predicted with reasonable certainty. Thus, in the case of 1,2-dimethylpyrrolidine hydrochloride [NMe doublets at τ 6·97 (more intense) and 7·27 in $CDCl_3$], the *trans* epimer (**24**) would be expected to be more stable than the *cis* form and hence the more intense lower field NMe signal is assigned to the *trans* epimer (Becconsall *et al.*,

1965). The same argument applies to salts of 1,2-dimethylpiperidine, the more intense N—Me and 2-Me doublets of the salt spectrum (Fig. 4.4) being allocated to the *trans* isomer (**25**). The N-methyl signal is the lower

25

field member and it appears to be generally true that *e*-NMe groups in 6-membered cyclic bases come to resonance at a lower magnetic field than axial N—Me groups (cf. the relative chemical shifts of *a* and *e* protons in

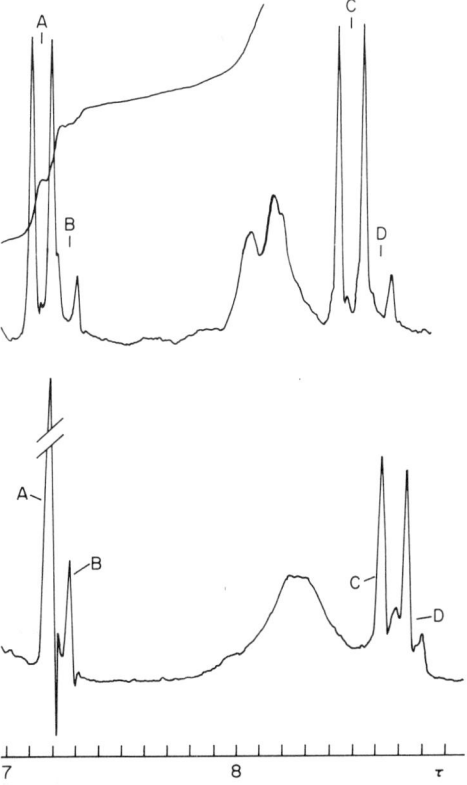

FIG. 4.4. Part of the 60 MHz PMR spectrum of 1,2-dimethylpiperidine hydrochloride in CDCl$_3$ (upper) and D$_2$O (lower figure). Signals A (HN<u>Me</u>) and C (2-Me) are assigned to the *trans* NMe/2-Me and B and D to the *cis* epimer.

cyclohexane derivatives, p. 118, and also quaternary salts discussed below†; exceptions arise in piperidine derivatives carrying an axial methyl substituent adjacent to nitrogen (p. 159). In the *cis* epimer, the N-Methyl group is unlikely to be entirely axial [as in **26**(a)] because the population of the 2-*a*-methyl conformer [**26**(b)] should also be significant. Distinct N—Me

(a)   (b)

26

and 2-Me signals are still seen in the spectrum of the salt in D_2O (Fig. 4.4) even though proton exchange is fast as established by the singlet nature of the N-methyl signals. Epimeric NCH$_2$Ph and 2-methyl signals are clear in the spectrum of 1-benzyl-2-methylpiperidine hydrochloride in CDCl$_3$. The major NCH$_2$Ph signal (Fig. 2.8, p. 64) forms the expected AB quartet, each component being further split into a doublet as a result of HCNH coupling.

27

coupling. The minor epimer, of probable conformation **27** (the bulk of the N-substituent makes the *a*-NR conformation unfavourable), gives rise to an NCH$_2$Ph *singlet* (in CDCl$_3$—D$_2$O) which overlaps the more intense higher field peak of the major methylene signal; it is a doublet when coupled to the $^+$NH proton and similarly overlaps part of the major signal, as is evident from the comparative intensities of the two halves of the composite methylene signal. These differences provide another example of conformational influence on benzylic methylene non-equivalence (see p. 139).

† Delpeuch and Deschamps (1970) consider this to be true only when an α-substituent is present. They detected N—Me signals due to the minor conjugate acid species of 1-methyl-4-*t*-butylpiperidine and 1,4-dimethylpiperidine in acid solvents, and in both cases these fell to *lower* field of the major signals due to the epimers with equatorial N-methyl groups.

Many piperidine salts fail to give PMR evidence of epimer formation even in solutions of low pH. Thus both 1-methyl-4-phenylpiperidine and 1-methyl-*trans*-decahydroquinoline display one N—Me singlet (or doublet) due, presumably, to one epimer being extensively favoured over the other (Becconsall *et al.*, 1965). Note, for example, that although the *e*-N—Me epimer (**25**) is more favoured than the *a*-N—Me form, it is even more favoured when no 2 (6) substituents are present (as in the 4-phenyl derivative **28**) as a result of the absence, in such cases, of a destabilizing *gauche* C—Me/N—Me interaction (see **29**). When two *gauche* Me/Me interactions operate, a higher proportion of the axial N-methyl epimer results as in the case of *cis*-1,2,6-trimethylpiperidine (see p. 153).

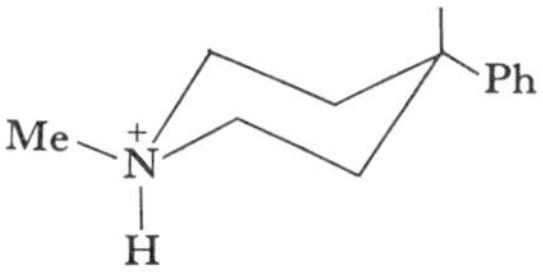

28

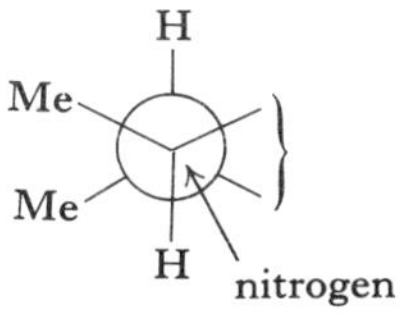

29

N-Epimers arise in the case of the piperidines **30**(b) and **30**(c), but the PMR spectra of their hydrohalide salts display two 3-Me doublets rather

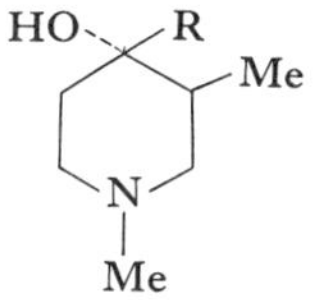

(a) R = CN
(b) R = CO_2H
(c) R = CO_2Et
(d) R = $C(OH)Ph_2$

(30)

than two N-methyl signals (Casy *et al.*, 1970) (Fig. 4.5). The favoured conformation of *cis* N—Me/3-Me epimers in these cases is clearly **31**. However, both conformers of the *trans* epimer (**32**) should be significantly populated because each entails two comparable 1,3-diaxial interactions involving ring substituents [note that the steric dimensions of OH and CO_2H or CO_2Et do not differ greatly (Eliel *et al.*, 1965)]. Axial 3-Me in **32**(b) is deshielded by the nearby charged nitrogen atom (see p. 203), hence

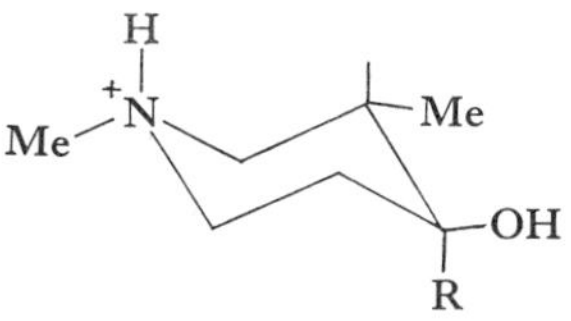

31

cis NMe/3-Me

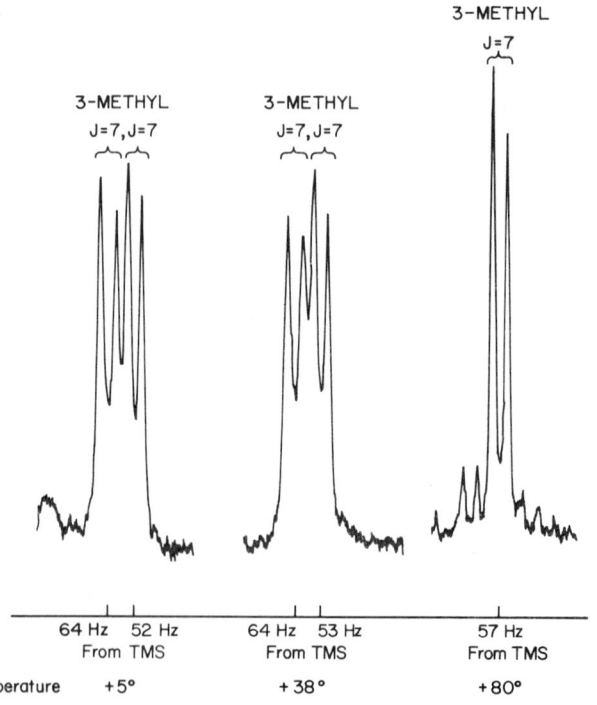

32(a)   **32(b)**

trans NMe/3-Me

the overall chemical shift of 3-Me in the *trans* N—Me/3-Me epimer will be lower field than that of 3-Me in the *cis* epimer (**31**), while chemical shift differences between the epimeric N-methyl groups are reduced to a point where separable signals are not observed. This interpretation is supported by temperature studies. At higher temperatures (faster rates of proton exchange) the duplicate 3-methyl signals merge to a single doublet of an intermediate chemical shift, while at a lower temperature (slower rate of proton exchange) the duplicate signals move further apart (Fig. 4.5).

FIG. 4.5. Part of the 60-MHz PMR spectra of *trans* 4-OH/3-Me 1,3-dimethyl-4-hydroxypiperidine-4-carboxylate hydrochloride in DMSO-d_6 at various temperatures. At 38° the higher field 3-methyl doublet falls at 9·12, and the lower field at 8·93τ.

The 3-methyl signal is not duplicated in spectra of the hydrochlorides **30**(a) and **30**(d). In the former, steric requirements of OH are substantially greater than the cyano function (Eliel *et al.*, 1965) hence populations of the *trans* epimer [**32**(b)], in which non-bonded interactions of this group (R) are relieved while those of OH are increased, will be low. When R is much larger than OH, as in **30**(d), the *trans* epimer (**32**) is favoured by a large factor and the PMR spectrum of this salt provides no evidence of

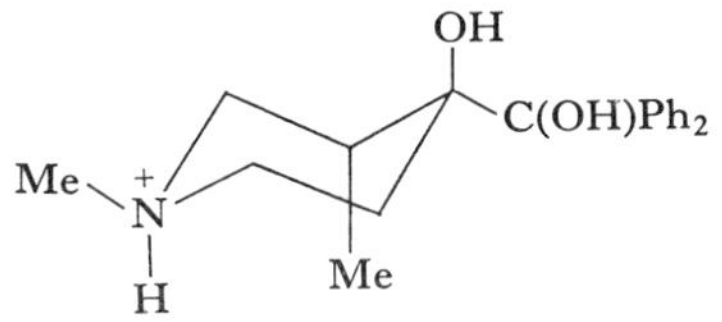

33

conjugate acid isomers. The conformation **33** is supported by the 3-methyl chemical shift of **30**(d) hydrochloride (τ 8·95) which is typical of an axially placed group in this series.

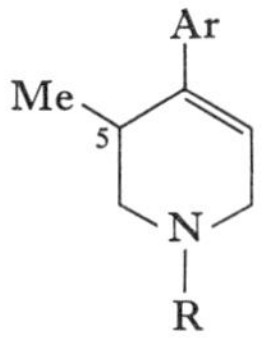

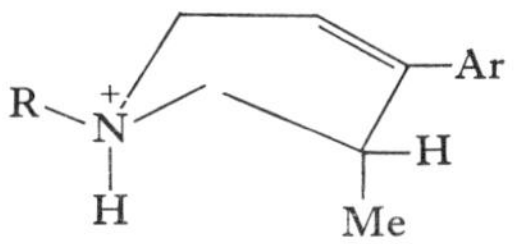

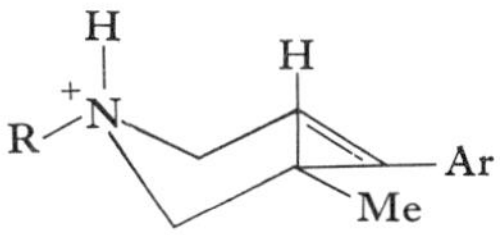

	trans NR/Me	*cis*
34	**35**	**36**

Epimer formation in the tetrahydropyridines (**34**) also manifests itself through the observation of duplicate C—Me rather than N—R signals (Casy *et al.*, 1965). Here preferred conformations for the *trans* and *cis* N—R/5-Me isomers are **35** and **36** respectively. An example is shown in Fig. 4.6; the lower field, less intense, 5-Me signal is assigned to the *trans* isomer (axial 5-Me as in **35**, hence close to and deshielded by the $^+$N function) and the higher field signal to the *cis* isomer (as **36**). The more intense O—Me and N—Me signals must also arise from the *cis* isomer. When the proton exchange rate is accelerated by the addition of D_2O, all duplicate signals coalesce to a point which is the weighted mean of the separate high- and low-field signals. The *cis* to *trans* ratio is almost 2:1 from the integrals of the 5-methyl signals, ratios close to 1:1 being obtained for 4-phenyl examples. The high populations of the axial methyl conformer observed is surprising, but it must be noted that only one 1,3-diaxial

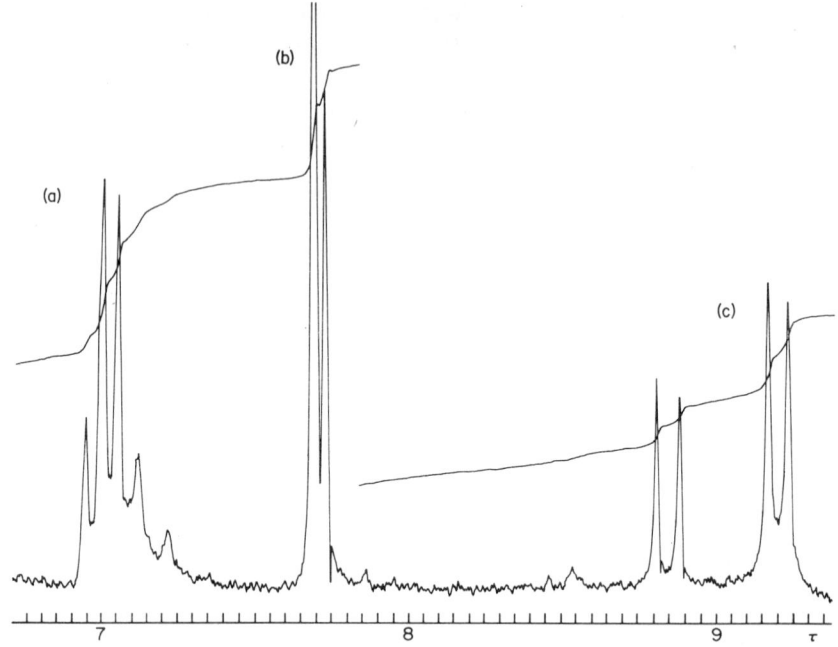

Fɪɢ. 4.6. Part of the 100-MHz PMR spectrum of 1,5-dimethyl-4-*o*-tolyl-1,2,5,6-tetrahydropyridine hydrochloride in CDCl₃: (a) N—Me (major and minor over-lapping doublets due to HNMe); (b) O—Me; (c), 5-Me signal (Iorio and Casy, 1970).

interaction is involved. There is, in addition, independent PMR evidence for the favoured nature of axial rather than equatorial substituents in 6-substituted-1-phenylcyclohexenes (**37**) (Garbisch, 1962), attributed to the

37   **38**   **39**

fact that *a*-X interferes less than *e*-X with the conjugated styrenoid unit of such derivatives. Garbisch measured the width at half height (W_H) of the 6-methine signal; this will be narrower in the *a*-X conformer (**38**) than in the *e*-X form (**39**) because only the latter's 6-H signal involves an axial–axial coupling contribution (in general, signal resolution was not good enough to allow measurement of the *J* values themselves). From results obtained in model compounds (e.g. the 4-proton resonance of *cis* and *trans* 4-*t*-butylcyclohexanol) a W_H range of 13–15 Hz was taken as evidence for a lack of conformational preference, $W_H > 15$ a preference for *e*-X and $W_H < 13$ a preference for *a*-X conformations. Bulky X substituents [e.g. *t*-Bu, Ph, $Me_2C(OH),NO_2$] all led to W_H values between 9 and 10 Hz showing axial conformers to be preferred. When an axial 4-substituent was present, however, as in **40**, the 6-H signals gave W_H values consistent

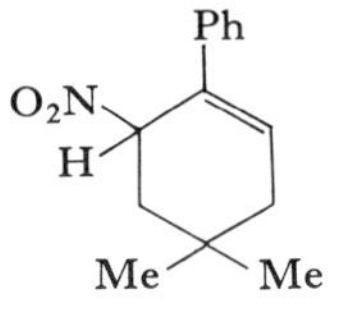

40 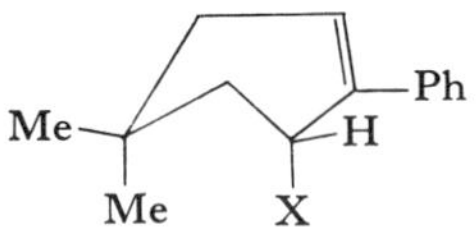   **41**

with a favoured 6*e*-X conformer; in these derivatives the energy of the 6*a*-X isomer (**41**) is raised by the 1,3 diaxial X/Me interaction. Similar data have been obtained in support of preferred *a*-X conformers for the tetra-hydropyridines **42** (Lyle and Krueger, 1967).

Vicinal coupling between a vinylic and adjacent *a* or *e* proton in cyclohexene derivatives is discussed on p. 259.

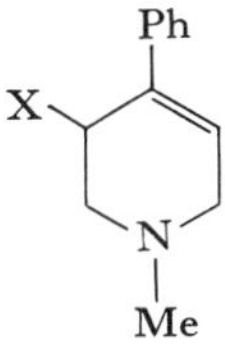

X = Br, OH, Cl, OAc
W_H 6–7·2 Hz

42

Epimer detection provides an alternative PMR method of differentiating between *cis* and *trans* 1,2,6-trimethylpiperidine. The hydrochloride of the

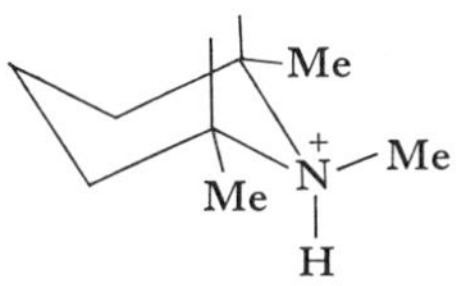

43 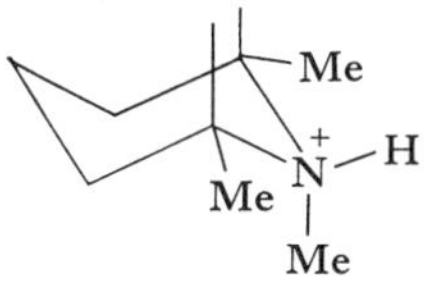   **44**

cis isomer shows two N—Me doublets in H_2O (two singlets in D_2O) with the lower field the more intense and assigned to the *e*-N—Me epimer (**43**); the 2,6-dimethyl signal is also duplicated since the responsible protons are *trans* to NMe in **43** and *cis* in **44** (Fig. 4.7). The epimer ratio, from integrals, is about 63 (*trans*):37 (*cis*). Epimers do not arise in the *trans* 2-Me/6-Me hydrochloride; the *a*-N—Me form (**47**) is a conformational, *but not a configurational*, isomer of the *e*-N—Me isomer (**45**). Hence **45** and **47** ≡ **46** will rapidly interconvert and a single averaged N—Me signal is seen. Two C—Me signals appear in CDCl₃ (not in H₂O) but this is due to the fact

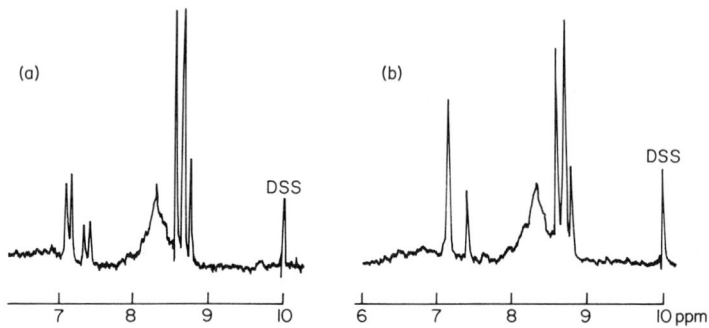

FIG. 4.7. The 60-MHz PMR spectrum of *cis*-1,2,6-trimethylpiperidine hydro-chloride (a) in H₂O and (b) in D₂O. Chemical shifts are: N—Me, τ7·42 and 7·17; C—Me, τ8·64 and 8·45 (Kawazoe *et al.*, 1967).

that one methyl group (Me_A) is *cis* and one (Me_B) *trans* to N—Me in both chair conformations.

45   **46**   **47**

If the epimer formed as the result of the equatorial protonation of a cyclic tertiary base enables the relief of non-bonded interactions obtaining in the product of axial approach (N-substituent equatorial), then significant populations of the two forms may be expected. Use of this principle in cases where epimer formation is restricted to one member of an isomeric series may provide useful corroboration of configurational assignments. This situation arises in the isomeric 1,2,5-trimethyl-4-phenyl-4-piperidinols (α,γ, -no epimers, β-epimers) and 1,2-dimethyl derivatives (α-epimers, β-no epimers) shown below (Casy and McErlane, unpublished results).

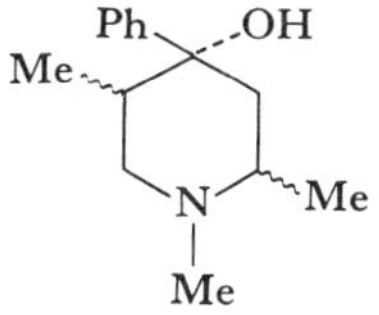

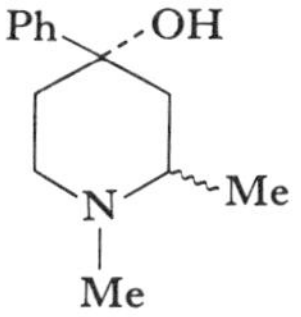

Ph .OH   Ph .OH
Me

N Me   N Me

Me   Me

In the former compounds, the γ- and β-isomers have been assigned the stereochemistry shown on the basis of other PMR evidence. Epimers are

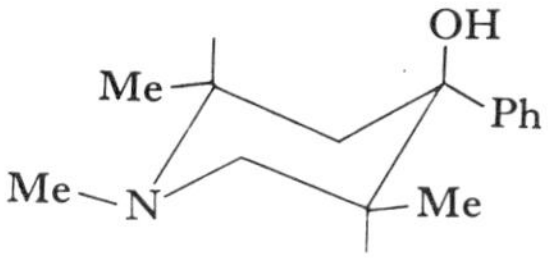

OH

Me Ph

Me—N Me

γ-isomer

not anticipated for the γ-form because the product of equatorial protonation (N—Me axial) merely introduces additional non-bonded interactions; these may not be relieved by a conformational change, because any departure from the *e*-4 Ph chair form must place the equatorial 2,5-dimethyl and 4-phenyl groups in less favourable orientations. In the β-isomer, however, a conformational change of the minor epimer, e.g. to a skew boat (as shown)

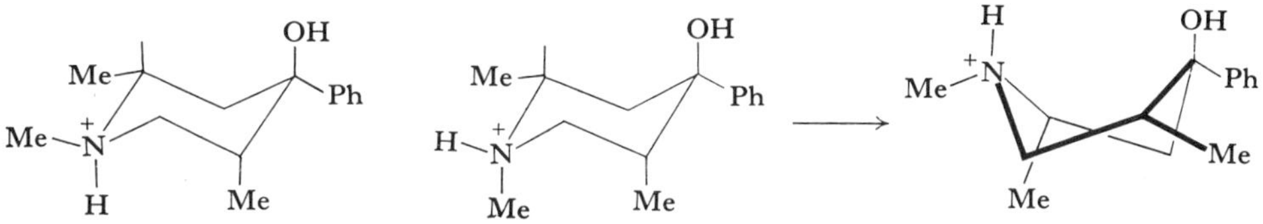

OH   OH   H OH
Me Ph   Me Ph   Me—N⁺ Ph
Me—N⁺   H—N⁺   Me
H Me   Me Me   Me

major   minor epimer

β-isomer

(epimeric ratio 3:1)

or inverted chair will relieve the non-bonded interactions of *two* axial substituents. Thus observation of epimers in the β-base is in accord with the proposed stereochemistry.

QUATERNARY SALTS

The PMR spectra of methiodides of N-methyl cyclic bases (such as piperidine) display duplicate N-methyl signals of equal intensity in cases where conformational preferences obtain. Signal separation is greatest when one conformer is greatly favoured over all others, e.g. **48**, and in these cases one signal represents an essentially axial N—Me and the other an equatorial N—Me environment (Stein *et al.*, 1968). There is good evidence that the higher field is due to *a*-Me and the lower field signal to *e*-Me (see later). In the absence of conformational preferences, N,N-dimethyl quaternary salts show a single N—Me signal which falls almost halfway between *a*- and *e*-NMe positions. Thus, the N—Me chemical shift of N,N-dimethylpiperidinium iodide (**49**) is about halfway between the two

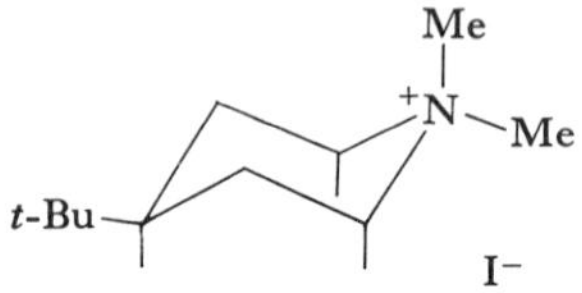
Me
+N
Me
t-Bu
I−

48

N—Me
188·5 Hz
196·8 Hz
from TMS in
$CDCl_3$-TFA (60 MHz)

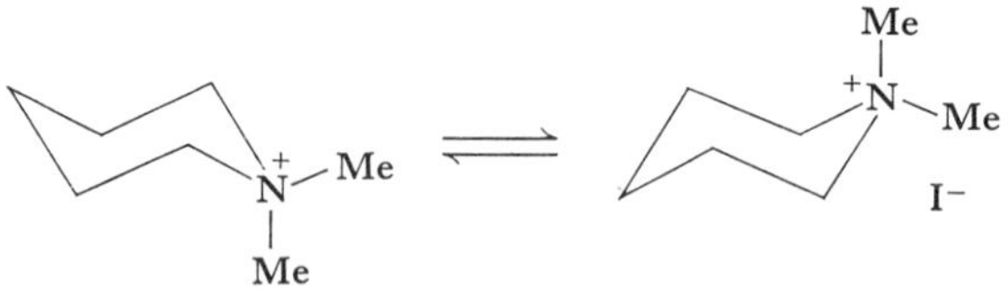
+N
Me
Me
Me
+N
Me
I−

49

N—Me
193·2 Hz
(same conditions)

N—Me signals of the rigid analogue (mean NMe 192·6 Hz); the small discrepancy is due to the influence of the *t*-butyl substituent and/or contributions from conformers additional to **48**.

The PMR characteristics of N,N-dimethyl salts may have stereochemical utility. Thus *cis* 1,2,6-trimethylpiperidine methiodide ("frozen" conformation **50**) is characterized by a spectrum that shows two N—Me

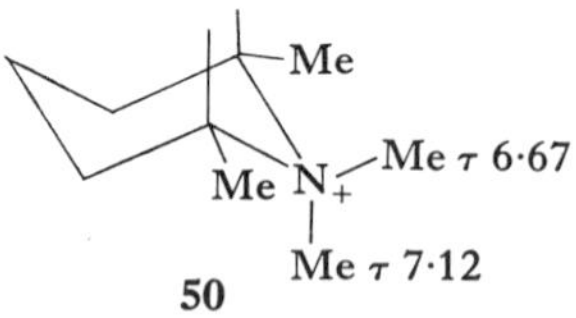
Me
Me τ 6·67
Me N+
Me τ 7·12

50

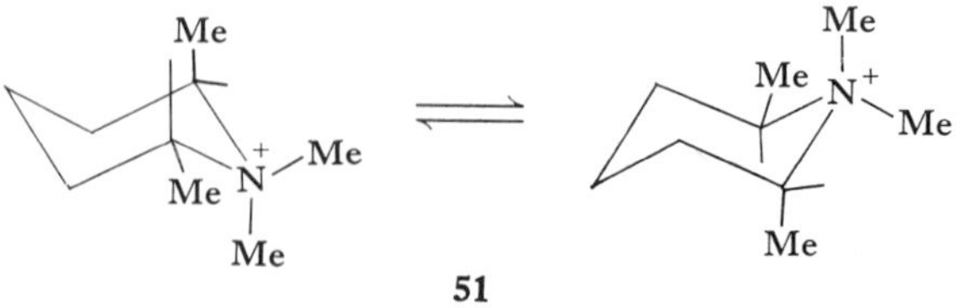
Me
+N
Me
Me
Me
Me
Me N+
Me
Me

51

signals while, in contrast, that of the *trans* methiodide (**51**) (no conformational preference) shows a single N—Me resonance (Kawazoe *et al.*, 1967).

To derive configurational information from substituted N,N-dimethyl-piperidinium salts carrying a 2-methyl substituent, it is first necessary to establish the effect of the 2-substituent upon the chemical shifts of the *a*- and *e*-N-methyl groups (Tsuda and Kawazoe, 1970).

2-*e*-Me: Comparison of data upon methiodides of 1,3- and 1,2-dimethyl-piperidine in CDCl$_3$ (in the 1,3 derivative the C-methyl group has a negligible effect upon the resonances of the N-methyl groups but serves to anchor the conformation), shows that the axial N-methyl protons are more

e N—Me δ 3·55	δ 3·45	Δ 0·10 ppm (shielding)
a N—Me δ 3·35	δ 3·15	Δ 0·20 ppm (shielding)

(Δ is the difference between the chemical shifts of equivalent 3-methyl and 2-methyl derivative N—Me signals.)

shielded than the equatorial by a factor of 2 in spite of the gauche N—Me/2-Me relationship concerned in either case. These conclusions depend on the assumption of higher and lower field N—Me signals being due to axial and equatorial groups respectively and evidence for this is discussed later.

2-*a*-Me: The methiodide of *cis*-1,2,3-trimethylpiperidine (the base is the exclusive product of the catalytic reduction of 2,3-dimethylpyridine methiodide) serves as a model for axial 2-methyl derivatives because the equatorial 2-Me conformer is unfavoured by a diaxial Me/Me interaction.

cis 2,3-dimethyl derivative

e N—Me δ 3·55	δ 3·41	Δ 0·14 (shielding)
a N—Me δ 3·35	δ 3·57	Δ −0·22 (deshielding)

This time the *lower* field signal is taken as due to the axial N-methyl protons (see p. 159) whence it follows that an axial 2-methyl group has a pronounced deshielding influence on axial N-methyl, but shields an equatorial N-methyl substituent. The average substituent effects in ppm (+ shielding, − de-shielding) of an α-C-methyl group on the chemical shifts of axial and equatorial N-methyl protons in piperidinium salts, derived from a series of derivatives are shown below. The shielding effects of *e*- and *a*-methyl upon

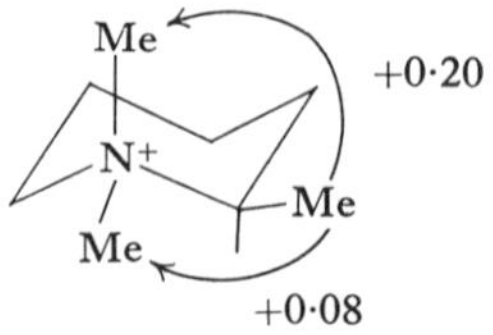

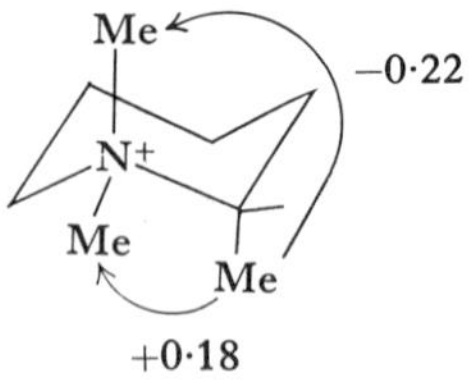

adjacent axial and equatorial protons differ in the same way (Booth, 1966).

Application of these data to the isomeric 1,2,5-trimethylpiperidines is now described; these are obtained in the ratio 65:35 by catalytic reduction of the corresponding dimethylpyridine methiodide. Calculated N—Me chemical shifts for the two isomers are obtained as follows:

trans

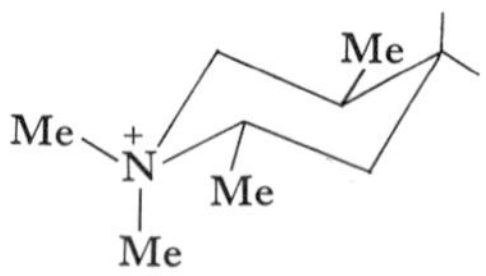

e N—Me   3·55* − 0·08 = δ 3·47
a N—Me   3·35* − 0·2 = δ 3·15

* Corresponding values for *e* and *a* N—Me in 1,3-dimethylpiperidine methiodide.

cis

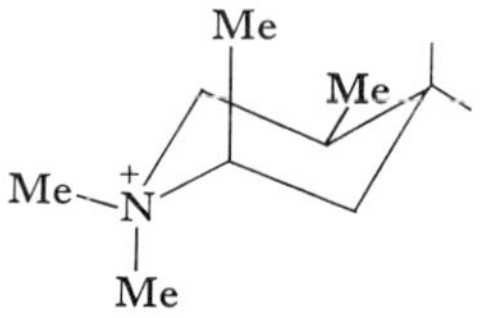

e N—Me   3·55 − 0·18 = δ 3·37
a N—Me   3·35 + 0·22 = δ 3·57

Comparison of the observed N—Me values (major δ 3·4 and 3·57, minor δ 3·48 and 3·18) with those calculated clearly show the major form to be the *cis*, and the minor the *trans*, isomer.

When an N-methyl base is quaternized with an alkyl (or arylalkyl) halide other than methyl, N-methyl-N-alkyl isomers may arise comparable in nature with N-protonated epimers (see p. 145). The spectra of total alkylation products obtained in such cases show two N-methyl signals but these usually differ in intensity because the formation of one isomer is usually favoured over that of the other (again compare N-protonated epimers). N-methyl signals of intensity ratio 8·3:1·7 occur, for example, in the spectrum of the ethylation product of N-methyl-4-*t*-butylpiperidine in acetone (see **52** and **53**) (Kawazoe and Tsuda, 1967).

The problem now arises of which isomer is responsible for the NMe signal of higher intensity. This may be solved if it is assumed that the *a*-N—Me group comes to resonance at higher field than *e*-N—Me (as mentioned previously); in the above case the more intense signal is also the *lower* field signal, thus the product of *axial* attack preponderates in this reaction (placing N—Me equatorial). Much interest has been aroused in the question of the preferred steric course of quaternization (see Brown *et al.*, 1967a and McKenna, 1970, for summaries of studies in this field) and it is generally concluded that axial approach of the quaternizing agent is the preferred reaction pathway provided the group already attached to nitrogen is not markedly smaller than the incoming group. A few aspects of PMR investigations of this problem are outlined below. Alkylations with tri-deuteromethyl iodide (CD_3I) provide information in situations where the two N-substituents are virtually equal in size (no information is forthcoming, of course, from N,N-dimethyl quaternary salts) and an example is shown in Fig. 4.8 (Kawazoe and Tsuda, 1967) which shows the spectrum of the N-quaternary products of 1,2-dimethylpiperidine and CD_3I. The greater intensity of the lower field N—Me signal (due to *e*-Me) shows that axial attack preponderates (see below).

lower field, more intense   higher field, less intense signal

Support for assignment of the *lower* field N-methyl signal to the axial substituent in N,N-dimethylpiperidinium salts that bear an axial 2-methyl substituent (see p. 157) is provided by the spectrum of the N-quaternary

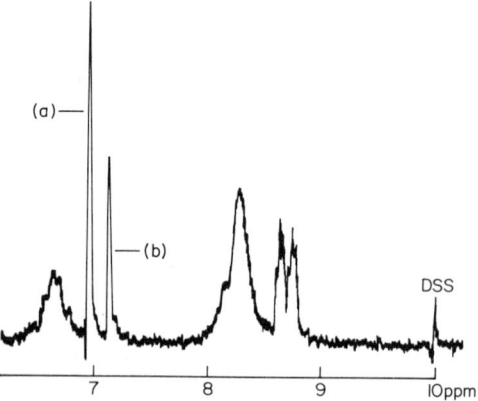

products of *cis*-1,2-dimethyl-3-phenylmorpholine and CD$_3$I (Casy and Hassan, unpublished results); in this case the *lower* field N—Me signal, due to axial N—Me, is the least intense.

FIG. 4.8. The 60-MHz PMR spectrum of the N-quaternization products of 1,2-dimethylpiperidine with CD$_3$I in acetone at 23° (Kawazoe and Tsuda, 1967). Signal (a) is assigned to the *e*-NMe isomer and (b) to the *a*-NMe isomer.

Two quaternary salts result when 1,2-dimethylpiperidine is treated with ethyl iodide as seen by the appearance of two N—Me signals in the total reaction product spectrum; the isomers are formed in unequal amounts, since one signal (that at lower field) is the more intense of the two. A qualitatively similar mixture is also obtained when 1-ethyl-2-methyl-piperidine is treated with methyl iodide (reverse quaternization). Here, however, the relative isomer proportions are reversed, the higher field signal being the more intense (Becconsall *et al.*, 1965). Axial and equatorial N—Me assignments, in this case, are made from comparison with the N-methyl signals of 1,2-dimethylpiperidine hydrochloride. It will be recalled that the spectrum of this salt shows two N—Me signals of different intensity, the more intense (lower field) being assigned to the equatorial group (see p. 147); by analogy the lower field signal of the N—Me, N—Et quaternary mixtures is taken as due to the *e*-N—Me group. Hence, the axial N—Et isomer is the major product of reaction A while the axial

major product
lower field NMe
more intense

Reaction A

----

major product
higher field NMe
more intense

Reaction B

N—Me isomer is the main product of reaction B, both results being consistent with a preferred axial "approach of quaternizing" agent. The equatorial 2-methyl group of these isomers must make a substantial contribution to the *a*- and *e*-N-methyl chemical shift difference through its differential shielding influence, as already discussed (p. 157).

The major products of methylation of 1-ethyl-4-phenylpiperidine and 1-benzyl-4-phenylpiperidine have now been shown to be the axial N-methyl salts by X-ray crystallography, confirming conclusions based on PMR evidence (Fedeli *et al.*, 1969; Brettle *et al.*, 1969). Fedeli *et al.* (1969) pointed out that the axial N-methyl signal of N-methylpiperidinium salts should be broader than that due to an equatorial group because of more effective long-range coupling between the N-methyl and α-axial hydrogen atoms (W-pathways link axial, but not equatorial, N-methyl protons to the α-protons, see Chapter 3, p. 91). In fact, the axial N-methyl signal of 1,1-dimethyl-4-phenylpiperidinium iodide, which may be assigned on the grounds of its higher field position and reduction in intensity in the spectrum of the corresponding trideuteromethyl iodide, proved to be slightly broader (W_H 2·17 Hz) than the corresponding equatorial signal (W_H 1·86 Hz). The width of the two signals did not differ significantly in the spectrum of the 2,2,6,6-tetradeuterio analogue, the values 1·85 Hz (low field) and 1·86 Hz (high field) being obtained [see also Jones *et al.* (1970a)].

The benziodide of N-benzyl-4-phenylpiperidine (**54**) has a PMR spectrum which displays two benzyl CH_2 singlets separated by about 30 Hz at 60 MHz in $CDCl_3$ (Brown *et al.*, 1967b). This is far greater than the separation seen for most *a*- and *e*-N-methyl signals and is attributed to differential aromatic shielding in terms of conformation **55**, and the equivalent one in which Ph and ···H of *a*-CH_2Ph are interchanged. In this, the

phenyl ring of the axial benzyl group would be expected to shield the methylene protons of the equatorial benzyl group, while phenyl of *e*-benzyl should have little influence upon the *a*-CH_2Ph group. The higher field CH_2 signal is therefore assigned to the *e*-CH_2Ph group; screening calculations based upon Johnson and Boveys' procedure (1958) give good agreement with the observed chemical shifts. In the analogue **56**, asymmetry resulting from the nearby 2-methyl group renders both *a*- and *e*-benzyl methylene groups non-equivalent (cf. p. 138) and a pair of AB multiplets are observed. When the 2-methylpiperidine base containing the group N—CD_2Ph is quaternized with benzyl iodide, the lower field CH_2Ph

signal of the total product spectrum is the more intense, a result in agreement with the preferred axial approach hypothesis (Brown *et al.*, 1966).

LONE-PAIR STEREOCHEMISTRY

The question of the position of equilibrium in cyclic bases such as derivatives of piperidine (**57**) has interested chemists for a number of years.

57

The problem may be regarded as an investigation of the steric requirements of the lone pair on nitrogen, i.e. the volume in space in close proximity to the nucleus in which there is the greatest probability of finding the electrons, although wording in these terms has been criticized (Allinger *et al.*, 1967; Riddell, 1967). Diametrically opposed conclusions have been drawn about the relative dimensions of the lone pair and hydrogen (based upon a variety of physical evidence), and the entire topic is still controversial. However, some interesting PMR studies have come out of this work and three of these are discussed here.

The approach of Lambert's group (1967) is based on the observation that the chemical shift difference (Δ_{ae}) between geminal protons adjacent (α) to nitrogen in rigid cyclic systems is enhanced to 50–60 Hz when the lone pair is axial (Hamlow *et al.*, 1964); the normal Δ_{ae} value in 6-membered rings is generally 25–30 Hz at 60 MHz (see p. 118). This enhancement is associated with a pronounced up-field shift of the axial proton which is *anti*-coplanar to the lone pair. Lambert argues that the α-methylene group of an isomer with an equatorially orientated lone pair should exhibit a near-normal Δ_{ae} value. This interpretation of the data obtained by analysis of the low temperature PMR spectra of the deuterated piperidines (**58**) (see p. 120 for details of one of the spectra) provides evidence that the lone

R	Δ_{ae} (Hz)	
	$\alpha-CH_2$	$\gamma-CH_2$
H	26·1	24·8
Me	56·5	31·1
t-Bu	59·8	31·7

58

(solvent: CD$_3$OD)

pair is equatorial in piperidine itself (normal Δ_{ae}), and axial in the N-alkyl derivatives (elevated Δ_{ae}). Note that the γ–Δ_{ae} enhancement is much smaller, the methylene protons in this case being well removed from possible lone pair influences. A normal Δ_{ae} value was also found for proton-ated piperidine in which the lone pair is, of necessity, absent.

These conclusions are invalid if factors other than the lone pair can be shown responsible for the Δ_{ae} increase, and the view has been expressed that the axial proton shielding is caused by the equatorial alkyl substituent on nitrogen rather than the lone pair (Robinson, 1968). Booth (1966) has drawn attention to the fact that an axial proton α- to equatorial methyl is shielded to a greater extent than is its geminal (equatorial) neighbour. In confirmation, Lambert and Keske (1969) found from analysis of the spectrum of the deuterated derivative (**59**) that equatorial methyl adds

59   **60**

16 Hz to the Δ_{ae} difference (in spite of the fact that the spacial Me/H$_a$ and Me/H$_e$ arrangements cannot differ greatly, see **60**). From this result, and the fact that only a 6 Hz increase in Δ_{ae} follows replacement of N—H by N—Me in the protonated piperidine anion (**61**), it was concluded that

61

Δ_{ae} 22 Hz   28 Hz

(in CD$_3$OD, gegenion Cl$^-$)

both lone pair and N-alkyl substituent shield the axial proton, but the former is the dominant contributor.

Booth's approach (1968) rests on the assumption that the rate of reaction of a base with D$^+$ is almost certainly greater than the rate of nitrogen inversion (Saunders and Yamada, 1963; Delpuech and Deschamps, 1967). If this is true,† the relative populations of the conjugate acids (iii) and (iv), obtained by treating pure dry *cis*-3,5-dimethylpiperidine with deuteriofluoroacetic acid (CF$_3$CO$_2$D) will reflect the populations of the equatorial and axial

---

† Its validity, particularly during the initial mixing process, has been questioned (McKenna and McKenna, 1969; Jones *et al.*, 1970b).

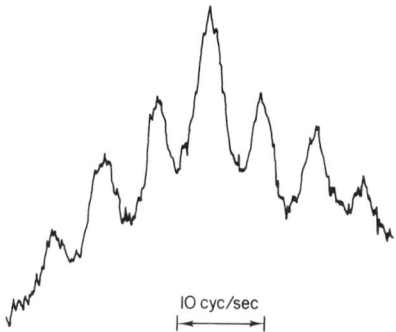

N—H forms of the free bases (i) and (ii) respectively (D$^+$ uptake without change of configuration must also be assumed). The spectrum of the base in CF$_3$CO$_2$D showed a 7-line multiplet at τ7·3 (Fig. 4.9) made up of a

10 cyc/sec

F$_{IG}$. 4.9. Resonance of 2,6-axial protons in the PMR spectrum of *cis*-3,5-dimethylpiperidine in CF$_3$CO$_2$D (Booth, 1968).

triplet and a quartet with identical chemical shifts. The triplet (separation ~12 Hz) was assigned to the identical 2,6-axial protons of the equatorial N—H form (iii), which are coupled about equally to the geminal 2,6-equatorial and the 3,5-diaxial protons. The quartet (separation also ~12 Hz) was assigned to the 2,6-axial protons of form (iv), which involves an additional large coupling to the axial N—H proton. All lines were broadened by small couplings to deuterium and/or equatorial N—H. The fact that a 7-line signal was observed which was unchanged after 2 days, shows that H$^+$ and D$^+$ exchange in the conjugate acids (iii) and (iv) is very slow. The relative areas of triplet and quartet, obtained by use of a DuPont 310 Curve Resolver, showed that the mixture contained 54% (iii) and 45% (iv). If the same ratio holds for the bases (i) and (ii), the free energy difference between the axial and equatorial forms is only about 0·1 kcal/mole.

The last method, due to Katritzky's group, involves comparing the equilibrium compositions of 2,6-dialkylcyclohexanone and 1-*t*-butyl-3,5-dialkyl-4-piperidones (Brignell *et al.*, 1968). In the former case, the *trans* isomer is destabilized by a 1,3-diaxial alkyl/H interaction (see **62**), whereas in the latter, the equivalent interaction is between an axial alkyl group and a

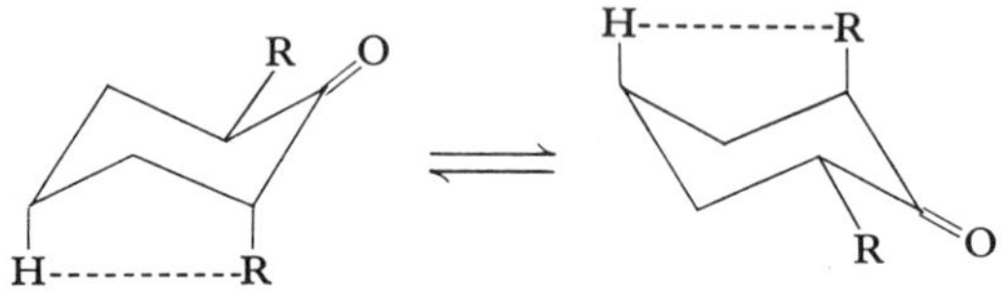

62

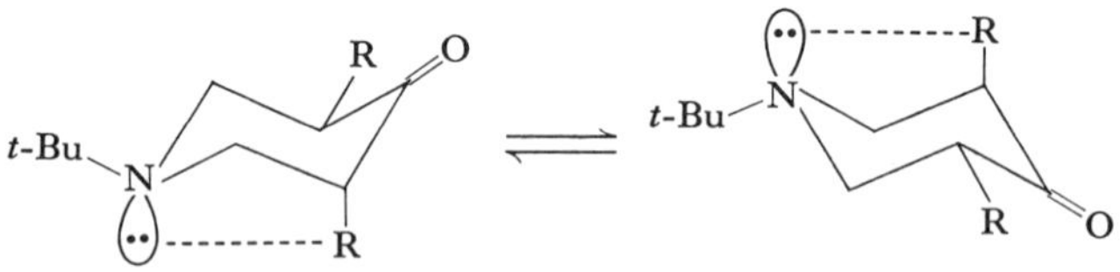

63

lone pair (see **63**). If the lone pair has smaller steric requirements than a hydrogen atom, the equilibrium population of the *trans* piperidone should exceed that of the cyclohexanone and vice versa if it occupies a greater space.

The PMR spectrum of a *cis–trans* mixture of 1-*t*-butyl-3,5-dideutero-3,5-dimethyl-4-piperidone (the normal ketone was equilibrated and deuterated simultaneously by treatment with CCl$_4$—Na—D$_2$O) shows two AB patterns in the ring methylene region (Fig. 4.10). One of these is derived from the conformationally fixed *cis* derivative (**64**), and the other to the *trans* dimethyl isomer (**65**) which gives rise to time-averaged spectra

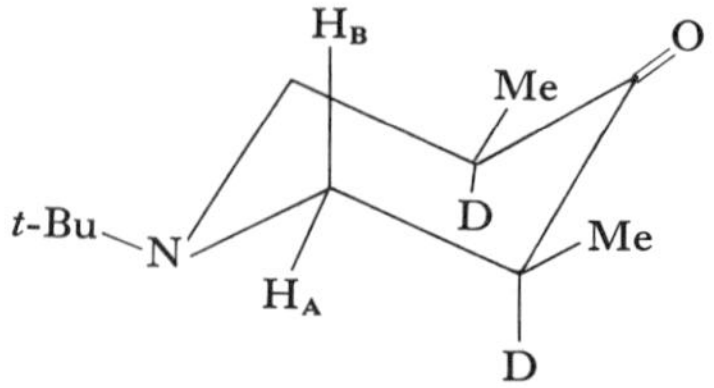

64

cis

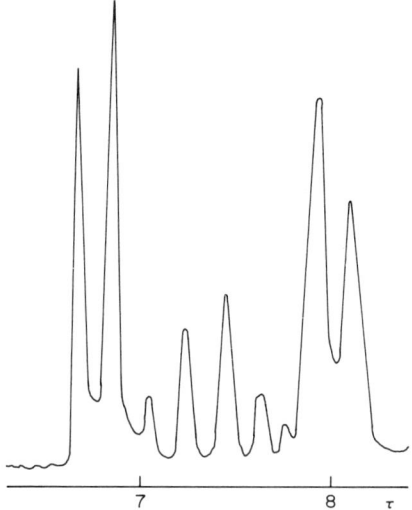

trans **65**

because of rapid interconversions between the equivalent conformations
65(a) and (b). The assignment of the two overlapping AB patterns is made
as follows (this analysis has much in common with that of 1,2,5-trimethyl-4-
piperidone, described on p. 140).

FIG. 4.10. 60-MHz PMR spectrum of the ring proton region of 3,5-dideutero-
3,5-dimethyl-1-*t*-butyl-4-piperidone in CCl$_4$ (Brignell *et al.*, 1968).

   (i) The AB chemical shift difference should be larger for the *cis* derivative,
as the AB pattern is derived from one axial and one equatorial proton,
whereas for the *trans* compound each proton is time-averaged to half axial
and half equatorial (note that the interconversion of **65**(a) and (b) does not
render the H$_A$ and H$_B$ protons equivalent; H$_A$ is *cis* and H$_B$ *trans* to 3-Me
in either form).

   (ii) The calculated H$_A$ and H$_B$ chemical shifts using the data of Booth
(1966) on methyl shielding effects and the ring methylene chemical shift

of the piperidone (**66**) (average of chemical shifts for axial and equatorial protons) are as follows:

$$H_A: 7\cdot15 + (0\cdot47 + 0\cdot40)/2 = \tau\ 7\cdot58$$
$$H_B: 7\cdot15 + (0.28 - 0.20)/2 = \tau\ 7\cdot19$$

These results agree better with the inner AB system (τ 7·52 and 7·11) than with the outer (τ 7·96 and 6·70), hence the outer pattern is assigned to the *cis*-piperidone.

(iii) The lines of the inner AB systems have similar half-widths, whereas the higher field lines of the outer AB are broader than the low-field lines, probably because of axial-axial H/D coupling in the higher field axial protons.

The AB systems of the 3,5-diethyl analogue were assigned similarly. The percentages of *cis* and *trans* forms (see below) were determined by integration of the pairs of AB spectra. Results show that, qualitatively, the proportion of axial alkyl for the piperidones are in each case larger than for

	% *cis*
2,6-dimethylcyclohexanone	91·6–92·9
1-*t*-butyl-3,5-dimethyl-4-piperidone	82·9
2,6-diethylcyclohexanone	79·4–85·8
1-*t*-butyl-3,5-diethyl-4-piperidone	56·6

the corresponding cyclohexanone (it is interesting that a larger alkyl substituent also favours the *trans* isomer). This demonstrates that repulsive interactions are smaller between an alkyl group and an (unsolvated) lone pair. It is assumed that the *t*-butyl substituent does not distort the ring and that the cyclohexanone and piperidone rings have the same geometry.

References

Allinger, N. L., Hirsch, J. A. and Miller, M. A. (1967). *Tetrahedron Lett.* 3729.
Becconsall, J. K., Jones, R. A. Y. and McKenna, J. (1965). *J. Chem. Soc.* 1726.
Booth, H. (1966). *Tetrahedron* **22**, 615.
Booth, H. (1968). *Chem. Commun.* 802.

Booth, H. and Gidley, G. C. (1965). *Tetrahedron* **21**, 3429.

Booth, H. and Little, J. H. (1968). *Tetrahedron* **24**, 279.

Brettle, R., Brown, D. R., McKenna, J. and Mason, R. (1969). *Chem. Commun.* 339.

Brignell, P. J., Katritzky, A. R. and Russel, P. L. (1968). *J. Chem. Soc.* (B), 1459.

Brown, D. R., Hutley, B. G., McKenna, J. and McKenna, J. M. (1966). *Chem. Commun.* 719.

Brown, D. R., Lygo, R., McKenna, J. and McKenna, J. M. (1967a). *J. Chem. Soc.* (B), 1184.

Brown, D. R., McKenna, J. and McKenna, J. M. (1967b). *J. Chem. Soc.* (B), 1195.

Casy, A. F., Beckett, A. H., Iorio, M. A. and Youssef, H. Z. (1965). *Tetrahedron* **21**, 3387.

Casy, A. F., Chatten, L. G. and Khullar, K. K. (1970). *Can. J. Chem.* **48**, 2372.

Closs, G. L. (1959). *J. Amer. Chem. Soc.* **81**, 5456.

Delpeuch, J. J. and Deschamps, M. N. (1967). *Chem. Commun.* 1188.

Delpeuch, J. J. and Deschamps, M. N. (1970). *Tetrahedron* **26**, 2723.

Eliel, E. L., Allinger, N. L., Angyal, S. J. and Morrison, G. A. (1965). "Conformational Analysis". Wiley, New York.

Fedeli, W., Jones, R. A. Y., Katritzky, A. R., Mazza, F., Mente, P. G. and Vaciago, A. (1969). *Accad. Naz. Lincei* **46**, 733.

Garbisch, E. W. (1962). *J. Org. Chem.* **27**, 4249.

Hamlow, H. P., Okuda, S. and Nakagawa, N. (1964). *Tetrahedron Lett.* 2553.

Hassan, M. M. A. and Casy, A. F. (1970). *Org. Magnetic Resonance*, **2**, 197.

Hill, R. K. and Chan, T-H. (1965). *Tetrahedron* **21**, 2015.

Iorio, M. A. and Casy, A. F. (1970). *J. Chem. Soc.* (C), 135.

Johnson, C. E. and Bovey, F. A. (1958). *J. Chem. Phys.* **29**, 1012.

Jones, R. A. Y., Katritzky, A. R. and Mente, P. G. (1970a). *J. Chem. Soc.* (B), 2110.

Jones, R. A. Y., Katritzky, A. R., Richards, A. C., Wyatt, R. J., Bishop, R. J. and Sutton, L. E. (1970b). *J. Chem. Soc.* (B), 127.

Kawazoe, Y. and Tsuda, M. (1967). *Chem. Pharm. Bull.* **15**, 1405.

Kawazoe, Y., Tsuda, M. and Ohniski, M. (1967). *Chem. Pharm. Bull.* **15**, 51.

Lambert, J. B. and Keske, R. G. (1969). *Tetrahedron Lett.* 2023.

Lambert, J. B., Keske, R. G., Carhart, R. E. and Jovanovich, A. P. (1967). *J. Amer. Chem. Soc.* **89**, 3761.

Lyle, R. E. and Krueger, W. E. (1967). *J. Org. Chem.* **32**, 3613.

Lyle, R. E. and Thomas, J. J. (1969). *Tetrahedron Lett.* 897.

McKenna, J. (1970). *In* "Topics in Stereochemistry" (E. L. Eliel and N. L. Allinger, eds), Vol. **5**. Wiley, New York.

McKenna, J. and McKenna, J. M. (1969). *J. Chem. Soc.* (B), 644.

Musher, J. I. and Corey, E. J. (1962). *Tetrahedron* **18**, 791.

Prostakov, N. S. and Mikheeva, N. N. (1962). *Russ. Chem. Rev.* **31**, 556.

Riddell, F. G. (1967). *Quart. Rev.* **31**, 304.

Robinson, M. J. T. (1968). *Tetrahedron Lett.* 1153.

Saunders, M. and Yamada, F. (1963). *J. Amer. Chem. Soc.* **85**, 1882.

Stein, M. L., Ottinger, R., Reisse, J. and Chiurdoglu, G. (1968). *Tetrahedron Lett.* 1521.

Tsuda, M. and Kawazoe, Y. (1970). *Chem. Pharm. Bull.*, **18**, 2499.

CHAPTER 5

PMR Studies of Optical Enantiomorphs

Over the past few years much interest has arisen in the differentiation of enantiomorphs and the determination of their optical purity by PMR. Raban and Mislow, the instigators of these investigations, have published a review of this topic (1967) and the first part of this chapter is based largely upon their account.

Consider the (R)- and (S)-enantiomorphs of α-phenylethylamine. The

$$
\underset{(R)\text{-}\mathbf{1}}{\overset{\text{Me}}{\underset{\text{H}}{\text{Ph}-\!\!\!\!\!-\!\!\!\!\!-\text{NH}_2}}} \equiv \underset{(R)\text{-}\mathbf{1}}{\overset{\text{H}}{\underset{\text{Me}}{\text{H}_2\text{N}-\!\!\!\!\!-\!\!\!\!\!-\text{Ph}}}} \qquad \underset{(S)\text{-}\mathbf{1}}{\overset{\text{Me}}{\underset{\text{H}}{\text{H}_2\text{N}-\!\!\!\!\!-\!\!\!\!\!-\text{Ph}}}}
$$

two methyl groups in the isomers are situated in enantiomorphic environments (i.e. those related as object to mirror image); the magnetic influences of these environments are equivalent, however, hence there is no difference in chemical shift between (R)-Me and (S)-Me PMR signals. It is, in fact, generally true that the NMR spectra of enantiomorphs are identical when examined in optically inactive (achiral) solvents (see later for results in optically active solvents). Reaction of (R)-1 and (S)-1 with (R)-O-methyl-mandelyl chloride 2 yields the amides (RR)-3 and (RS)-3 respectively. These derivatives bear a diastereoisomeric, rather than enantiomorphic relationship and may therefore differ in their PMR characteristics, in

$$
\underset{2}{\overset{\text{OMe}}{\underset{\text{H}}{\text{Ph}-\text{C}-\text{COCl}}}} \qquad \underset{(R)\quad(R)\text{-}\mathbf{3}}{\overset{\text{OMe}\quad\quad\text{H}}{\underset{\text{H}\;\;\text{O}\quad\quad\text{Me}}{\text{Ph}-\text{C}-\text{C}-\text{NH}-\text{C}-\text{Ph}}}}
$$

$$
\underset{(R)\quad(S)\text{-}\mathbf{3}}{\overset{\text{OMe}\quad\quad\text{Me}}{\underset{\text{H}\;\;\text{O}\quad\quad\text{H}}{\text{Ph}-\text{C}-\text{C}-\text{NH}-\text{C}-\text{Ph}}}}
$$

170

greater or lesser degree depending on differences in conformer populations and the effect of highly anisotropic groups (see p. 111). In the present

H
|
R—C—Me
|
H

protons with an enantiomorphic
environment

H
|
R—C—R*
|
H

protons with a diastereoisomeric
environment

(R is a symmetric and R* a dissymmetric group)

example, chemical shift differences of the diastereoisomeric methine (a), methoxy (b) and C-methyl (c) groups are clearly differentiated, as shown in Fig. 5.1. If the optically pure (R)-acid chloride **2** reacts with partially

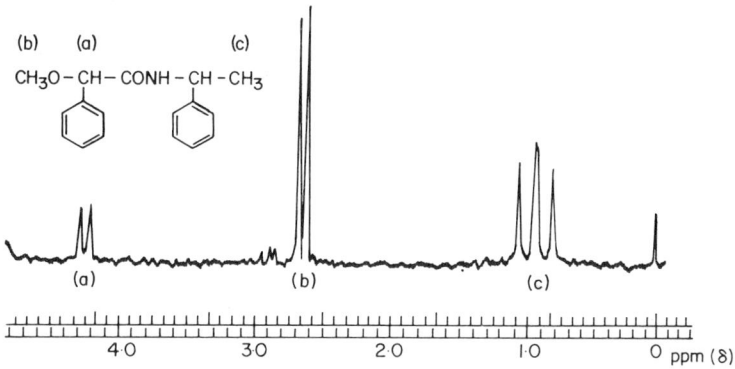

Fig. 5.1. Part of the 60 MHz PMR spectrum of a sample of amide prepared from racemic α-phenylethylamine and O-methylmandelyl chloride in CDCl$_3$. The spectrum of this mixture is identical with that of a 50:50 mixture of (RS)- and (RR)-diastereoisomers (Raban and Mislow, 1967).

resolved 1-phenethylamine the same pair of amides (RR)- and (RS)-**3** results but in unequal amounts, and the relative proportions of the two isomers may be gauged by the integrated intensities of one or more of the signals a, b and c in the spectrum of the *total* product. Resolution of the amine may be followed by converting the amine to the amide at each stage (assuming that amide formation induces no further resolution) and observing the diminution of one set of abc signals; when one set is completely absent, an optically pure product will have been obtained.† In principle, therefore,

† This resolution has been adapted as an advanced undergraduate laboratory experiment (Jacobus and Raban, 1969).

optical antipodes may be differentiated for the purposes of identification and resolution by linking them to the optically pure enantiomorph of a suitable reagent and examining the PMR spectrum of the *total* diastereo-isomeric mixture. For the method to be successful, chemical shift differences between diastereoisomers must be sufficiently pronounced (signals which overlap are useless while those that are very close cannot be integrated separately) and care must be taken to ensure: (i) the quantitative reaction of both enantiomorphs (so that the diastereoisomer ratio exactly reflects the enantiomeric ratio) and, (ii) that no racemization or equilibration processes occur (see later for this last point). Methylmandelic acid is a particularly useful reagent for the PMR differentiation of enantiomorphic amines and alcohols for the following reasons:

1. It contains two groups which give rise to singlet resonance signals (a and b in Fig. 5.1); note that signals due to either component of the diastereo-isomer may serve for differentiation purposes.

2. It contains a highly magnetically anisotropic group (phenyl) which increases the probability of chemical shift non-equivalence between diastereoisomers being large enough to permit the separate integration of related signals.

The most convenient and labour-saving application of the PMR pro-cedure to a resolution problem occurs when both the physical separation and spectral differentiation of a diastereoisomeric pair are possible. An example concerning the separation of the racemic *cis* and *trans* cyclo-

(Ar = *o*-MeC₆H₄)

hexanols **4** by the aid of (−)-menthoxyacetyl chloride **5**, is given below (Galpin and Huitric, 1968). The *trans*-(±)-alcohol **4** and the acid chloride gave a crude, solid mixture of esters which on fractional crystallization yielded one pure diastereoisomer; the second isomer was obtained by column chromatography of the mother liquor residues. The *cis*-(±)alcohol **4** formed a crude ester mixture which could not be solidified. It was separated by chromatography into one pure and one partially pure diastereoisomer. Separation progress was monitored by PMR spectroscopy, the signals of importance being those due to the non-equivalent methylene protons of the esters **6**. These had similar chemical shifts in isomeric pairs but the separations of the central doublets of the two AB spectra differed. In the *trans* mixture the two lower field components of these doublets overlapped and a three peak signal resulted (Fig. 5.2), resolution being followed by the progressive disappearance of the central signal and attainment of the symmetrical (a) or (b) pattern.

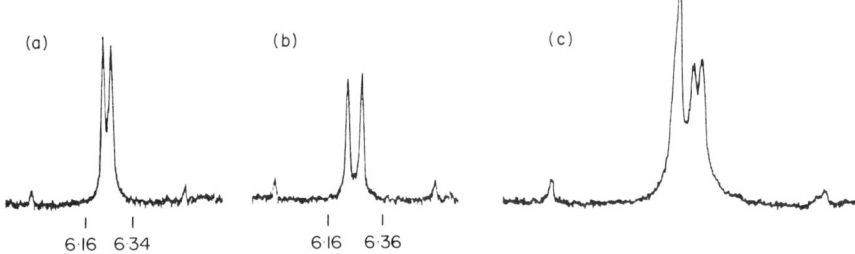

FIG. 5.2. The 60 MHz PMR signals of the OCH$_2$CO methylene protons of (a) (+)-*trans*-2-*o*-tolylcyclohexyl(−)-menthoxyacetate 6; (b) the (−)-*trans* diastereoisomer; (c) a mixture of the (+)- and (−)-*trans* diastereoisomers (Galpin and Huitric, 1968). The scale of c is larger than that of a and b.

Dale and Mosher (1968) tested the accuracy of the PMR method of enantiomorphic purity assay in cases where optical purities were known by other means. Agreement between polarimetric and PMR analyses were good for (S)-O-methylmandelic acid of 90·9% enantiomorphic purity by optical rotation (see Scheme I) and in other instances (Table 5.1).

OMe
|
Ph—C—CO$_2$H $\xrightarrow[\substack{\text{benzene} \\ \text{reflux 1 hr.}}]{\text{SOCl}_2}$ Ph—C—COCl $\xrightarrow[\substack{\text{pyridine-} \\ \text{benzene}}]{\substack{\text{4 mols} \\ (-)\text{-menthol}}}$
| |
H   H

OMe

(S)-acid 95·5 : 4·5 enantiomer ratio by optical rotation

unpurified

Me

OMe
|
Ph—C—CO$_2$—⟨cyclohexyl⟩
|
H$_B$ CHMe$_2$

Product extracted with H$_2$O, then H$_2$O—HCl; ester isolated by glpc. Ratio of H$_B$ PMR singlets 95·8 : 4·2

Scheme I

The NMR spectrum of *t*-butyltrifluoromethylcarbinyl O-methyl-mandelate shows duplicate signals for the fluorine resonances and for all proton groups save aromatic (Fig. 5.3); however, signal intensities for related groups are not identical in spite of the fact that the ester was made from racemic materials (cf. spectrum of the amide of Fig. 5.1, also made from racemic reagents). Further to this observation, the PMR spectra of

TABLE 5.1

Comparison of alcohol enantiomorphic purities as determined by NMR and other methods[a]

$$
\begin{array}{c}
\quad\ \ \text{MeO}\ \ \ \text{O}\qquad\ \text{R}\\
\quad\ \ \ |\qquad\ \ \|\qquad\ |\\
\text{Ph—C—C—O—C—H}\\
\quad\ \ \ |\qquad\qquad\quad\ |\\
\quad\ \ \ \text{H}\qquad\qquad\ \ \text{R}'
\end{array}
$$

R	R'	Polarimetry	Glpc	NMR (100-MHz)
Me	C_6H_{11} [b]	$2 \cdot 0 \pm 1 \cdot 0$	—	$3 \cdot 0 \pm 2 \cdot 0$
Me	$n\text{-}C_6H_{13}$	$96 \cdot 1 \pm 0 \cdot 3$	—	$97 \cdot 5 \pm 2 \cdot 0$
Me	CF_3	—	$62 \cdot 3 \pm 0 \cdot 5$	$62 \cdot 5 \pm 0 \cdot 3$
Ph	CF_3	$100 \cdot 0 \pm 0 \cdot 1$	$100 \cdot 0 \pm 0 \cdot 5$	$100 \cdot 0 \pm 0 \cdot 5$
Me	$i\text{-}Pr$	$100 \cdot 0 \pm 1 \cdot 0$	—	$100 \cdot 0 \pm 0 \cdot 5$

[a] Values are per cent enantiomorphic purity.
[b] C_6H_{11} represents cyclohexyl.

diastereoisomers prepared from the *t*-butyl secondary alcohol **7a** and O-methylmandelyl chloride are identical whether prepared from optically

$$
\begin{array}{c}
\quad\ \text{OH}\\
\quad\ \ |\\
\text{R—C—}t\,\text{Bu}\qquad \text{R} = \text{(a) Ph; (b) Et; (c) Me}\\
\quad\ \ |\\
\quad\ \text{H}\\
\quad\ \ \textbf{7}
\end{array}
$$

active or racemic reagents. Similar results are obtained in the case of the same esters of **7b** and **7c**, spectra repeatedly consisting of unequal signals for epimerically related groups. Ester preparation, in these examples, must therefore involve an equilibration process at the α-position of the acid moiety induced, most probably, by a base-catalysed carbanion mechanism.

$$
\underset{\substack{\uparrow\\ \text{Base}}}{\underset{\text{H}}{>}}\!\!\text{C—C}\!\!\begin{array}{c}\nearrow\text{O}\\ \searrow\text{OR}\end{array}
\ \underset{\longleftarrow}{\overset{-\text{H}^+}{\rightleftharpoons}}\
>\!\bar{\text{C}}\text{—C}\!\!\begin{array}{c}\nearrow\text{O}\\ \searrow\text{OR}\end{array}
\ \overset{+\text{H}^+}{\rightleftharpoons}\
\underset{}{\overset{\text{H}}{>}}\!\text{C—C}\!\!\begin{array}{c}\nearrow\text{O}\\ \searrow\text{OR}\end{array}
$$

(S)-ester   carbanion   (R)-ester

In support, the lithium aluminium hydride reduction of *t*-butylphenyl-

$$\underset{\mathbf{8}}{\overset{\overset{\displaystyle CF_3}{|}}{\underset{\underset{\displaystyle H}{|}}{PhCCO_2}}-\overset{\overset{\displaystyle CMe_3}{|}}{\underset{\underset{\displaystyle Ph}{|}}{C}}-H} \quad \xrightarrow{\text{LiAlH}_4} \quad \underset{\mathbf{9}}{\overset{\overset{\displaystyle CF_3}{|}}{\underset{\underset{\displaystyle H}{|}}{PhCCH_2OH}}} \quad + \quad \underset{\mathbf{10}}{HO-\overset{\overset{\displaystyle CMe_3}{|}}{\underset{\underset{\displaystyle Ph}{|}}{C}}-H}$$

carbinyl α-trifluoromethylphenylacetate **8**, prepared from optically active reagents gave the partially racemic **9** and optically pure **10** alcohols showing

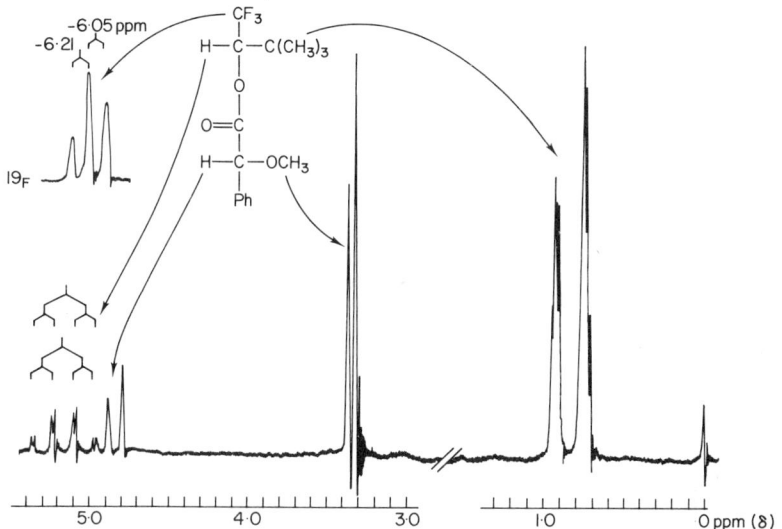

FIG. 5.3. 60 MHz PMR and 56·4 MHz fluorine NMR spectra of *t*-butyltri-fluoromethylcarbinyl O-methylmandelate, prepared from racemic acid and carbinol. Solvent: α,α,α-trifluorotoluene (Dale and Mosher, 1968).

that only the acid moiety had been racemized. The equilibration of dia-stereoisomers which may occur during the synthetic procedure is therefore a serious drawback to the general applicability of the NMR method to the determination of the optical purity of secondary carbinols and allowance should be made for equilibration complications in any application of this method. In these examples, equilibration was seen only in esters of *t*-butylcarbinols; it did not occur in esters formed from optically active methylisopropylcarbinol or methyl-*n*-hexylcarbinol and in these cases NMR spectra accurately reflected the enantiomorphic composition known from polarimetric data (Table 5.1). Hence ester equilibration must be dependent upon steric factors. One way of avoiding the ester equilibration difficulty would be to use an acid reagent which lacks an α-hydrogen atom. One such agent, namely α-methoxy-α-trifluoromethylphenylacetic acid (MTPA, **11**), has now been developed (Dale *et al.*, 1969) and is available

$$
\begin{array}{c}
\text{OMe} \\
| \\
\text{Ph—C—CO}_2\text{H} \\
| \\
\text{CF}_3
\end{array}
$$

11

from the Aldrich Chemical Company in either enantiomorphic form. Advantages claimed for MTPA are (i) marked stability toward race-mization, (ii) good separation of both proton and fluorine NMR signals of the diastereoisomers, with the fluorine signals in an uncongested spectral region, (iii) its application with both amines and alcohols and (iv) the fact that this absolute method (i.e. independent of optical rotation) can be used accurately on samples as small as 20 mg.

Another example of the NMR analysis of diastereoisomers which are obtained when enantiomorphs are converted to suitable derivatives, is the

$$
\begin{array}{ccc}
\text{Ph} & & \text{Ph} \\
\diagdown & \xrightarrow{\text{(+)-RCH}_2\text{Br}} & \diagdown^+ \\
n\text{-Pr—P} & & n\text{-Pr—P—CH}_2\text{CH(OMe)Ph} \quad \bar{\text{Br}} \\
\diagup & & \diagup \\
\text{Me} & & \text{Me}
\end{array}
$$

[R = PhCH(OMe)]

12   **13**

case of the optically active phosphine **12** and related compounds (Casey *et al.*, 1969). The PMR spectrum of the quaternary salt mixture **13** obtained when the phosphine is treated with (+)-2-phenyl-2-methoxyethyl bromide [a reagent derived from methyl (S)-O-methylmandelate], displays duplicate OMe (separation 4·2 Hz) and PMe (doublets due to PCH coupling, separa-tion 9·2 Hz) signals in D_2O at 60 MHz. Relative intensities may be used to judge optical purity, given that the reaction is essentially quantitative and stereospecific. There is evidence that such is the case.

Enantiomorphs which owe their asymmetry to the replacement of hydrogen by deuterium form diastereoisomers which differ little in physical properties and are therefore very difficult to separate. The NMR method for the determination of optical purity is thus very useful in such cases because the method requires no physical separation.

$$
\begin{array}{cc}
\text{H} & \text{H} \quad \text{O} \\
\text{Me} \text{--} \diagup & \text{Me} \text{--} \diagup \quad \| \\
\text{Me} \diagup \diagdown \text{OH} & \text{Me} \diagup \diagdown \text{OC—R}^* \\
\textbf{14} & \textbf{15}
\end{array}
$$

Consider the case of isopropanol; the two methyl groups of **14** reside in enantiomorphic, but identical, environments and this gives rise to coincident PMR signals. Reaction of isopropanol with a chiral (i.e., optically active) reagent, such as (+)-O-methylmandelic acid chloride gives the ester **15** in which the methyl groups now have diastereoisomeric environments and thus each gives rise to a unique signal (cf. previous discussion) [Fig. 5.4(a)].

(a)

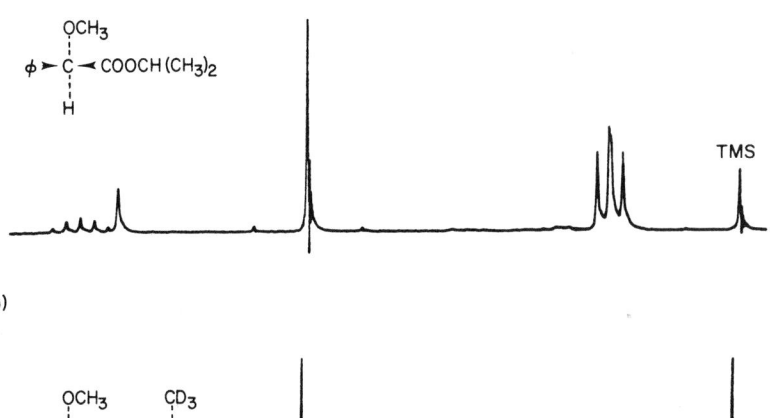

(b)

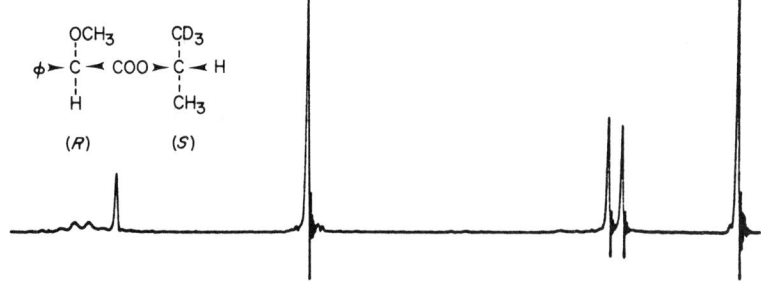

FIG. 5.4. 60 MHz PMR spectra in benzene of (a) isopropyl-(R)-O-methyl-mandelate and (b) the (R)-O-methylmandelate of a sample of propan-2-ol-1-d$_3$ (Raban and Mislow, 1966).

By the same argument, the enantiomorphic (and identical) environments of methyl in antipodal forms of propan-2-ol-1-d$_3$ (**16**) become non-equivalent in the (+)-O-methylmandelate esters. If the CD$_3$ derivative is optically pure,

16

one of the *sec*-Me signals of the *iso*propyl ester **15** will be absent (the deuterium isotope effect upon the chemical shift is very small). If the deuterated compound is not pure, the same ester will show two *sec*-Me signals and their relative intensities will be a measure of optical purity. Thus, a sample prepared from optically pure lactic acid by a route which did not affect bonds linked to the asymmetric centre, was shown to be 98–100% pure by this method [Fig. 5.4(b)] (Raban and Mislow, 1966).

Gerlach (1966) has solved the problem of both absolute configuration and optical purity of a sample of (+)-α-deuterated benzylamine by PMR spectroscopy. The method requires the conversion of benzylamine into 3-benzyl-4-phenyl-oxazolidin-2-thione (see Scheme II). The PMR spectrum

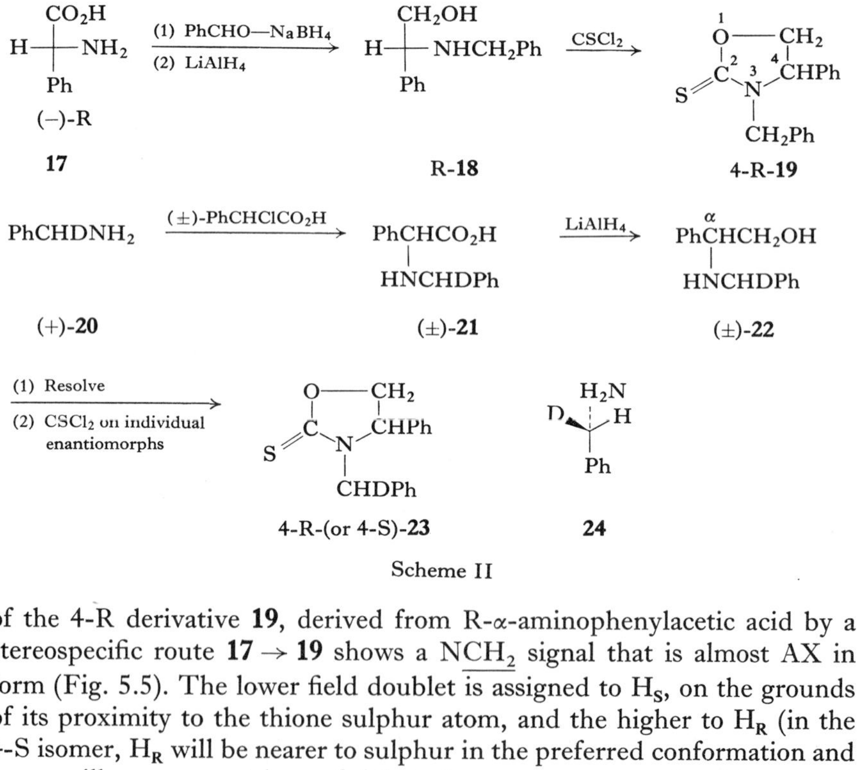

Scheme II

of the 4-R derivative **19**, derived from R-α-aminophenylacetic acid by a stereospecific route **17** → **19** shows a NCH_2 signal that is almost AX in form (Fig. 5.5). The lower field doublet is assigned to H_S, on the grounds of its proximity to the thione sulphur atom, and the higher to H_R (in the 4-S isomer, H_R will be nearer to sulphur in the preferred conformation and hence will resonance at lower field). A dextrorotatory sample of deuterated benzylamine **20** was then converted to the intermediate **22** via **21**. This substituted alcohol, racemic with respect to the α-carbon atom, was then resolved. Two pure enantiomorphs were obtained and one of them shown to correspond with the isomer **18**; hence its configuration at the α-carbon centre was established as R. Cyclization of the R-isomer gave the corresponding oxazolidin-2-thione **23**. The NCHD signal of this product displayed a major high field singlet (broad due to D—C—H coupling) corresponding with the doublet assigned to H_R, and a minor singlet corresponding with the H_S signal (Fig. 5.5). The relative intensities of the signals were 80 (H_R):20 (H_S), and these intensities were reversed in the spectrum of the thione derived from the S-isomer of **23**. The results establish that the (+)-deuterated benzylamine sample contains 80% of the enantiomorph **24**.

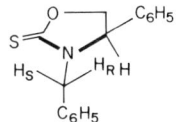

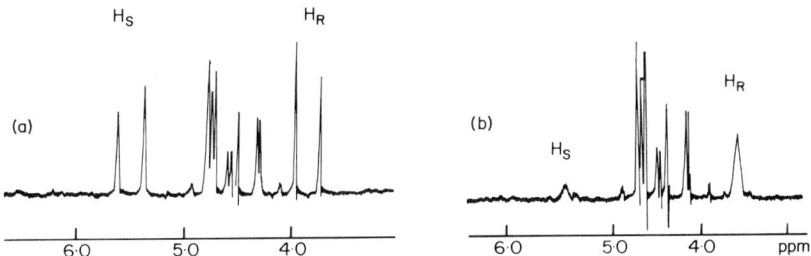

FIG. 5.5. Part of the 60-MHz PMR spectrum of (a) 4R-3-benzyl-4-phenyl-oxazolidin-2-thione, and (b) 4R-3-(α-$^2H_1$-benzyl)-4-phenyl-oxazolidin-2-thione in CDCl₃ (Gerlach, 1966).

The study of an asymmetric synthesis which has been aided by application of similar principles is outlined below. Reaction between racemic *p*-toluenesulphinyl chloride and (S)-pinacolyl alcohol (96% optical purity) gives an unequal mixture of the two possible diastereoisomeric esters **25**. The PMR spectrum of the mixed esters shows duplicate methine, *sec*-Me

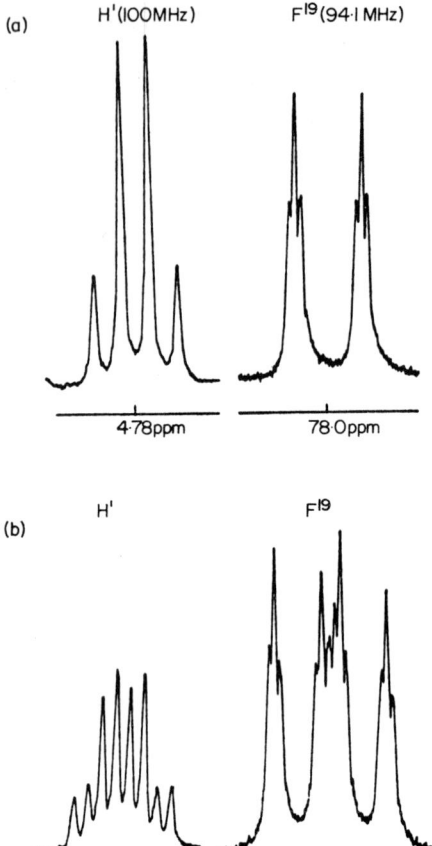

Fig. 5.6. Part of the NMR spectra of partially resolved 2,2,2-trifluorophenyl ethanol in (a) racemic and (b) optically active α-(1-naphthyl)ethylamine. The left hand signals are due to the methine proton, and the right hand to the fluorine atoms of the fluoroalcohol (Pirkle and Beare, 1967).

and CMe$_3$ signals in the ratio 60·5 : 39·5 (Mislow and Raban, 1967); thus the diastereoisomeric purity of these esters and the enantiomorphic purity† of the derived methyl *p*-tolyl sulphoxides (**26**) is 21%.

Pirkle has now reported several examples in which enantiomorphic pairs themselves may be differentiated directly by NMR spectroscopy when

† Defined as $\dfrac{\text{mols levo isomer} - \text{mols dextro isomer}}{\text{total mols}} \times 100$, assuming the major isomer to be the levo form.

examined as solutes in optically active (chiral) solvents (Pirkle, 1966). Differing diastereoisomeric environments again form the basis for differences in NMR characteristics but these arise as a result of solute-solvent interactions (e.g. by hydrogen bonding or dipolar attraction) rather than covalent bonding. Thus the fluorine resonance of optically inactive 2,2,2-

$$Ph-\underset{\underset{\text{H}}{|}}{\overset{\overset{\text{OH}}{|}}{C}}-CF_3 \qquad\qquad PhCHMeNH_2$$

27   **28**

trifluoro-1-phenylethanol **27**, a doublet of triplets in pyridine (long range coupling of the fluorine and aromatic protons occurs), appears as two sets of doublets (each showing fine structure) in optically active α-phenethylamine **28**; the two sets are equally intense as expected when the racemic alcohol is used as solute. In the racemic solvent, the fluorine resonance of the racemic alcohol **27** reduces to one doublet (plus fine structure). In this case exchange between solvent partners [e.g. R (solute) R (solvent) + S (solvent) ⇌ R (solute) S (solvent) + R (solvent)] must be so fast that all the fluoroalcohol molecules experience the same average environment (note that exchange when the solvent is optically active does not alter the diastereoisomeric environments). The ^{19}F resonance signals of the (±) alcohol **27** differ similarly in racemic and (+)-α-(1-naphthyl)ethylamine (Fig. 5.6); at 100 MHz the two methine proton signals may be differentiated in the latter solvent (they overlap to give an octet). The partially resolved fluoroalcohol **27** shows two sets of doublets in (−)-α-phenethylamine with the higher field set the more intense. The lower field set becomes more intense when the dextrorotatory amine **28** is used as solvent and relative intensities of the two signals may be used to assess optical purity as usual.

Another proton example is the racemic alcohol **29** (Burlingame and Pirkle, 1966). The two doublets coalesce in racemic solvent and are closer

$$Ph-\underset{\underset{\underline{\text{H}}}{|}}{\overset{\overset{\text{OH}}{|}}{C}}-CHMe_2$$

29

methine H $\underline{\underline{\ }}\begin{cases} \text{doublet in } CFCl_3 \\ \text{2 doublets in } (+)\text{-}\alpha\text{-}(1\text{-naphthyl})\text{ethylamine} \\ \text{1·6 Hz separation at 60 MHz} \\ \text{2·5 Hz separation at 100 MHz} \end{cases}$

together when the solvent is not completely optically pure. The ^{19}F resonance signal of the O-methyl ether of the (±)-fluoroalcohol **27** forms a single doublet (plus fine structure) in optically active α-(1-naphthyl)ethyl-amine (Pirkle and Burlingame, 1967) and this result indicates that hydrogen bonding is responsible for the solute-solvent associations observed for the alcohol itself. The separations of the two ^{19}F signals of the alcohols **30b** and **30c** are larger and smaller respectively than that found for the phenyl alcohol **30a** (Pirkle and Burlingame, 1967); hydrogen bonding in **30b**

$$\text{(±)} \quad R{-}\underset{\underset{\text{H}}{|}}{\overset{\overset{\text{OH}}{|}}{C}}{-}CF_3$$

30   Δδ at 56·4 Hz

(a) R = Ph   3·3 Hz

(b) R = 2,4-diNO$_2$.C$_6$H$_3$ 5·8 Hz

(c) R = α-pyridyl   0·4 Hz

should be augmented by π-complexation whereas the α-pyridyl nitrogen of **30c** can compete with OH as a hydrogen bonding site and produce a different type of complex.

Pirkle and Beare (1967) examined a series of optically active alcohols, RCH(OH)Ar, of known configuration in (+)-α-(1-naphthyl)ethylamine. These alcohols contained minor amounts of their optical antipodes and the usual signal duplication was observed. For alcohols of absolute configuration **31**, the more intense methine multiplet appeared at higher field

$$\underset{\underset{\displaystyle R \qquad Ar}{}}{\overset{\overset{\displaystyle OH}{|}}{\diagup\!\!\diagup}}\text{--H}$$

31

while, in cases where R = CF$_3$, the more intense ^{19}F doublet fell at lower field. These results enable the absolute configuration of an enantiomorphic alcohol of type RCH(OH)Ar of unknown stereochemistry to be deduced from the relative positions of its major and minor C—H and (when appropriate) CF$_3$ signals (a small amount of the second enantiomorph must, of course, be present). Spectral differences between enantiomorphs of these alcohols are explained in terms of the models **32a** and b which present a probable conformation of the complex formed between one of the alcohol enantiomorphs and the optically active solvent (stabilized by hydrogen bonding and weak charge-transfer interactions between the aromatic

32a   32b

systems). In **32a** the methine proton of the fluoroalcohol is screened by the naphthyl ring because it lies close to it and above its plane; in the complex involving the antipodal form of this alcohol **32b**, the methine proton takes the position occupied by CF_3 in **32a** and thus, being less subject to aromatic screening, comes to resonance at lower field. In contrast, aromatic screening of the CF_3 group will be greater in the enantiomorph depicted in **32b**. Thus the higher field methine signal is always associated with the lower field CF_3 signal and vice versa.

The same principles have been used to establish the absolute configuration of $(+)$-α-hydroxy-α-trifluoromethyl-phenylacetic acid (Pirkle

33   34

(a) R = H
(b) R = Me
(c) R = CCl_3
(d) R = CF_3

and Beare, 1968a). Thus the carbomethoxy resonances of enantiomorphs of the esters **33** a, b and c (of known configuration) have observable different chemical shifts in $(+)$-α-(1-naphthyl)ethylamine with the (R)-signal at higher field than the (S) in all three cases. Partially resolved **33d** (dextrorotatory) displays two $CO_2\underline{Me}$ resonance signals with the major at *lower* field, hence the $(+)$ ester is assigned the configuration **34** (it is argued that since replacement of Me by CCl_3 maintains the sense of the $CO_2\underline{Me}$ signal non-equivalence, cf. **33** b and c, it is reasonable to assume the same of its replacement by CF_3). The major ${}^{19}F$ resonance signal of the mixture **33d** is *upfield* of the minor signal, hence the explanation of relative chemical shifts of the two 'H and ${}^{19}F$ signals of the alcohols may be explained by models similar to **32**.

Pirkle has further extended this work to the determination of optical purity and absolute configuration of amines and amino acids (Pirkle *et al.*, 1968; Pirkle and Beare, 1969). For the latter, the procedure is to convert the amino acid sample to the corresponding methyl ester hydrochloride, liberate the free base by a technique which minimizes the extent of racemization, and to examine the PMR spectrum of the basic ester as solute in optically active 2,2,2-trifluoro-1-phenylethanol (a convenient resolution

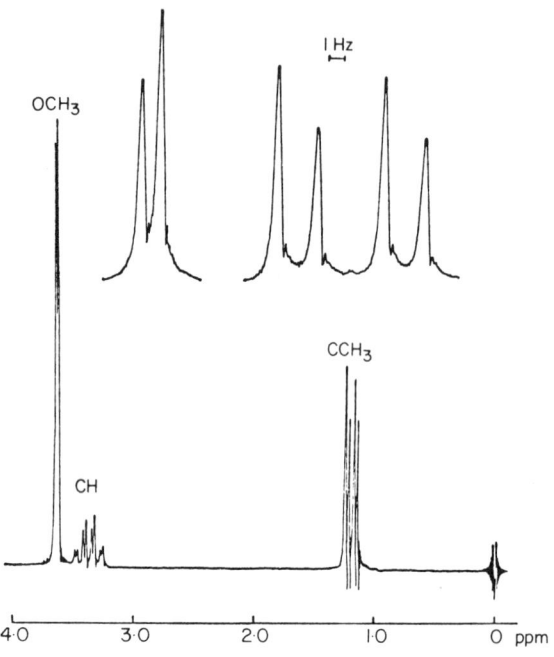

FIG. 5.7. Part of the 100 MHz PMR spectrum of partially resolved (S)-methyl alaninate in (−)-2,2,2-trifluoro-1-phenylethanol. The upper traces are scale expansions of the O—Me and C—Me resonances (Pirkle and Beare, 1969).

of the racemic fluoroalcohol is reported by the same group, Pirkle *et al.*, 1969a). The spectrum of partially resolved (S)-methyl alaninate, shown in Fig. 5.7 is typical of the results obtained. The CO_2Me, α-C—H, and C—Me resonances are all duplicated with the more intense CO_2Me and C—H signals at higher and the C—Me signal at lower field. The relative peak heights (or peak areas) of the expanded enantiomorphic C—Me and CO_2Me resonances gave an isomeric S:R ratio which represented an optical purity of 17·8% (cf. the initial value of 20% calculated from weights of racemic and optically pure alanine used).

When the (−)-fluoroalcohol **27** was used as solvent, it was generally true that the α-C—H and α-CO$_2$Me signals of (S)-amino acid esters (XCHNH$_2$CO$_2$Me) fell at higher and the X signals at lower field than corresponding signals of (R)-isomers; this observation provides a method for establishing the configuration of amino acids and is rationalized in terms of solute-solvent models similar to those already described. An objection to the use of methyl esters for these purposes is the fact of their ready base-catalysed racemization and their conversion to diketopiperazines on storage. These processes are very slow when the esters are solutes in the fluoroalcohol **27**, possible as a result of the feeble acidic properties of this solvent (pK_a 11·9).

In recent papers of this series, similar studies of asymmetric sulphur derivatives (e.g. sulphoxides), phosphine oxides and amine oxides are reported (Pirkle and Beare, 1968b; Pirkle *et al.*, 1969b). In all these cases spectral non-equivalence between enantiomorphs is thought to arise by the formation of short-lived hydrogen-bonded diastereoisomeric solvates with the optically active alcoholic solvent. Similar interactions may occur in solutions of equimolar proportions of an optically active sulphoxide and (±)-alcohol, e.g. **35** and **36**, in achiral solvents such as CCl$_4$ or CDCl$_3$ (Anet *et al.*, 1968). The methine signals of these alcohols form two over-

35   36

lapping quartets (coupling to OH is excluded because the O<u>H</u> signal formed a single peak) but normal signals are obtained in DMSO or when benzyl alcohol is added (these changes break up $\rangle$S → O···H—O— hydrogen bonds).

One common interaction between optically active molecules which is so important in resolution procedures, namely the dipolar attraction between anions and protonated bases, has also been shown to result in the duplication of PMR signals. Horeau and Guetté (1968) report separations ($\Delta\delta$) of enantiomorphic methine resonances in the 60-MHz spectra of the phenylacetic acids **37** in solvent (+)-1-(α-naphthyl)ethylamine, and have used relative intensities of the signals to determine optical purity.

R	$\varDelta\delta$ Hz
Me	9
Et	18
iso-Pr	38

PhCHRCO$_2$H

37

A recent paper has drawn attention to differences in the PMR spectra of natural (−)-dihydroquinine and synthetic racemic material (Williams *et al.*, 1969) in the achiral solvent CDCl$_3$; spectra of mixtures of the two materials showed clear signal duplication, especially in the aromatic region. It is well known that the physical properties of optically inactive modifications of asymmetric molecules which exist as racemic compounds differ significantly from those of the corresponding enantiomorphs, and it is probable that the above result is a further extension of this generality. It was also found that spectral differences diminished on dilution and this result supports the racemic compound interpretation since formation of the latter would be less favoured as the solute concentration falls.

Mention of NMR shift reagents containing paramagnetic ions has already been made (p. 36). Whitesides and Lewis (1970) have prepared a chiral reagent of this type (**38**) from europium (III) trichloride and *t*-butyl-

38

hydroxymethylene-*d*-camphor as a tool for determining enantiomorphic purity. They illustrate the power of the reagent in separating the resonance frequencies of R and S-α-phenylethylamine. Large downfield shifts of all resonances were seen in a carbon tetrachloride solution of **38** ($\sim$ 0·15 M) (the spectrum fell between 8 and 18 ppm), the expected result of pseudo-contact interaction between the europium (III) ion and a rapidly exchanging mixture of coordinated and free amine. More significant, however, were the frequency differences in the resonance of corresponding protons of R and S isomers which varied from $\sim$ 0·5 ppm for the CHNH$_2$ proton to about 0·07 ppm for the para hydrogen of the aromatic ring. The authors conclude that the frequency shifts probably reflect differences in the stability constants for the diastereoisomeric complexes formed between R and S-α-phenethylamine and **38**.

References

Anet, F. A. L., Sweeting, L. M., Whitney, T. A. and Cram, D. J. (1968). *Tetrahedron Lett.* 2617.

Burlingame, T. G. and Pirkle, W. H. (1966). *J. Amer. Chem. Soc.* **88**, 4294.

Casey, J. P., Lewis, R. A. and Mislow, K. (1969). *J. Amer. Chem. Soc.* **91**, 2789.

Dale, J. A. and Mosher, H. S. (1968). *J. Amer. Chem. Soc.* **90**, 3732.

Dale, J. A., Dull, D. L. and Mosher, H. S. (1969). *J. Org. Chem.* **34**, 2543.

Galpin, D. R. and Huitric, A. C. (1968). *J. Org. Chem.* **33**, 921.

Gerlach, H. (1966). *Helv. Chim. Acta.* **49**, 2481.

Horeau, A. and Guetté, J-P. (1968). *Compt. Rend.* **267**, 257.

Jacobus, J. and Raban, M. (1969). *J. Chem. Educ.* **46**, 351.

Mislow, K. and Raban, M. (1967). *In* "Topics in Stereochemistry" (N. L. Allinger and E. Eliel, eds), Vol. II, pp. 199. Interscience, New York.

Pirkle, W. H. (1966). *J. Amer. Chem. Soc.* **88**, 1837.

Pirkle, W. H. and Beare, S. D. (1967). *J. Amer. Chem. Soc.* **89**, 5485.

Pirkle, W. H. and Beare, S. D. (1968a). *Tetrahedron Lett.* 2579.

Pirkle, W. H. and Beare, S. D. (1968b). *J. Amer. Chem. Soc.* **90**, 6250.

Pirkle, W. H. and Beare, S. D. (1969). *J. Amer. Chem. Soc.* **91**, 5150.

Pirkle, W. H. and Burlingame, T. G. (1967). *Tetrahedron Lett.* 4039.

Pirkle, W. H., Burlingame, T. G. and Beare, S. D. (1968). *Tetrahedron Lett.* 5849.

Pirkle, W. H., Beare, S. D. and Burlingame, T. G. (1969a). *J. Org. Chem.* **34**, 470.

Pirkle, W. H., Beare, S. D. and Muntz, R. L. (1969b). *J. Amer. Chem. Soc.* **91**, 4575.

Raban, M. and Mislow, K. (1966). *Tetrahedron Lett.* 3961.

Whitesides, G. M. and Lewis, D. W. (1970). *J. Amer. Chem. Soc.*, **92**, 6979.

Williams, T., Pitcher, R. G., Bommer, P., Gutzwiller, J. and Uskoković, M. (1969). *J. Amer. Chem. Soc.* **91**, 1871.

CHAPTER 6

PMR Studies of Compounds of Pharmacological Interest

INTRODUCTION

PMR spectroscopy has been extensively applied in the field of medicinal chemistry both from analytical and structural aspects. Examples of its use in the quantitative analysis of pharmaceuticals have been given in Chapter 1. It is also a very powerful tool in qualitative analysis and·is of special value in the identification of structurally related groups of drugs. Warren *et al.* (1966), for example, have reported the PMR spectra of over 20 pheno-thiazine derivatives in clinical use and in many cases the spectra display distinctive features useful for diagnostic purposes. For instance, com-pounds with branched 1-aminoalkyl substituents, such as promethazine

1

(a) $R = H$; $R' = CH_2CHMeNMe_2$
(b) $R = Cl$; $R' = (CH_2)_3NMe_2$

(**1a**) are readily distinguished from analogues with straight chain 1-substituents such as chlorpromazine (**1b**) by the presence of a secondary methyl doublet near δ 1·0.

The great value of PMR spectroscopy in this field, however, lies in the detailed structural information it can provide, not only that concerning gross features but also fine points of molecular geometry. In the examples which follow (Chapters 6 and 7), stress is laid upon stereochemical aspects and many of the studies described throw light upon the configuration and preferred conformation (this often under physiological conditions) of pharmacologically active agents. Most of the principles employed have been outlined in previous chapters but several others are introduced here for the first time in this book.

NARCOTIC ANALGESICS

Morphine and Related Compounds

There have been several reports upon the PMR features of morphine and its derivatives (Yamaguchi *et al.*, 1963; Rull and Gagnaire, 1963). A study of 6,14-*endo*-ethenotetrahydrothebaines (Fulmor *et al.*, 1967) is of particular interest to medicinal chemists because of the extremely high potencies of many members of this series (Bentley and Hardy, 1963). These analgesics are derived from Diels–Alder adducts formed from thebaine

thebaine   adduct **2**

(the diene) and various electrophilic alkenes. NMR data confirms previous conclusions that the R substituent of the adduct **2** is at C-7 rather than C-8 and also the *endo* disposition of the 6,14 etheno bridge "inside" the tetrahydrothebaine system (see **3**). The adduct (**2**, R = CN) is obtained in

3

epimeric forms about C-7 and the spectrum of the α-form is shown in Fig. 6.1. The complexity of the spectrum is typical of the series and analysis involves extensive spin–spin decoupling experiments. The signal of the vinylic protons at C-17 and C-18 (see **3**) form a typical AB pattern ($J_{17,18}$ 9 Hz) with H-18 being weakly coupled in addition to both H-5β ($J \sim 2$ Hz) and H-7β ($J \sim 1$ Hz). A model shows that both H-5β and H-7β together with H-18 and the intervening carbon atoms approximate to a plane in which the connecting bonds resemble a "W"; this arrangement satisfies the requirements for long range coupling (see p. 91) and the C-7 cyano substituent is assigned an α-configuration on these grounds. If this is correct, H-18 in the β-isomer should not be coupled to H-7 since this proton is not correctly orientated for long range coupling; this was in fact confirmed, H-18 of the second isomer (β) forming part of a three rather than a four spin system. The H-8α and H-8β protons both form a doublet of doublets but the β-signal is downfield of the α-signal by about 100 Hz. This large difference in chemical shift is attributed to the deshielding influence of the tertiary nitrogen atom which is held in proximity to H-8β

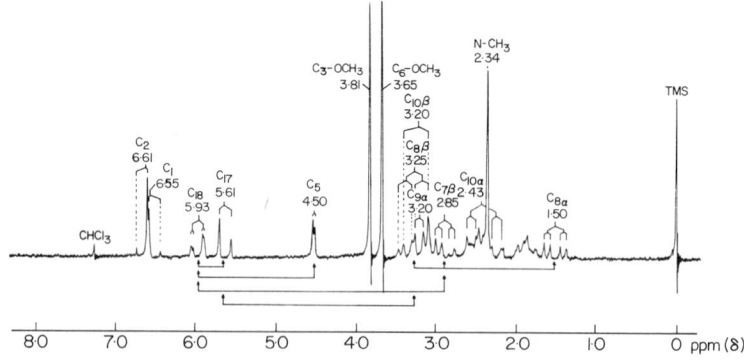

FIG. 6.1. 60 MHz PMR spectrum of the 7α-cyano epimer **3** in CDCl₃. The arrows indicate the spin systems established by decoupling experiments (Fulmor *et al.*, 1967).

(separation 2·9 Å) by the rigid ring system. This effect would not be seen if the double bond bridge occupied an *exo*-position.

When the C-7 substituent in **2** is an asymmetric tertiary alcohol function (RR′COH), diastereoisomeric alcohols result and potency differences between isomers have been reported (Lister, 1964). Assignment of configuration to the diastereoisomeric phenylmethyl *t*-carbinols **4** and **5** proved possible from PMR data. The vinylic protons at C-18 and C-17 in one alcohol are upfield of corresponding signals due to the second isomer ($\Delta\delta$ 33 Hz for H-17 and 55 Hz for H-18). These upfield shifts are attributed to aromatic screening (in **5** the vinylic protons lie close to and above the plane of the aromatic ring), the greater upfield shift of H-18 over H-17 being consistent with the closer proximity of H-18 to phenyl; hence the isomer with the higher field vinylic signals must have phenyl placed as in **5**. Intramolecular hydrogen bonding was evident in both isomers from IR (concentration independent band at 3425 cm^{-1}) and PMR (low field OH signals in DMSO-d$_6$) evidence which must involve the 6-methoxy group as shown in **4** and **5**. The configurations of the two isomers are therefore established.

Benzomorphan derivatives

In analgesics based on benzomorphan **6**, ring C of morphine is reduced to C-5 and C-9 methyl substituents; the latter may be *cis* or *trans* and

6

diastereoisomers (α- and β-) are encountered, the β-isomers being the more potent (Eddy and May, 1966). While the 5-methyl signals have similar resonance positions in the spectra of an isomer pair (lower field than usual for a *t*-Me group because both lie near the plane of, and fairly close to, the 6,7-fused benzene ring), the α-9-Me signal is about 0·5 ppm higher field than the β-signal, the latter being nearly coincident with the β-5-Me resonance (Fig. 6.2). May's group (Fullerton *et al.*, 1962) interpreted these differences in terms of differential aromatic screening, α-Me being shielded because it lies within the aromatic screening zone, whereas the β-group is too far removed to be affected by phenyl (see **7**). The proximity of the

α-**7** β-**7**

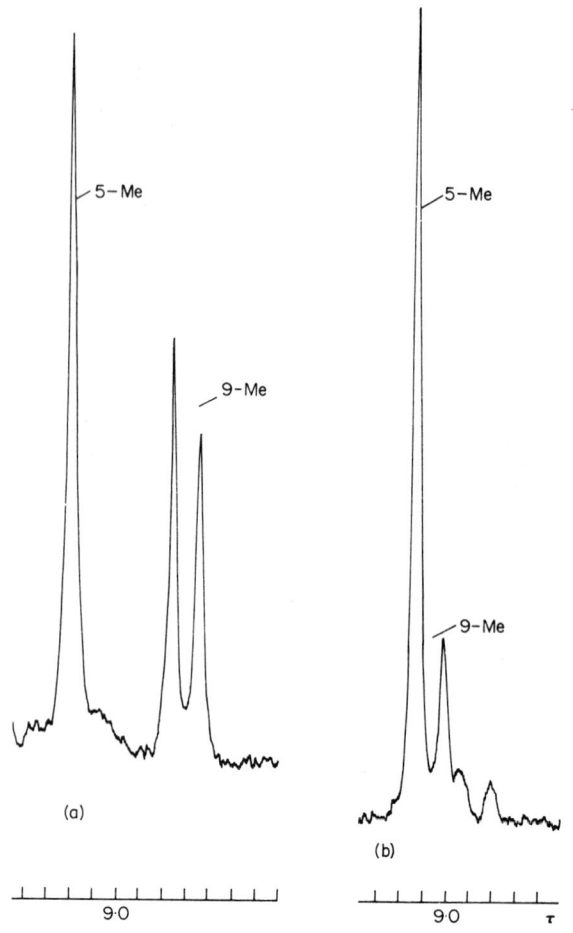

Fig. 6.2. 60 MHz PMR 5- and 9-methyl signals of isomeric 2,5,9-trimethyl-2′-hydroxybenzomorphans **6** (R = Me) in DMSO-d$_6$. (a) α-isomer; (b) β-isomer.

nitrogen lone-pair to the 9-Me group in the β-base **7** is probably also of significance in regard to the α/β 9-Me chemical shift difference in view of demonstration of the greater deshielding influence of the lone-pair orbital upon axial than upon equatorial 3-Me substituents in *trans* quinolizidines (**8**) (Moyneham *et al.*, 1962). These PMR data form part of the evidence for the original assignments of α-*cis* and β-*trans* 5,9-dimethyl configurations (with reference to the hydroaromatic, not the piperidine, ring of **6**), later confirmed for the α-N-allyl analogue by X-ray crystallography (Fedeli *et al.*, 1966).

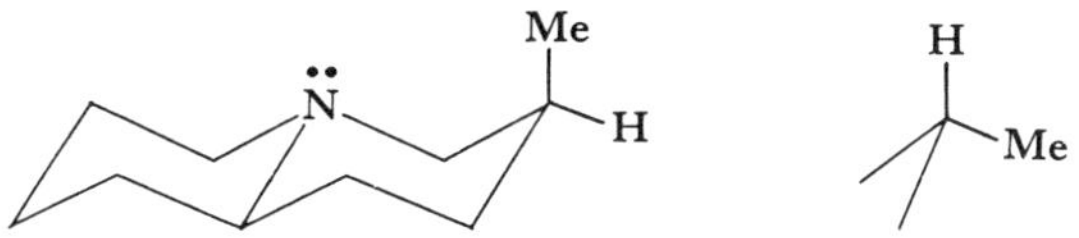

8

a-Me downfield of *e*-Me signal by 0·27 ppm

A quaternization study of α- and β-metazocine (**6**, R = Me), investigated by PMR spectroscopy, provides some information about the conformation of the piperidine ring of these derivatives in the polar solvent, DMSO-d$_6$ (Casy and Parulkar, 1969b). Interest in conformation arises because of a possible correlation between isomeric potency and stereochemical differences. The α-isomer is readily methylated with methyl iodide and the PMR spectrum of the product displays distinct, equally intense, N—Me singlets at 202 and 194·5 Hz (Fig. 6.3a), (Chemical shifts are given in Hz from TMS, the operating frequency being 60 MHz). In the corresponding salt prepared from the α-isomer and trideuteromethyl iodide, only one N—Me signal is present [Fig. 6.3(b)]; hence, assuming a preferred axial approach of the quaternizing agent (CD$_3$I) (see p. 159), the N—Me signal of the α-N—Me, N—CD$_3$ salt is probably due to an equatorial group. This is the *higher* field member of the N,N-dimethyl quaternary salt [Fig. 6.3 (a) and (b)], thus the lower field signal (202 Hz) is assigned to the axial N-methyl. In monocyclic N,N-dimethylpiperidinium salts, the axial N—Me group generally has the higher field position (p. 156) and anisotropic effects of the fused hydroaromatic system, which will influence *e*- rather than *a*-N-methyl, are most probably responsible for the reversal noted in the α-benzomorphan case. The results of methylating α-**6** (R = Et) (Reaction A) and ethylating (α-**6**, R = Me) support this interpretation.

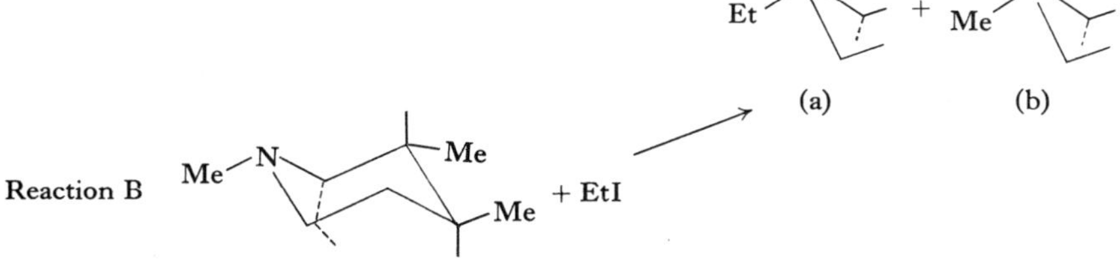

Reaction A

+ MeI

(a)   (b)

Reaction B

+ EtI

Two isomers may result in each case, (a) and (b). Only one N—Me signal (201 Hz) was resolvable in the total product of Reaction A and the minor N—Me signal (185 Hz) was identified in spectra of recrystallized samples. The same N—Me signals were found in the total product of Reaction B but minor and major values were reversed. Hence, assuming axial approach of quaternizing reagent, the major product of Reaction A is the axial N—Me isomer (a) (lower field NMe) while that of reaction B is the equatorial isomer (b) (higher field NMe).

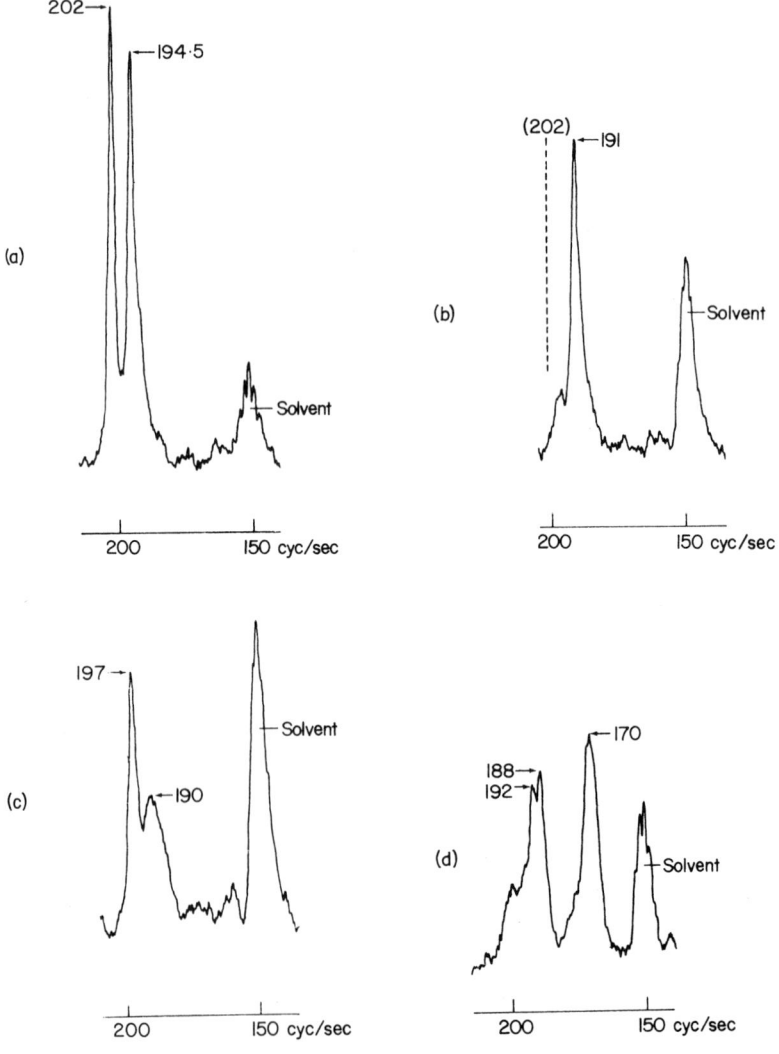

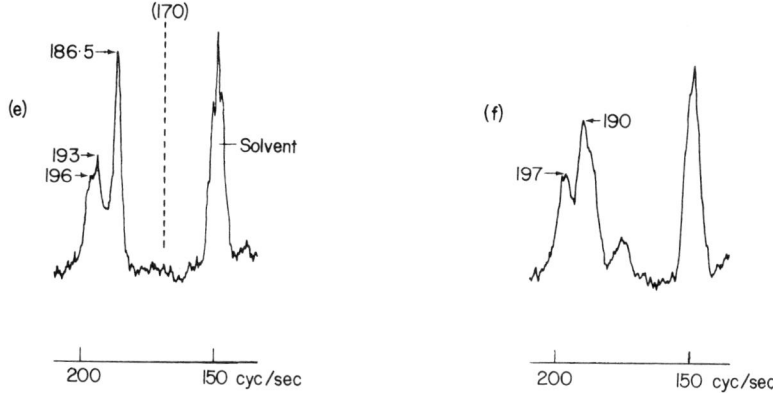

Fɪɢ. 6.3. Part of the 60 MHz PMR spectra of (a), N,N-di-Me; (b), N—Me, N—CD$_3$; (c) N,N-di-CD$_3$ iodide salts of α-5,9-dimethyl-2′-hydroxy-6,7-benzomorphan in DMSO-d$_6$. (d), (e) and (f) are spectra of the corresponding salts of the β-isomer. Spectral amplitudes are 12·5 (a), 25 (d), 32 (b) and (e), and 80 (c) and (f). Relative signal intensities may be gauged by comparison with solvent signals near 150 Hz. Chemical shifts are in Hz from TMS at a 60 MHz frequency (Casy and Parulkar, 1969b).

The β-methiodide of (**6**, R = Me), prepared after a much longer reaction period, gave a spectrum in which identification of the N,N dimethyl signals is equivocal because more than two signals fall in the N—Me resonance region [Fig. 6.3(d)]. One must be the unusually broad 170 Hz band because of its complete absence in the mono N—CD$_3$ compound [Fig. 6.3(e)], while the other is the 188 Hz signal, as both this and the 170 Hz band are missing from the spectrum of the N,N-di-CD$_3$ derivative [Fig. 6.3(f)]. Note that the *lower* field N-methyl signal of the β-salt remains in the N—Me, N—CD$_3$ derivative, while the higher field signal persists in the analogous α-N—CD$_3$ salt. Signals remaining in α- and β-N,N-di-CD$_3$ derivatives must be due to ring protons [Fig. 6.3(c) and (f)]. The greater separation of the β-N-methyl signals (18 compared with 7·5 Hz for α-) and the unusually high field position of one of the signals cannot be explained in terms of the chair conformation β-**9** but is understandable if the quaternary salt adopts the skew-boat conformation **10** in which 1,3 diaxial interactions are relieved. In this conformation one N—Me group falls well within the aromatic shielding zone, while the other has an environment similar to that of *e*-Me in the salt α-**9**. In the chair conformer β-**9**, the *e*-NMe signal should have a similar chemical shift to *e*-NMe in α-**9**, but *a*-NMe of the β-isomer would be expected to be lower field than the comparable signal of the α-form on the grounds of the known deshielding influence of an axial methyl substituent upon an axially placed 3-hydrogen

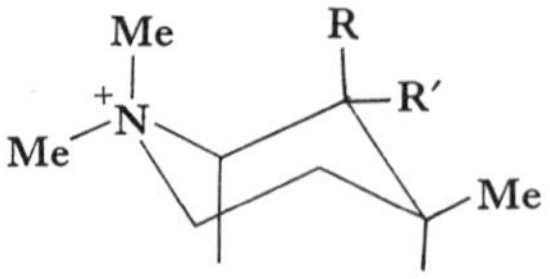

9

α, R = H; R′ = Me
β, R = Me, R′ = H

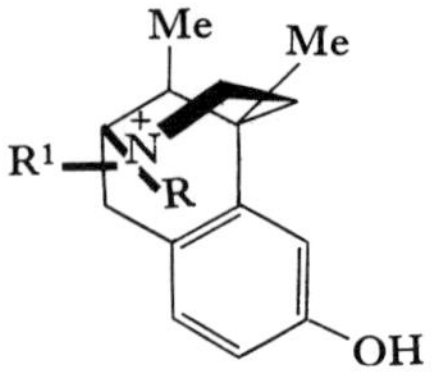

10

MeI (R = R′ = Me)
HCl (R = R′ = H)

atom (Booth, 1966). Taking the skew-boat **10** as the favoured β-methiodide conformation, two reaction pathways for quaternization of the β-base with CD_3I may be formulated on the basis of axial approach of reagent:

1. axial approach of CD_3I upon the chair conformation **11** and subsequent conformation change of **12** to **13**.

2. *pseudo*-axial approach of CD_3I upon the skew-boat conformation of the β-base **14** with direct formation of a favoured quaternary salt conformation **15**.

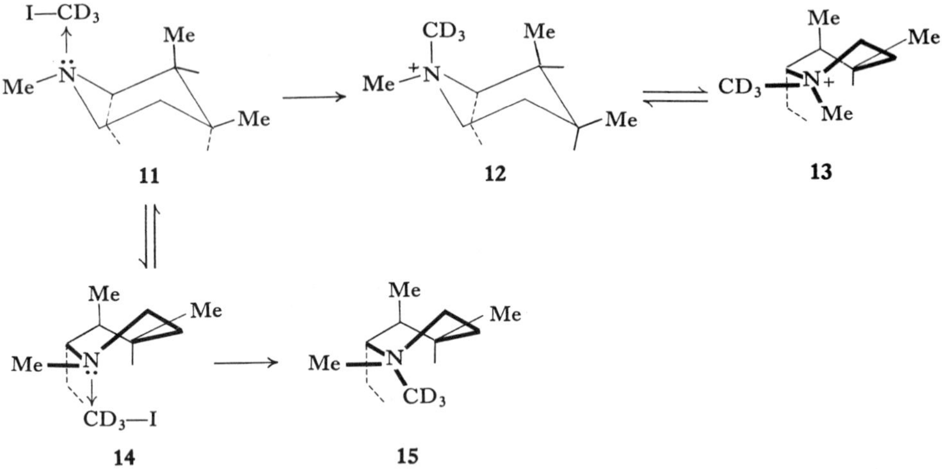

11   **12**   **13**

14   **15**

In the former case, the higher field NMe signal will remain, whereas in the latter, the lower field signal; results obtained [Fig. 6.3 (d) and (e)] thus support pathway 2.

There is evidence that skew-boat conformers of hydrochloride salts of the β-benzomorphans are also favoured in polar solvents. If skew-boat populations are significant in the β-secondary base hydrochloride (**16**), the difference between the chemical shifts of the two +NH protons would

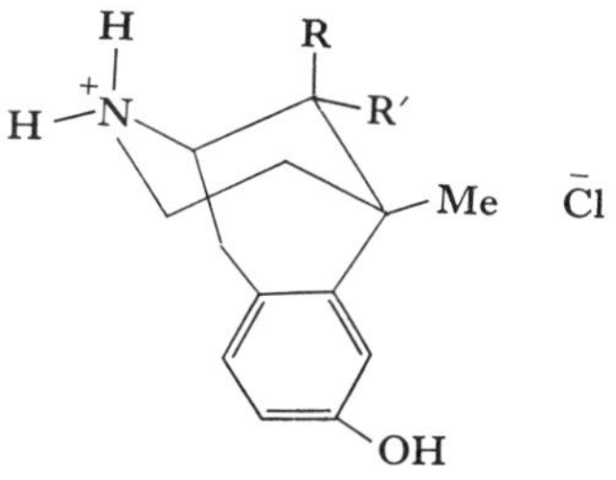

16

α, R = H; R′ = Me
β, R = Me; R′ = H

be expected to be greater than that between the α-+NH$_2$ pair since one of the β-+NH protons (R = H in **10**) falls close to and above the plan of the aromatic ring in the skew-boat (**10**, R = R′ = H). In fact; while the α-+NH$_2$ signal formed a two proton multiplet centered at 570 Hz, the β-salt showed two distinct +NH signals at 571 Hz and 525 Hz, the latter, higher field signal value, being consistent with the aromatic screening of one of the protons as anticipated for (**10**, R = R′ = H) [Fig. 6.4 (a) and (b)]. Note that the +NH$_2$ chemical shifts are unlikely to be averaged through proton exchange because the rate of this process is slow in the hydrogen bonding solvent DMSO-d$_6$ (Chapman and King, 1964). The broad nature of these signals (α-W_H 14, β-24 and 26 Hz) differentiated them from the sharper phenolic OH resonances. The latter fell within the aromatic multiplet in base and hydrohalide spectra as was confirmed by comparing spectra of salts of α-**16** and the corresponding O-methyl ether [Fig. 6.4 (a) and (c)] but were moved downfield in methiodides [forming isolated singlets, see Fig. 6.4 (d)] possibly as a result of hydrogen bonding of OH to the iodide anion. In β-N-methyl salts (**6**), the unfavoured nature of chair conformers in DMSO-d$_6$ is suggested by the negligible downfield shifts of the 9-methyl signal following protonation of the base (see p. 203) but clear evidence could not be obtained because the salts were insoluble in CDCl$_3$, the best solvent for observing this effect.

PMR spectroscopy has been used to help elucidate the structure of the

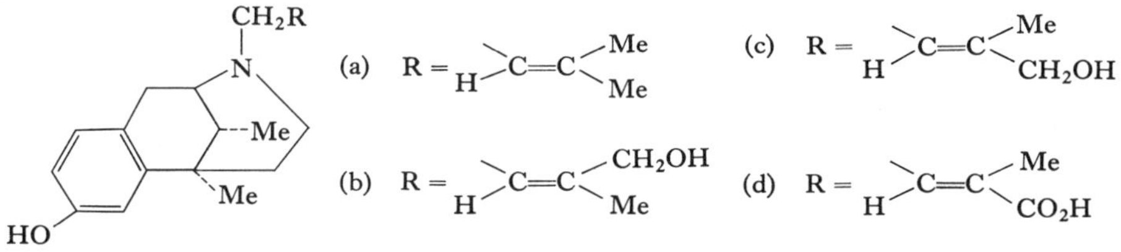

17

(a) R = $_H$>C=C<$^{Me}_{Me}$

(b) R = $_H$>C=C<$^{CH_2OH}_{Me}$

(c) R = $_H$>C=C<$^{Me}_{CH_2OH}$

(d) R = $_H$>C=C<$^{Me}_{CO_2H}$

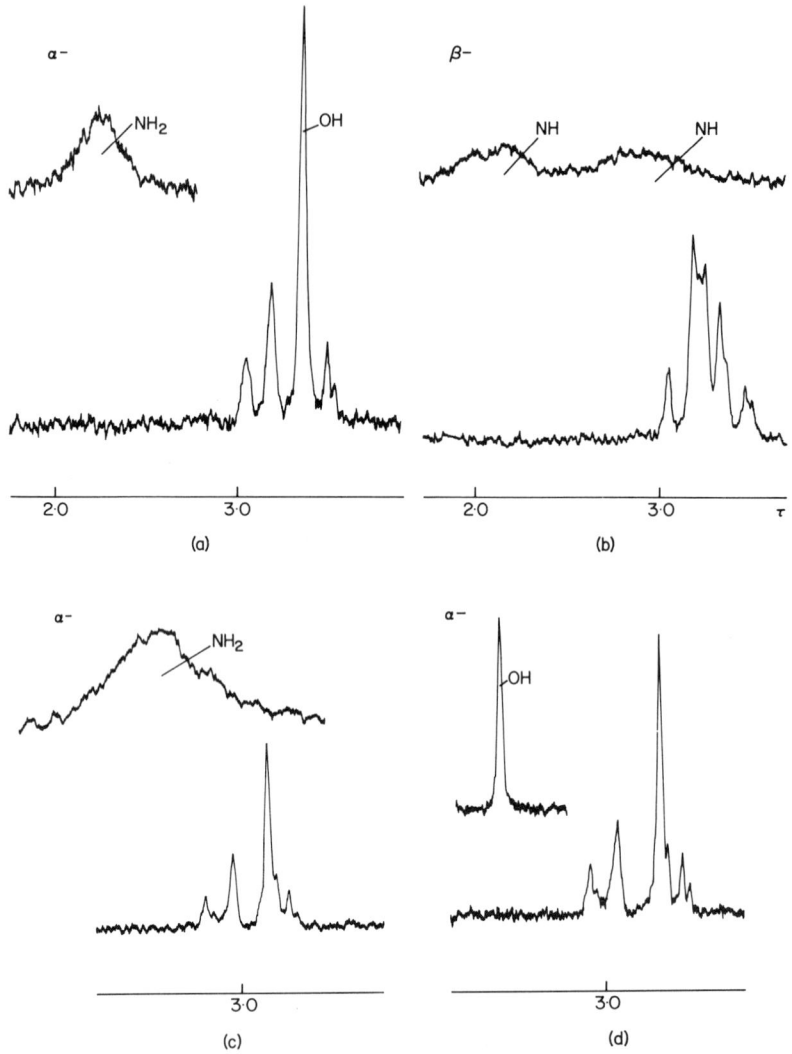

Fɪɢ. 6.4. Part of the 60 MHz spectra of the isomeric secondary base hydrochlorides **16** and the related α-O-methoxy derivative. (a) α-**16**: NH₂ signal is offset
100 Hz, phenolic OH coincides with part of the aromatic signal near τ 3·4 [cf. (a)
and (c)]; (b) β-**16**: NH₂ signals are offset 100 Hz; (c) O-methoxy derivative of
α-**16** hydrochloride: NH₂ signal is offset 130 Hz; (d) methiodide of α-**6** (R = Me):
OH signal is offset 100 Hz.

metabolic products of pentazocine (**17a**), a benzomorphan analgesic that is claimed to be non-addicting (Pittman *et al.*, 1969). Four products were recovered from incubates of liver preparations and the drug, and from the urine of monkeys given pentazocine. Three, identified by chromatographic and infra-red comparisons with authentic materials, proved to be the unchanged drug, and the oxidized forms **17c** and d respectively. The PMR spectrum of the fourth metabolite was very similar to, but not

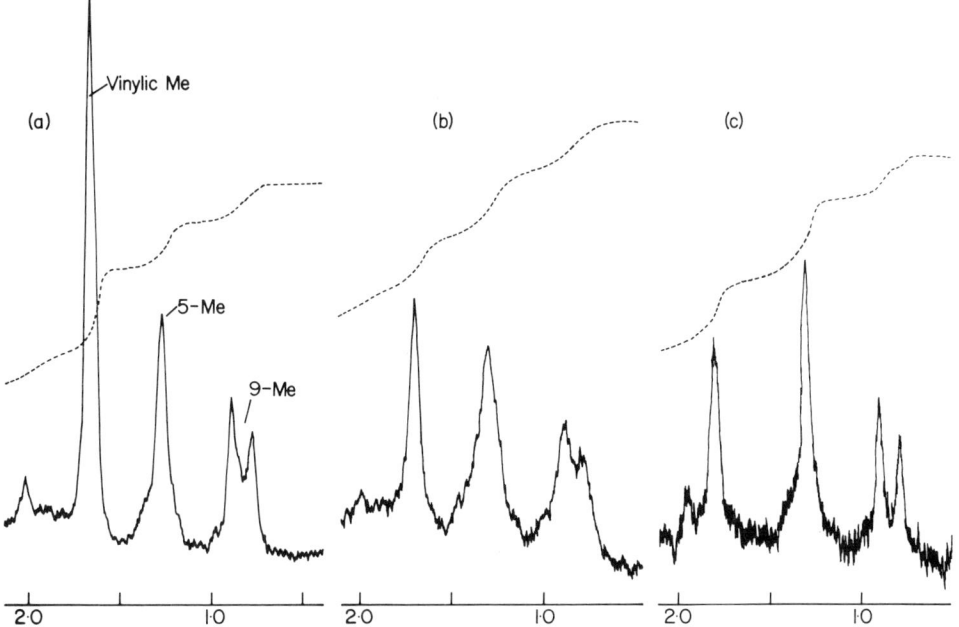

FIG. 6.5. The methyl resonance region of the 60 MHz spectra of (a), pentazocine (**17a**); (b) the *trans* alkene **17c**; and (c) the fourth metabolite **17b**. Solvent is CDCl₃ (Pittman, 1969).

identical with, that of the *trans* H/Me alkene (**17c**). Part of the spectra of authentic pentazocine (**17a**), the *trans* alkene (**17c**) and the fourth metabolite (from monkey urine) are shown in Fig. 6.5. Only a small quantity of the metabolite could be isolated (about 11 mg); nevertheless the single span spectrum of this material, dissolved in the minimum volume of CDCl₃ and examined in a microtube, provided clear information about structure (Pittman, 1969). The critical points in identifying the metabolite as (**17b**) by spectral comparison with pentazocine and the *trans* alkene were: (i) the presence of unchanged signals due to protons on the aromatic nucleus (near δ 7) and to protons due to the methyl groups at ring positions 5 and 9 (singlet δ 1·3 and doublet 0·85 respectively), (ii) the approximate halving of the relative peak height of the signal due to protons on the methyl group(s) attached to the allyl side chain (near δ 1·7), and (iii) the appearance of a new signal at the expected position for a methylene group attached to hydroxyl (near δ 4).

4-Phenylpiperidines

The best known analgesics of this class are pethidine **18** (R = H) and the reversed ester of pethidine **19** (R = H). The analogues with R = Me have

been isolated in diastereoisomeric forms (Ziering and Lee, 1947; Casy *et al.*, 1969) and, in each case, the isomer obtained in minor yield (β-) is distinctly more potent than the major form (α-). The configurations of the reversed esters **19** (R = Me) (termed α- and β-prodine) are known from X-ray crystallography to be *trans* 3-Me/4-Ph (α-) and *cis* (β-), and PMR studies have provided evidence of the solute conformation of these isomers (Casy, 1966a, 1968). PMR differences between α- and β-prodinol (**20**),

(R = H)

the alcohols from which the prodine esters are derived, are consistent with α-(*trans*) and β-(*cis*) 3-Me/4-Ph configurations and support structures (**21**) as the preferred base and hydrochloride conformations in non-polar solvents. Interpretations of the PMR data are given below.

The 4-phenyl signal

In α-**21** the preferred orientation of the 4-phenyl group will be a plane approximately at right angles to that of the piperidine ring, the equatorial 3-methyl/*ortho* aromatic hydrogen interactions being minimum in this conformation. In β-**21** however, the same orientation is not favoured since it would bring aromatic protons in close proximity to an axial methyl group, and the preferred orientation of the 4-phenyl group will be when it is approximately coplanar with the piperidine ring. Averaged environments will be experienced by all protons as a result of rapid rotation about the bond linking the two rings. In both isomers, however, it is probable that the populations of conformers akin to α- and β-**21** will be higher than all others at any one time. In α-conformers the environments of *ortho*

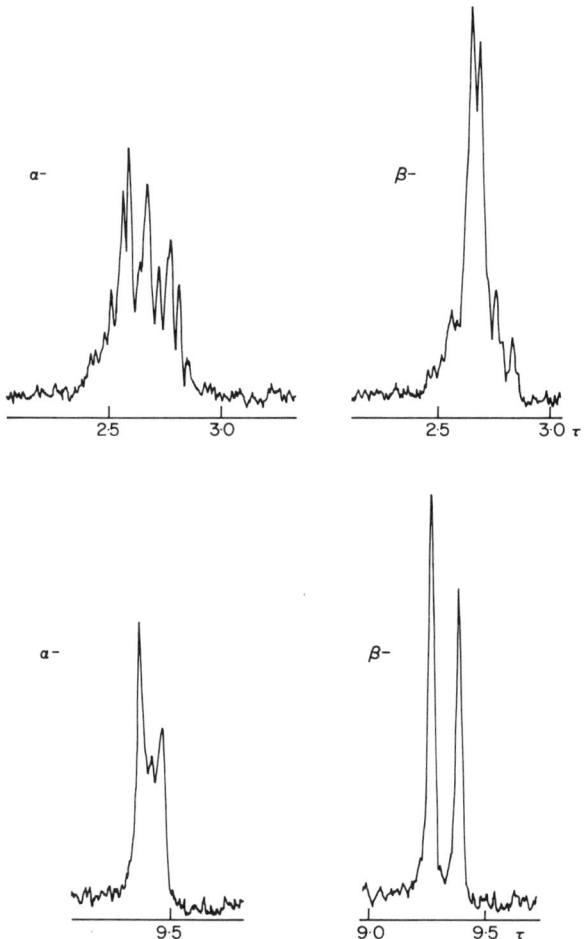

FIG. 6.6. Part of the 60 MHz PMR spectra of the isomer 1,3-dimethyl-4-phenyl-piperidin-4-ols in CCl₄. Upper: aromatic signals; Lower: 3-methyl signals (Casy, 1966a).

aromatic hydrogens should differ markedly from the corresponding *meta* and *para* atoms because an *ortho* proton is close to the electronegative oxygen of the 4-hydroxyl group in the preferred conformation α-**21** and the equivalent form in which the *ortho* protons are interchanged. In β-conformers, the *ortho* protons are further removed from the hydroxyl group and hence their chemical shift will be closer to those of the other aryl hydrogens. Thus chemical shift differences among the aromatic protons are expected to be more pronounced in the *trans* isomer with the result that the *trans* aromatic signal should be more complex than that of the corresponding *cis* isomer. This conclusion is confirmed experimentally, the aromatic signal of the α-isomer being markedly broader than the corresponding β-signal (Fig. 6.6).

The 3-methyl signal

Chemical shifts. When the 4-phenyl group adopts a near perpendicular orientation with respect to the plane of the piperidine ring (as in α-**21**), the equatorial 3-Me substituent is judged to fall just within the diamagnetic screening zone of the aromatic nucleus, while the axial 3-Me group falls on the periphery of the same zone when phenyl is orientated as in β-**21** (these conclusions were reached by inspection of Dreiding models and application of the Johnson and Bovey (1958) screening data for benzene). Hence 3-methyl substituents should be screened in similar degree from the applied magnetic field in both isomers. In fact, the 3-methyl signals of the two isomers are upfield relative to those of 3-methyl in cyclic analogues in which a phenyl group is not adjacent to methyl, cf. **22** (Garbish, 1963). However, the α-3-methyl signal is slightly higher field than the corresponding β-signal (Fig. 6.6). Other factors may contribute to this chemical shift difference, e.g. influence of the nitrogen lone pair, the hydroxyl group, and the C—C bonds of the piperidine ring which render analysis in terms

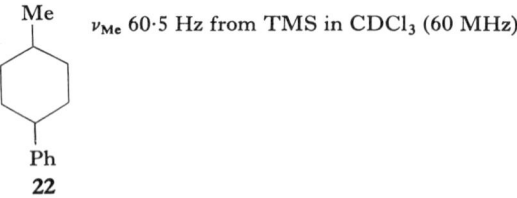

ν_{Me} 60·5 Hz from TMS in CDCl$_3$ (60 MHz)

22

of α- and β-**21** difficult. The separation of the α- and β-3-methyl signals increases from 4·5 to 10 Hz when DMSO-d$_6$ replaces CCl$_4$ as solvent; the stereochemical significance of this result, which may be related to the geometry of solute-solvent hydrogen bonded complexes of the two alcohols, is discussed in the Appendix (p. 387).

Effect of protonation of the basic signal. When protonated forms (hydro-chlorides) of the diastereoisomers (**20**) are examined the α-3-methyl signal suffers a relatively small downfield shift (2 Hz) whereas the corresponding β-signal is moved downfield by 14 Hz (base and salt comparisons both made in CDCl$_3$). It is to be noted that 3-methyl is closer to positively charged nitrogen in the chairs (**21**) when it is axial, and would therefore be expected to be more deshielded than an equatorial group by this feature.

Evidence for the influence of proton uptake at a basic centre two carbons removed from a methyl group upon the chemical shift of the latter is well demonstrated by the following examples. When 1-benzyl-3-methyl-3-phenylpiperidine **23** in CDCl$_3$ is protonated the 3-methyl resonance (a singlet) moves downfield by 25 Hz (Casy *et al.*, 1969). There is little doubt that 3-methyl is axially orientated in the preferred conformation of this compound. Hence, it will be in close proximity, bearing a 1,3-diaxial relationship, to the +NH proton of the conjugate acid. However, when 3-methylpiperidine **24** is protonated in CDCl$_3$, there is a smaller difference (8 Hz) between the chemical shifts of the 3-methyl group in the free base

23   24

and hydrochloride (unpublished results). Here the 3-methyl group is equatorially orientated in the preferred conformation. Evidence that protonated nitrogen deshields protons bearing a 1,3-diaxial relationship to +NH more than it deshields 3-equatorial protons is provided by data upon *trans* 2,6-dimethylpiperazine **25**. Sudmeier (1968) analysed the PMR

25

spectra of *trans* 2,6-dimethylpiperazine in base, monoprotonated, and diprotonated forms as ABCY$_3$ systems, and the downfield shifts which follow successive protonation are shown in Table 6.1. It is seen that the greater shifts are suffered by protons A and C that are axial to +NH. These results establish that protonated nitrogen in 6-membered cyclic bases has a differential deshielding influence upon axial and equatorial proton groups in the 3-position of the ring. Hence differences in the protonation shifts observed for α- and β-prodinol support an equatorial conformation for α-3-methyl as in α-**21**, and an axial orientation for β-3-methyl, as in β-**21**, (cf. chemical shift assignments to 3-methyl in epimeric 1-alkyl-3-methyl-4-phenyl-1,2,3,6-tetrahydropyridines, p. 151). The mechanism of deshielding by protonated nitrogen has not been established but results with α-prodinol show that the inductive withdrawal of electrons by positively charged nitrogen makes only a small contribution, hence magnetic anisotropic effects induced by the positively charged centre must be involved.

Appearance of the 3-methyl signal. While the 3-methyl signal of the β-isomer (base in $CDCl_3$ and CCl_4) is a near-symmetrical doublet (J 7–7·5 cyc/sec), that of the α-isomer is a narrower (peak separation 5–5·6 cyc/sec), non-symmetrical doublet and shows evidence of a third peak midway between the main peaks, most clearly apparent in CCl_4 (Fig. 6.6). Distortion of a methyl doublet is a typical result of virtual long-range

TABLE 6.1

**Downfield chemical shift (ppm)
in aqueous buffers**

Proton	HP+	H_2P++
A	0·46	0·59
B	0·39	0·48
C	0·41	0·54

(P denotes a piperazine nitrogen atom.)

coupling† and its occurrence in the α- rather than the β-isomer, is interpreted in terms of conformations **21** as follows:

$$\begin{array}{c} \text{Me} \\ | \quad | \\ -C-C- \\ | \quad | \\ H_A \ H_B \end{array}$$

26

$$\begin{array}{c} H_A \ H_B \ H_X \\ | \quad | \quad | \end{array}$$

† First order treatment of the system C—C—C predicts a sharp doublet for the X signal. When H_A and H_B are strongly coupled, however (i.e. when J_{AB} is of the same order of magnitude or exceeds the chemical shift difference between H_A and H_B), a first order analysis is not valid. The correct signal, exactly derived by a rigorous ABX calculation, consists of 6 lines with the two most intense having the same separation as the lines of the 1st order doublet. In these circumstances the X proton appears to behave as if it were coupled to both H_A and H_B, whereas in fact it is coupled to only one of the two. This effect, termed virtual coupling, is not to be confused with the long-range variety which involves *real* coupling (Musher and Corey, 1962; Becker, 1965).

In the system (**26**) virtual coupling between methyl and H_A will occur when the coupling constant between H_A and H_B is large and of the same order as the chemical shift difference between the two protons. These conditions are more likely to prevail in α-**21** than in β-**21**. The former contains an *axial* proton at C-3 which will be strongly coupled to the C-2 *axial proton*; the latter contains an equatorial proton at C-3 which is only weakly coupled to the methylene protons at C-2 ($J_{aa} > J_{ae} \sim J_{ee}$). It is significant that virtual coupling effects are not seen in the α-base hydrochloride, a result probably due to the fact that conditions for virtual coupling no longer obtain in α-**21** when a strong deshielding influence (the protonated nitrogen atom), which affects C-2 protons more than the proton at C-3, is introduced into the molecule. This PMR analysis, especially in regard to differing α/β-3-methyl

Ph CN

~Me

N

Me

27

protonation shifts, is advanced as a convenient method of configurational assignment of general applicability to isomeric 1-substituted-4-aryl-3-methylpiperidines. Thus, the PMR spectra of the isomeric cyanides **27**, precursors of α- and β-3-methylpethidine (**18**, R = Me) differ respecting their aromatic signals (α- broader than β-) and 3-methyl signal protonation shifts (β 21·5 Hz, α-3·5 Hz) and configurations are assigned accordingly (Casy *et al.*, 1969).

The α- and β-3-methyl signals of the prodine esters (**19**, R = Me), themselves differ in the same way as those of the parent piperidinols; the α-3-Me signal (a deformed doublet, 41 Hz) of the base appears at 44 Hz in the hydrochloride spectrum, corresponding values for the β-signal being 44 Hz (base) and 62 Hz (HCl). The esters also differ in the positions of their CO.CH₂Me and CO.CH₂Me signals, the former being triplets near 73·5 Hz (α) and 65 Hz (β) in the bases, and the latter quartets near 157 Hz (α) and 143 Hz (β) in the salts (all 60 MHz spectra in CDCl₃). The higher field values of the β-signals are consistent with the *cis*-ester having a preferred conformation analogous to structure β-**21**, because the ester CO.CH₂Me group will spend some of its time above the plane of the phenyl group (i.e. well within the aromatic shielding zone) if the plane of the C—O bond and the aromatic ring are approximately perpendicular. A similar difference in the α/β ester signals is seen in the isomeric pethidine esters (**28**), a result in accord with their configurational assignments. While the chemical shifts of α- and β-3-methyl in 4-piperidinol (**20**) and 4-cyano

	α	β
OC$\underline{H}_2$Me	258	251
OCH$_2$$\underline{Me}$	76	68
3-Me	64	45

(bases in CDCl$_3$ at 60-MHz; chemical shifts in Hz from TMS)

28

(27) bases are similar (β-signals higher field by a few Hz), the α-3-methyl signals of the 4-ethoxycarbonyl (28) and 4-propionyl (29) derivatives are markedly lower field (17–19 Hz for esters, 25 Hz for the ketones) than

α-3-Me　67 Hz

β-3-Me　42 Hz

(bases in CDCl$_3$)

29

the corresponding β-signals. These differences confirm the stereochemical assignments since an equatorial 3-methyl group should be more subject to the magnetic anisotropic effects (Karabatsos *et al.*, 1967) of the adjacent carbonyl function than an axial group. The carbonyl function of the 4-propionyloxy group in the reversed esters **19** is further removed from the 3-methyl group and appears to have little influence upon its chemical shift (α 41, β 44 Hz, bases in CDCl$_3$).

Firmest evidence upon the conformation of cyclic derivatives derives from the chemical shifts and particularly the coupling constants of the ring proton signals. Unfortunately such information is not readily available in the cases of the pethidine stereoisomers so far discussed because of spectral complexity. Some of the ring proton signals of the isomeric 1,2,5-trimethyl-4-phenyl-4-piperidinols (the propionate ester of the γ-form is promedol, an analgesic used clinically in the U.S.S.R.) are well resolved,

deuterated analogue

however, and provide part of the PMR evidence for their stereochemistry (McErlane, 1971). Multiplets near δ 2 in spectra of the three available isomers (γ, β and α) are assigned to one or both of the 3-methylene protons because the signals are absent in spectra of the corresponding deuterated forms obtained from the α,α'-deuterated 4-piperidone precursor (see formulae). Information available from 100 MHz spectra and some preliminary conclusions are summarized below.

γ-*isomer*. A two proton doublet of quartets anticipated for the coupled system 3-H_a, 3-H_e, 2-H is present near δ 2. First order analysis gives J_{gem} 13·5 (identified by its occurrence four times within the 8-line signal). and J_{vic} 11·0 and 3·5 Hz. The vicinal values are typical of a/a and e/a (or e/e) couplings respectively, thus the arrangement shown below must occur

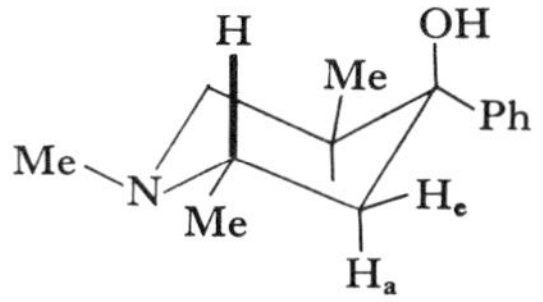

(evidence that 5-Me is also equatorial is not given here)

in the γ-isomer, the 2-methyl substituent being equatorial in the preferred conformation.

β-*isomer*. One of the 3-methylene signals is clearly resolved as a quartet near δ 2, J_{gem} 13·5 Hz and J_{vic} 10·0 Hz. The latter is, again, a typical axial-axial coupling, thus 2-methyl must also have an equatorial orientation in

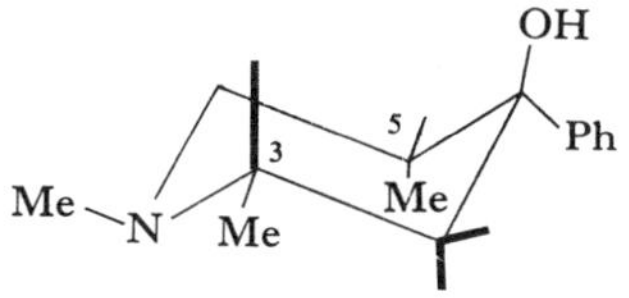

β-isomer

this isomer. The PMR characteristics of the 5-methyl signal are almost identical with those of the 3-methyl (axial) signal of β-prodinol (see p. 202), and these results together with deductions drawn from the detection of protonated epimers (Chapter 4, p. 155) allow the β-isomer to be assigned the preferred conformation shown above.

α-*isomer*. One complete 3-H signal (quartet J_{gem} 14·5, J_{vic} 4·5 Hz) near δ 2 and half of the other just upfield of the N-methyl signal, giving the second J_{vic} value of 6 Hz, are resolvable. If assignments to the γ- and β-isomers are correct, the α- form may be represented by either of the two structures shown below. The small 3-H J_{vic} value (4·5 Hz) agrees with both but the

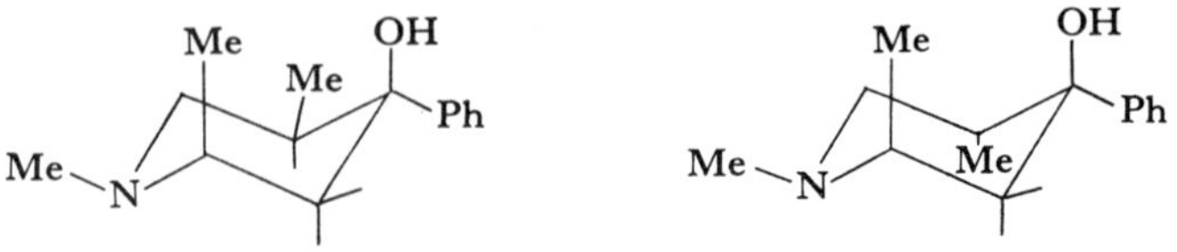

α-isomer (possible structures)

second value (6 Hz) is too large for a gauche coupling yet too small for a diaxial interaction. It is probable, therefore, that the compound has a preferred non-chair structure or exists as an equilibrium mixture of conformers.

Diphenylpropylamines

The PMR characteristics of the open-chain analgesic methadone **30** are noteworthy in two respects. In the first case, the $^+HN\underline{Me}_2$ signal of the

$$Me_2N\overset{6}{\underset{\substack{|\\ Me}}{C}}HCH_2CPh_2COEt \qquad Me_2NCH_2\overset{*}{\underset{\substack{|\\ Me}}{C}}HCPh_2COEt \qquad Me_2NCH_2CH(Me)\overset{\substack{Ph\\ |}}{\underset{\substack{|\\ OCOMe}}{C}}CH_2Ph$$

30 **31** **32**

hydrochloride salt in $CDCl_3$ is a triplet made up of two overlapping doublets (J ~5 Hz) (Fig. 6.7), typical of HNCH coupling; this arises because the two NMe groups are rendered non-equivalent in the protonated form by the adjacent asymmetric centre (C-6). When proton exchange is accelerated, e.g. in solvent D_2O, or in methadone free base, the N—Me groups are equivalent and the N—Me signal is a sharp singlet. Centres of asymmetry β- to a dimethylamino group also lead to magnetic non-equivalence as is evident from the $^+HNMe_2$ signal of salts of isomethadone **31** and the amino-ester **32**. The degree of resolution of separate NMe signals is highly sensitive to the proton exchange rate as is evident from results with isomethadone hydrochloride in $CDCl_3$ (a broad unresolved multiplet is seen), isomethadone base in $CDCl_3$-TFA (two broad bands appear) and the base in $CDCl_3$ plus a large excess of TFA (a clear pair of doublets, J 5 Hz, are resolved) [Fig. 6.7 (b), (d) and (e)].

Non-equivalence of geminal nuclei or proton groups (e.g. methylene or *gem* dimethyl groups) can always be expected if there is present somewhere in the molecule a carbon atom with three different substituents because the environments of one group (or nucleus) can never be exactly the same as that of the other, the effect being termed intrinsic asymmetry (Whitesides *et al.*, 1962). Environmental differences are magnified, however, in cases of conformational preference, and at ordinary temperatures this factor is normally more important than the intrinsic asymmetry effect. In the case of methadone hydrochloride, for example, conformer **34** is probably favoured over the other two staggered conformations **35** and **36** in which all four bulky substituents interact with one another; in **34** only one NMe

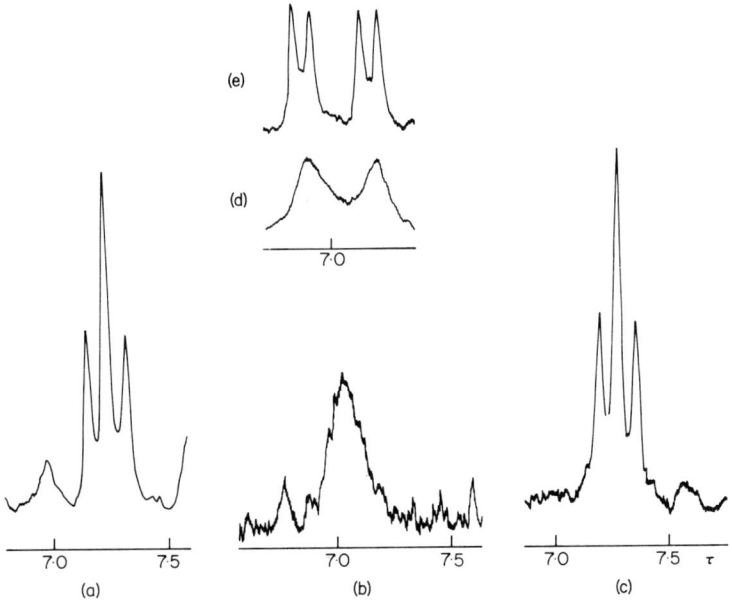

34   **35**   **36**

R = CPh₂COEt (front atom ⌖ is nitrogen)

group is influenced by the R substituent, hence environmental differences between the two N-methyl groups are to be anticipated. The carbonyl rather than aromatic part of the R group appears more important in this respect, because the HNMe₂ signal of the corresponding hydrocarbon (R = CHPh₂ in **34** etc.) forms a sharp doublet with J about 5 Hz due to HNCH coupling. This result suggests that the carbonyl function is directed

Fɪɢ. 6.7. 60 MHz PMR dimethylamino signals of: (a) methadone hydrochloride (**30**); (b) isomethadone hydrochloride (**31**); (c) the acetate analogue of propoxyphene hydrochloride (**32**) in CDCl₃; (d) isomethadone base in CDCl₃-TFA; (e) iso-methadone base in CDCl₃ plus a large excess of TFA.

towards the basic group in the preferred conformation of methadone hydro-chloride as solute in CDCl$_3$ (see **38**).

A clear account of factors responsible for magnetic non-equivalence is given in a recent review (van Gorkum and Hall, 1968).

A second distinctive feature of the PMR spectrum of methadone base is the unusually high field position of its secondary methyl signal. This falls at 29 Hz from TMS in CDCl$_3$ (60 MHz spectrum) whereas the *sec*-Me signals of the related derivatives **37** a–c are lower field by over 20 Hz (Casy, 1966b). The spectrum of related amino-ketones, e.g. heptalgin (**30**),

Me
|
Me$_2$NCHCH$_2$CPh$_2$R

37

	Sec-Me chemical shift (Hz from TMS in CDCl$_3$)	
a: R = H	54·5	
b: R = OH	52	
c: R = CN	54	
d: R = CO$_2$Et	38	
e: R = CH(OCOMe)Et	27	(α-isomer)

NMe$_2$ replaced by —N⟨ ⟩O also display high field *sec*-Me signals. When the bases **30** and **37** a–c are protonated the *sec*-Me signals move downfield as a result of the increased deshielding influence of the nitrogen centre; however, downfield shifts are lower (13 Hz compared with 20–25 Hz for the bases **37** a–c) for the amino-ketones. These results are taken to indicate that in favoured conformations of methadone and related ketones, the 6-methyl group lies close to and above the plane of one of the aromatic

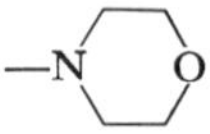

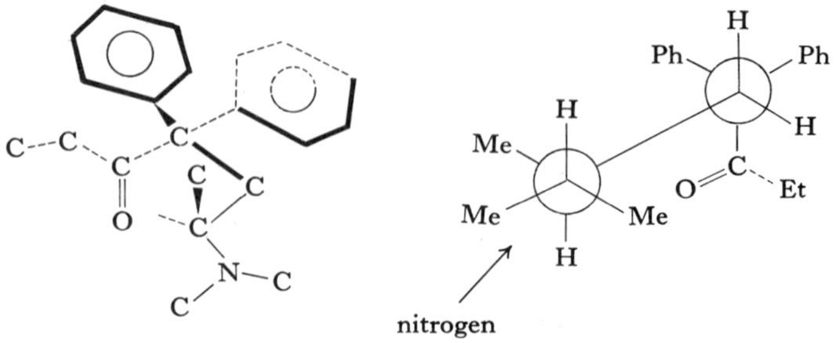

38

(heavy lines lie above and dotted lines below the plane of the paper) rings in the molecule; this orientation occurs in crystalline methadone hydrobromide as shown by X-ray crystallography (Hanson and Ahmed, 1958). The model **38**, based upon the structure of the crystalline salt, is therefore taken to represent a favoured conformation of methadone hydrochloride as solute in $CDCl_3$ or water (an unusually high field *sec*-Me signal is also observed when D_2O is solvent). High field *sec*-Me signals are also seen in the spectra of the analgesically active ester analogue of methadone (**37**d) and α-acetylmethadol (**37**e); the spectra of salts of these compounds in $CDCl_3$ also show non-equivalent $H\overset{+}{N}Me_2$ signals (Hassan, 1967).

Reduction of methadone gives two diastereoisomeric aminoalcohols (α- and β-methadol **39**) which differ in their analgesic potencies (Eddy *et al.*,

$$Me_2NCH(Me)CH_2CPh_2\overset{3}{C}H(OH)Et$$

39

1952). Indirect evidence regarding the configuration of the newly created C-3 asymmetric centre in **39** is derived from the fact that reaction of α- and β-methadol with cyanogen bromide gives two different, stereochemically pure, tetrahydrofuran derivatives (**40**) (Casy and Hassan, 1967). If the configurations of these 2-ethyl-5-methyl-3,3-diphenyltetrahydrofurans

40

are known (*cis* or *trans* 2-Et/5-Me), it is possible to deduce the relative configurations of the precursor methadols, provided the assumption is made that cyclization proceeds with inversion (see **41**) and there is good evidence that such is the case. Hence the methadol shown in **41** will give

41

the *cis* isomer whereas the methadol with the C-6 configuration reversed will yield the *trans* isomer. The configurations of the tetrahydrofurans may be assigned from PMR data upon the two isomers and related cyclic derivations (Table 6.2). The two C-4 methylene and the C-5 methine protons of the 5-methyl cyclic derivatives form spin–spin coupled systems ranging from AMX to ABX type. Analysis of the AM (or AB) signal is possible in most cases, but resolution of the C-5 proton signal (X) is hampered by its additional coupling to the 5-Me protons. When the three protons approach an AMX system, the high field methylene signal is near 150 Hz, and the low field 180 Hz, (Table 6.2 and Fig. 6.8) both being 4-line

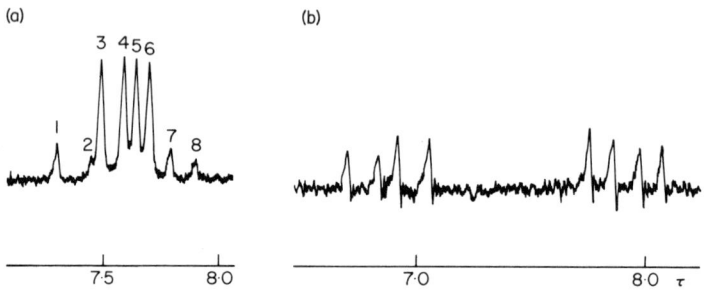

Fig. 6.8. 60 MHz PMR 4-methylene signals of isomeric 2-ethyl-5-methyl-3,3-diphenyltetrahydrofurans in $CDCl_3$. (a) α-isomer signal (subquartets composed of lines 1, 3, 4, 7 and 2, 5, 6, 8 respectively); (b) β-isomer signal (Casy and Hassan, 1967).

signals with J_{gem} 12–13 Hz. The higher field signal may be assigned to the proton H_β and that at lower field to H_α (see **42**) by analogy with the greater

(low field) H_α H_x

(high field) H_β Me

42

shielding of protons *cis* to methyl than those *trans*, as established for cyclohexane derivatives (Booth, 1966). The two methylene signals are most widely separated and show the greatest uniformity in line height in the case of the β-tetrahydrofuran **40**. In the α-isomer, however, the two signals overlap, the inner being much more intense than the outer lines (Fig. 6.8), and the spectral analysis (carried out by the same method as described in detail for the spectrum of β-methylacetylcholine, p. 221) shows that one proton has a normal and the other an abnormally high field position. Aromatic shielding effects based upon the predicted favoured conformation of the two isomers allow a decision upon the relative field positions of the methylene protons to be made. In the *cis* 5-Me/2-Et isomer, the conformation of phenyl *cis* to 2-ethyl will be influenced largely by the bulky flanking

TABLE 6.2

No.	Structure	4-Methylene proton data[a]		
		Centre of quartet[b]	J_{gem} (Hz)	J_{vic} (Hz)
1	Ph, Ph—⟨NH, NMe, Me HCl⟩	188 155·5	13·5 13·5	6·0 9·0
2	Ph, Ph— NR, Me R = (CH₂)₂O(CH₂)Cl	168 139	12·5 12·5	5·5 9·0
3	Ph, Ph—⟨NH, O, Me⟩	174 158	12·5 12·5	5·0 10·0
4	Ph, Ph—⟨O, O, Me⟩	186 153	12·5 12·5	5·0 10·5
5	Ph, Ph— α- ⟨Et, O, Me⟩	151 134	12 12	9·5 5·5
6	Ph, Ph— β- ⟨Et, O, Me⟩	187 125	13 13	8·0 6·0

[a] Hz from TMS at 60 MHz in CDCl₃.

[b] Resonance positions do not represent the true chemical shifts. All coupling constants, except the J_{vic} values for No. 5, were obtained by first order analysis.

substituent, while that of *trans* Ph (not adjacent to a bulky group) will be determined by the *gem*-Ph group. Dreiding models indicate that a preferred anti-planar *cis* Ph/heterocyclic ring orientation (in which *cis* Ph/2-Et interactions are a minimum) makes the same orientation for *trans* phenyl unfavourable because of *o*-hydrogen interactions. A favoured conformation for the latter group (in which nonbonded interactions involving the *trans* *o*-hydrogen protons and both the *gem*-Ph and 2-Et groups are a minimum) is shown in Fig. 6.9. This places the α-C-4 methylene proton within the screening zone of the adjacent Ph group; in consequence, its resonance position is moved up-field and the chemical shift difference between H_α and H_β decreases, and it is likely that H_α comes to resonance at the higher field position. In another interpretation, based upon a somewhat different model conformation (Portoghese and Williams, 1969), the chief aromatic

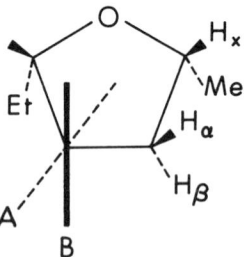

FIG. 6.9. Diagram of Dreiding model of *cis* (2-Et/5-Me) **40** viewed from above. Heavy lines lie above and dotted lines below, the plane of the tetrahydrofuran ring; A and B denote the planes of the aromatic rings *cis* and *trans* to 2-Et respectively.

influence is considered to be deshielding of H_β by the phenyl group *cis* to 2-ethyl, but the conclusion is the same. In the β-isomer (*trans* 2-Et/5-Me) interactions between 2-Et and a flanking Ph group will likewise be an important factor in determining the preferred configuration of Ph *trans* to the Et group. Here, however, the methylene proton *cis* to 5-Me (H_β) will be screened and, as a result, the H_α—H_β chemical shift difference will increase; in this case, aromatic screening effects augment the chemical shift difference induced by the 5-methyl substituent. These arguments account for the well-separated H_α and H_β signals and the unusually high field position of H_β in the β-isomer. Models of *cis* and *trans* (**40**) with preferred phenyl conformations as in Fig. 6.9 show the methylene protons of the 2-Et group of both isomers and the 5-Me group of the β-isomer to lie in an aromatic screening zone, while the 5-methine proton (β-isomer) falls approximately in the plane of the *cis* aromatic ring. These observations are in accord with the nature (α- a broad singlet, β- a deformed triplet) and similar high-field resonance positions of the two 2-Et signals, the higher field position of the β- 5-Me and the lower position of the β-5-methine proton, further supporting the configurational assignments.

Coupling constant data must be applied with caution to the configurational assignment of 5-membered ring isomers because of uncertainties of the preferred conformation of such cyclic systems (Chapter 3, p. 121). In the present examples, the higher J_{vic} values obtained for the methylene signals near 150 Hz (assigned to H_β) in the derivatives 1–5 (Table 6.2) are evidence that these derivatives adopt a preferred half-chair conformation as shown, for example, in (**43**a) for α-(**40**) with 5-methyl groups *pseudo*-equatorial.

43a   **43**b

(3,3-diphenyl substituents omitted)

In β-**40**, however, both vicinal coupling constants have values close to 6 Hz indicative of a lack of conformational preference; in this compound, an axial 5-methyl orientation does not generate a severe Me—Et interaction as arises when 5-methyl is *pseudo* axial in the α-isomer (see **43**b).

The *cis* geometry of the tetrahydrofuran derived from α-methadol has now been confirmed by an X-ray study (Singh and Ahmed, 1969).

CHOLINERGIC AGONISTS

The Conformation of Acetylcholine (Ach)

Many molecules of pharmacological interest are 1,2 disubstituted ethanes, e.g. acetylcholine, histamine and serotonin. The analysis of the PMR spectra of such molecules is of importance because it provides information about conformation which, in turn, may shed light upon the way in which the agonist interacts with its receptor (Gill, 1965). There are three possible staggered conformers for molecules of type XCH_2CH_2Y, one *trans* (**44**) and two equivalent *gauche* forms (**45** and **46**). In each conformer at least two distinct vicinal coupling constants are involved; thus in **44** there is one *J trans* and one *J gauche*, and in **45** one *J trans* and three *J gauche* couplings. Rapid interconversion of conformers by rotation about the C—C bond renders each geminal pair of protons chemically equivalent and the four proton system is described as AA′XX′ or AA′BB′ depending on whether the chemical shift difference between the methylene pairs is

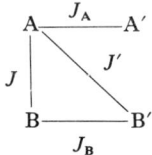

	trans	gauche	gauche
	44	**45**	**46**

large or small; the designation A_2X_2 or A_2B_2, also advocated for these systems (Hirst and Grant, 1964), is strictly reserved for the rare cases in which both A nuclei are coupled to an equal extent with the two X or B nuclei. The same rotation does not, however, lead to a single, averaged vicinal coupling constant except when the conformer populations are equal (see later). Hence, spectra obtained are complex and cannot be analysed by a first order treatment. The detailed analysis of these systems (Pople *et al.*, 1959; Wiberg and Nist, 1962; Emsley *et al.*, 1965; Bovey, 1969) is outside the scope of this book but certain features are summarized below.† The AA′BB′ (and AA′XX′) system gives patterns which are symmetrical about their midpoints; one half is due to the AA′ and the other to the BB′ (XX′) group and either may be used for spectral analysis. Each half of an AA′XX′ spectrum has a centre of symmetry (i.e. is centrosymmetric) additional to that of the group as a whole. In the most general case four

$$A \overset{J_A}{\rule{2cm}{0.4pt}} A'$$

coupling constants are involved. These can not be derived directly from the spectrum but may be calculated in terms of the parameters:

$$K = J_A + J_B; \quad L = J - J'$$
$$M = J_A - J_B; \quad N = J + J'$$

In many cases $J_A = J_B$ whence $M = 0$.

In principle, an AA′BB′ system gives rise to 12 transitions per group (i.e. 24 transitions in all) but in practice, fewer lines are generally observed [Fig. 6.10(a) and (b)]. A strong doublet is usually discernible formed from

† Garbisch (1968) has described a second order perturbation approach for approximate hand analyses of complex PMR spectra arising from 2, 3 and 4 spin systems. Such analyses are often a necessary preliminary to computer calculations of the iterative type.

lines 1, 2 and 3, 4 and the distance between these lines gives the sum of the vicinal coupling constants (N). In certain cases, the distance between lines 11 and 10 gives an approximate value for the magnitude (but not the sign) of the difference between J and J' (L) (Garbisch, 1968). If N and L are known, the values J and J' follow since $N + L = 2J$. When J and J' are equal ($L = 0$), a triplet is obtained [Fig. 6.10(c)].

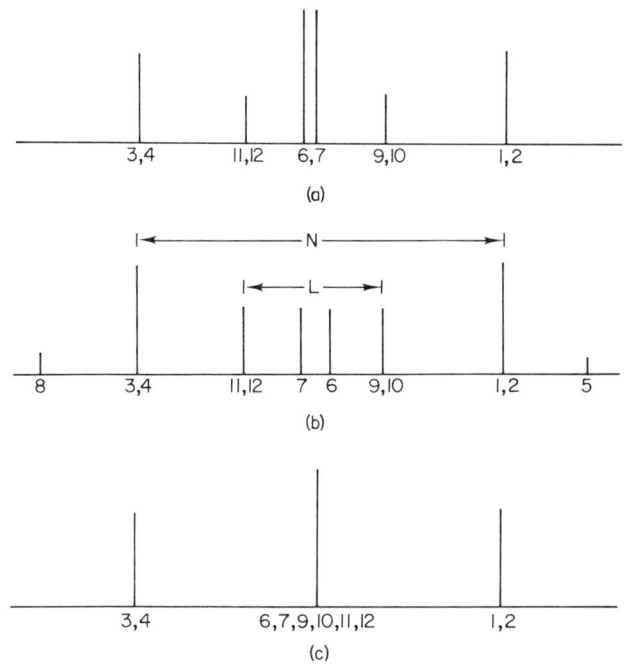

FIG. 6.10. Diagrammatic representation of the A_2 portion of A_2B_2 PMR spectra with $M = 0$ and (a) $L \neq 0$, $N = 3L$, $K \gg L$ or N; (b) $L \neq 0$, $N = K = 3L$; (c) $L = 0$ (Sheppard and Turner, 1959).

Culvenor and Hams' treatment (1966) of the spectrum of acetylcholine may now be discussed. Part of the 60 MHz spectrum of Ach chloride in D_2O is shown in Fig. 6.11; the methylene proton region may be classified as an AA'XX' type because the chemical shift difference ($\sim$50 Hz at 60 MHz) is almost certainly larger than 6 times J_{AX} (Bible, 1965). The lower field multiplet, due to the OCH_2 protons is broader than the $\overset{+}{N}CH_2$ multiplet because of an additional coupling of about 2 Hz between the OCH_2 protons and the positively charged nitrogen atom (nuclear spin 1). In quaternary nitrogen salts the 3-bond NCCH coupling is known to be larger ($1\cdot5$–2 Hz) than the 2-bond NCH coupling (0–$0\cdot6$ Hz) (Kawazoe *et al.*, 1967). N and L

values were estimated as shown from the high field multiplet [Fig. 6.11(b)] and refined by means of an iterative calculation and computer programme,† giving N, 9·49 Hz and L, 4·43 Hz; thus $2J = 13·92$ (N plus L), whence $J = 6·96$ and $J' = 2·53$ Hz (take as 7 and 2·5 Hz). These results are now compared with values anticipated if (i) *trans* and (ii) *gauche* conformers be favoured.

(i) *trans* conformer favoured

$$J_{11}' = J_g$$
$$J_{12}' = J_t$$

In this case the smaller coupling (2·5 Hz) could correspond with J_g, but the larger value (7·0 Hz) is abnormally low for J_t even allowing for substituent effects.

(ii) *gauche* conformers favoured

$$J_{11}' = 1/2(J_g + J_t)$$
$$J_{12}' = 1/2(J_g + J_g) = J_g$$

Here the 2·5 Hz coupling is consistent with the J_g values whereas $J = 7·0$ Hz approaches the expected value of $1/2 (J_t + J_g)$, e.g. about $(10 + 3)/2 = 6·5$ Hz. On this basis J_t has a value (11·5 Hz) within the normal range.

The methylene PMR signals of dioxane and N-methylmorpholine both closely resemble that of NCH$_2$ in the spectrum of Ach (Fig. 6.12). These compounds are constrained to *gauche* geometry by their cyclic nature and hence serve as models for the XCCY conformation. It appears, therefore, that the signal appearance of the methylene resonances shown in Figs 6.11 and 6.12 is characteristic of a *gauche* XCCY conformation.

† In this procedure judicious estimates of chemical shifts and spin–spin coupling constants are inserted in the theoretical equations which describe NMR transition energies and relative transition probabilities, and the results used to plot a calculated line spectrum. This spectrum is compared with the experimental one and any differences are used as a basis for readjustment of the initial estimates of NMR parameters. This technique has been reviewed (Pan and Rogers, 1968).

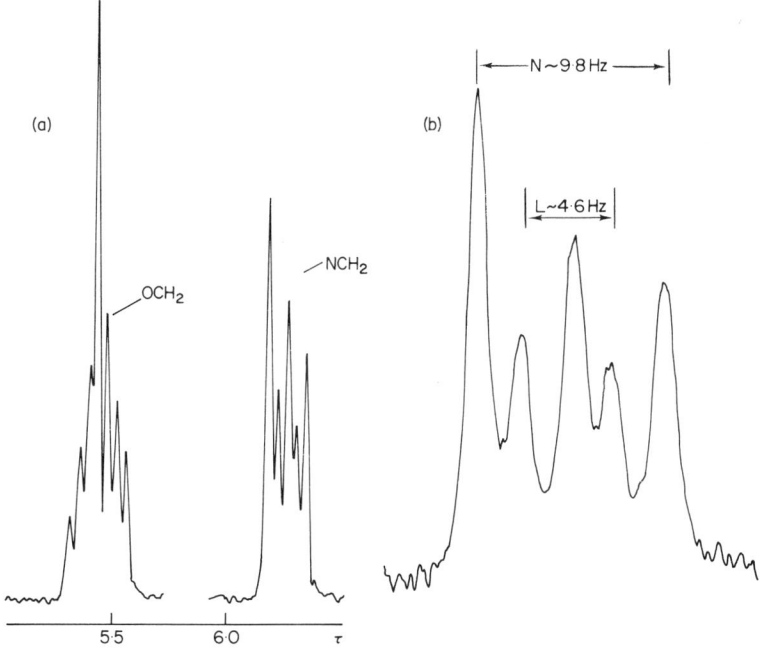

Fɪɢ. 6.11. Part of the 60 MHz PMR spectrum of acetylcholine chloride in D_2O. (a) OCH_2 and NCH_2 signals (sweep width 500 Hz); (b) N-CH_2 signal (sweep width 100). Approximate N and L values are marked.

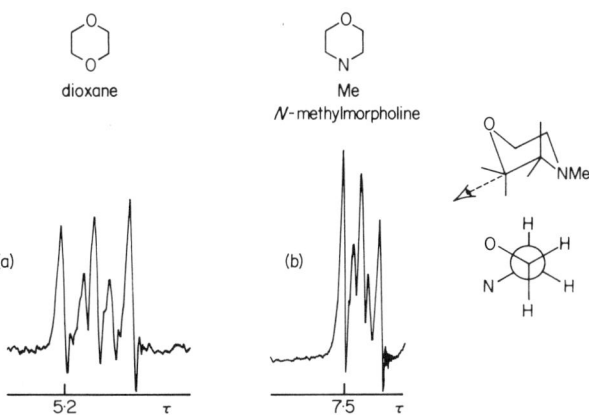

Fɪɢ. 6.12. 60-MHz PMR methylene signals of (a) dioxane (neat), lower field ^{13}C satellite signal, sweep width 250 Hz; (b) N-methylmorpholine (in D_2O), N-CH_2 signal, sweep width 500 Hz.

The dioxane case illustrates the use of the ^{13}C satellite signal for investigating coupling between identical protons (Sheppard and Turner, 1959). The proton signal due to the fragment **47** of dioxane is a singlet because all

$$
\begin{array}{cc}
\text{H H} & \text{H H} \\
^{12}|\ ^{12}| & ^{13}|\ ^{12}| \\
\text{O—C—C—O} & \text{O—C—C—O} \\
|\ \ | & |\ \ | \\
\text{H H} & \text{H H} \\
\mathbf{47} & \mathbf{48}
\end{array}
$$

protons are chemically equivalent. However, in the fragment **48** which incorporates carbon-13 (present in 1% natural abundance), protons

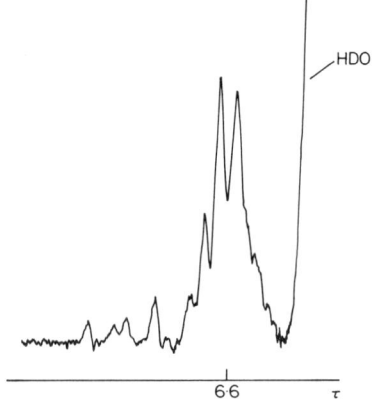

Fɪɢ. 6.13. 60 MHz PMR methylene signal of thioacetylcholine chloride in D_2O (sweep width 250 Hz).

attached to ^{13}C have a different environment to those attached to ^{12}C, and thus the two sets will couple. The signal appears at about 70 Hz on either side of the principal ^{12}C methylene signal as a result of ^{13}C—H geminal coupling, and the expansion of one of these signals is shown in Fig. 6.12(a).

Thus PMR evidence favours a high proportion of an Ach conformation which is *gauche* about the $\overset{+}{N}$CCO bond.

The spectrum of thioacetylcholine (thio Ach) is markedly different from that of Ach, the NCH_2 and SCH_2 signals being much closer together and forming a complex multiplet approximately symmetrical about its midpoint and obscured on the high field side by the HDO signal (Fig. 6.13). No assignment of L and N values is possible by inspection. This result is evidence that conformational preferences in the thio derivative are different from those in Ach and it is significant in this respect that X-ray analysis of thio-Ach bromide shows that the crystalline salt has a *trans* N/S conformation (Mautner, 1969). Recent PMR analyses confirm a *trans* conformation for this compound in solution (Cushley and Mautner, 1970; Culvenor and Ham, 1970).

PMR evidence of the conformation of β-methylacetylcholine (β-Me-Ach) requires analysis of the ABX system formed from two methylene (A,B) and one methine (X) proton (**49**). The non-first order analysis described here follows the procedure of Bible (1965). Ideally data from both AB and

$$\text{Me}_3\overset{+}{\text{N}}-\underset{\underset{\text{H}_B}{|}}{\overset{\overset{\text{H}_A}{|}}{\text{C}}}-\underset{\underset{\text{H}_X}{|}}{\overset{\overset{\text{Me}}{|}}{\text{C}}}-\text{OCOMe} \quad \text{Cl}^-$$

49

X portions of the spectrum should be used, but in the present case the methine signal is so broad (as a result of additional coupling with the

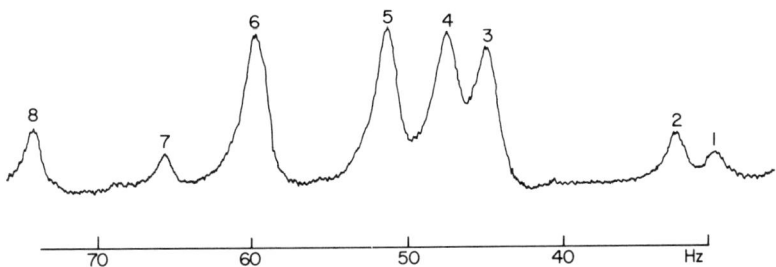

Fig. 6.14. 100 MHz PMR methylene signal of β-methylacetylcholine chloride in D_2O; sweep width 100 Hz and sweep offset 370 Hz. Lines 8, 6, 4 and 2, and 7, 5, 3 and 1, form the two AB quartets respectively (Casy *et al.*, 1971).

β-methyl protons and with the ^{14}N quaternary atom) that it cannot be analysed. Each methylene signal involves a geminal (J_{AB}) and vicinal (J_{AX} or J_{BX}) coupling; hence the AB signal is composed of two quartets. These partially overlap and each is identified by the separation 14·4 Hz (J_{AB}) which occurs four times within the 8-line signal (Fig. 6.14). The distance between the centres of the two quartets is 5·6 Hz, whence $|(J_{AX} + J_{BX})| = 11\cdot2$ Hz (the modulus bars denote that only the magnitude, but not the sign, of the quantity within the bracket is known). A pair of right angled triangles (a) and (b) are now constructed having J_{AB} as bases and with $2D_+$ as the hypotenuse of one and $2D_-$ as the hypotenuse of the other triangle. $2D_+$ and $2D_-$ are the distances between the first and third lines of the two AB quartets. From these triangles, two alternative solutions for the parameters of the ABX system are derived. In (a), $|(J_{AX} - J_{BX})|$

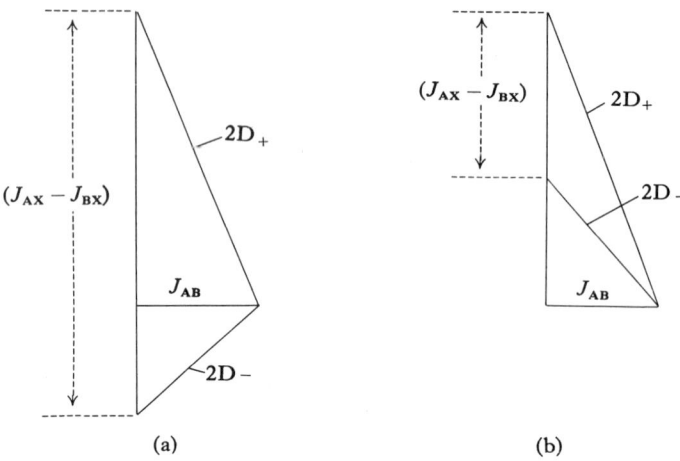

(a)   (b)

is equal to the sum of the two vertical sides, while in (b) this quantity is equal to the difference between the two sides. The halfway point on the combined vertical side (a) or the point halfway between the vertical apexes of (b) is marked. The distance from these marks to the base gives the two possible values for the chemical shift difference, $|\Delta v_{AB}|$, between the A and B protons. Following this procedure gives a $|\Delta v_{AB}|$ value of 4 or 19 Hz and a $|(J_{AX} - J_{BX})|$ value of 37·2 or 8·2 Hz. The correct alternatives are normally selected from the intensity ratio of peaks in the X signal. Since this was not available in the present case, the values $|\Delta v_{AB}| = 19$ and $|(J_{AX} - J_{BX})| = 8·2$ Hz were chosen as the more probable, and on the basis of comparisons with calculated spectra (Bovey, 1969). Combination of the quantities $|(J_{AX} - J_{BX})| = 8·2$ Hz, and $|(J_{AX} + J_{BX})| = 11·2$ Hz, gives the values $J_{AX} = 9·7$ and $J_{BX} = 1·5$ Hz. From the order of magnitude of these values (p. 88), it is clear that the preferred rotamer(s) of β-Me-Ach involve *trans* and *gauche* vicinal coupling constants; this conclusion supports either **50** or **51** as the preferred conformation. The fact of a chemical shift difference between H_A and H_B of 19 Hz further indicates that β-Me-Ach displays a marked conformational preference but does not aid choice of the preferred conformation because environments of the two protons differ in both conformers. The morpholino derivatives **52a** and **b** serve as rigid models for the *gauche* $^+N/O$ conformer **51**. Data upon their spectra generally support the latter as a preferred conformation in that vicinal coupling constants (for **52a** and **b**) of similar orders of magnitude as those found for β-Me-Ach,

52

a: $R = H$, $\overline{X} = \overline{Cl}$
b: $R = Me$, $\overline{X} = \overline{I}$

are derived from their spectra. Thus, analysis of the X multiplet of (**52a**) gives the value $|(J_{AX} + J_{BX})| = 13\cdot3$ Hz whence $J_{AX} = 11\cdot2$ and $J_{BX} = 2\cdot1$ Hz by first order treatment, following Booth and Gidley (1965) who previously examined the 2,6-dimethylmorpholine base (p. 135). Similar values are obtained by the same analysis of the methine signal of **52b** (the AB and X signals of the two salts virtually overlap). Chemical shift data upon the secondary methyl group of β-Me-Ach and the model **52** also supports the preferred nature of conformer **51**. Although the two values differ (ν_{Me}; **49** δ 1·33, **52b** δ 1·23 Hz), allowance for the acylation shift (ν_{Me} in **53** is δ 1·24) shows that β-methyl has a similar environment in both **49** and **52b**.

$$\overset{+}{Me_3N}CH_2CHMeOH \quad \overline{I}$$

53

The small, but significant, differences observed between the J_{AX} and J_{BX} values of the model **52b** and those of β-Me-Ach are best accounted for in terms of a *gauche* $^+N/O$ conformer in which the NCCO dihedral angle is increased from 60° (as in **51**) to about 90° (as in **54**); dihedral angles of this magnitude occur in solid state conformations of Ach, α-, and β-Me-Ach (Chotia and Pauling, 1969) and are probably the result of a compromise between steric and electronic interactions of the nitrogen and oxygen functions. In **54**, the H_XCCH$_A$ and H_XCCH$_B$ dihedral angles exceed and fall below, respectively, corresponding angles in the fully staggered conformation **51**. On the basis of the Karplus $\cos^2\phi/J$ relationship (see p. 88) both these changes will lead to a reduction in vicinal coupling constants; hence values lower than those in models such as (**52b**), in which dihedral angle deviations from 60° are less likely on account of constraints imposed by the cyclic structure, are to be anticipated. A significant population of the sterically unfavourable all-*gauche* conformer **55** would also account for the lower J_{AX} and J_{BX} values of **49** compared with those of **52b**, but the

54   **55**

β-methyl chemical shift values would not be expected to correspond so closely because methyl is deshielded by charged nitrogen in **55**. The population of the *trans* $^{+}$N/O conformer **50** is likely to be small since its contribution, although providing a lower J_{BX} coupling (J_{BX} is *trans* in **51** but *gauche* in **50**), would enhance the J_{AX} value.

In the case of α-methylacetylcholine (α-Me-Ach), knowledge of the J_{AX} and J_{BX} coupling constant magnitudes is obtained by analysis of the H_X signal of the ABXK$_3$ system (see **56**) since this exhibits negligible coupling with ^{14}N (cf. H_X signal of the β-isomer). The AB signal (a broad multiplet low field of the methine signal) is unsuitable for analysis because of coupling with ^{14}N and its overlap with the HDO peak; ^{14}N coupling effects are particularly clear in the α-methyl signal which forms a sharp

(α-Me protons are designated H_K)

56

doublet of triplets (J 7·0 and 2·0 Hz) (Fig. 6.15). Identification of the four XK$_3$ quartets amongst the 12 line methylene signal (Fig. 6.15) by Booth and Gidleys' procedure (1965), taking $J_{XK} = 7\cdot0$ Hz from the α-methyl signal, gives the quantities ($J_{AX} + J_{BX}$) = 8·8 Hz and the approximate values, $J_{AX} = 4\cdot2$ and $J_{BX} = 4\cdot6$ Hz; the latter values cannot be refined by data from the AB portion of the spectrum because this signal is poorly resolved. The first order values are of sufficient accuracy, however, for the present purpose. The geminal H_A and H_B protons formed a multiplet of five broad lines (partially obscured by the HDO peak) about 1·1 ppm downfield of the methine signal. The overall narrowness of the signal ($W_H \sim 10$ Hz) shows that the methylene protons have very similar chemical shifts, and the appearance of the ABX spectrum obtained closely approaches that calculated for the values $\Delta\nu_{AB} = 2$, $J_{AB} = 10$, and $J_{AX} = J_{BX} = 4$ Hz (Bovey, 1969).

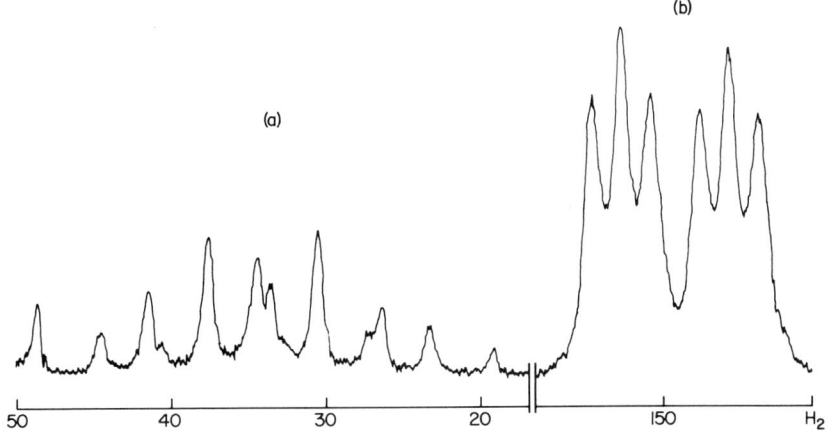

Fig. 6.15. 100 MHz PMR methine (a) and secondary methyl (b) signals of α-methylacetylcholine iodide in D_2O; sweep width 100 Hz and sweep offset 420 Hz for (a) and zero for (b) (Casy *et al.*, 1971).

The near-coincidence of the vicinal coupling constants derived by the above analysis indicate that conformational preferences among α-Me-Ach conformers are low. *Gauche* and *trans* coupling constants for the conformers **57–59** are unknown but values predicted on the basis of data upon β-Me-Ach, and the models **52a**, **52b**, **60** (the methiodide of phendimetrazine) and **61** (derived from ephedrine) give reasonable estimates of their orders

57

$J_{AX} \sim 1\cdot5$
$J_{BX} < 1\cdot5$†

58

$\sim 1\cdot5$
~ 10

59

~ 10 Hz
$\sim 1\cdot5$ Hz

$J_{AX} = 10\cdot4$ Hz (*trans*)

60

$J_{AX} = 1\cdot5$–$2\cdot0$ Hz (*gauche*)

61

† This value is judged the lowest because both coupling protons are *trans* to highly electronegative groups (Booth, 1965).

of magnitude; probable values are listed beneath each conformer. Similar populations of **57**, **58** and **59** should lead to average J_{AX} and J_{BX} coupling constants of just over 4 Hz, e.g. $(1\cdot5 + 1\cdot5 + 10)/3 = 4\cdot3$ Hz, as obtained by analysis. Conformer distributions of this order would also result in similar averaged environments (and hence like chemical shifts) for the H_A and H_B protons, again in agreement with spectral observations.

The α-methyl chemical shift (δ 1·5) is lower field than that of methyl in the model **62** (δ 1·33) in D_2O; this result is evidence for significant populations of **57** and **59** in which methyl is close to the deshielding oxygen function.

62

The solute conformations of the isomeric methylacetylcholines deduced from PMR evidence and infra-red spectroscopy (Casy, Hassan and Wu, 1971), correlate well with the solid state conformations of these compounds established by X-ray studies (Chotia and Pauling, 1969); that of the β-isomer is similar to **54** and the α-analogue exists in two forms that are related to **57** and **59** respectively. Inch and others (1970) have reported a similar PMR study of methylacetylcholines.

The preferred conformations in D_2O of the cyclic Ach analogues *cis*(**63**) and *trans*(**64**)2-dimethylaminocyclohexyl acetate methiodide follow from analysis of their 1- and 2-methine PMR signals (Wu, 1970). The 1-methine signal of the *cis* isomer (W_H 8·5 Hz) is much narrower than that of the *trans* derivative (W_H 16·5 Hz); two axial-axial couplings contribute to signal width in the latter case (see p. 88). The broad appearance of the 2-methine signal seen in the spectra of both isomers (W_H *cis* 16; *trans* 19 Hz) establish the axial orientation of the 2-methine proton in each case. Use of the acylation shift (p. 30) assists identification of the 1- and 2-methine signals

cis **63**   *trans* **64**

since only the former suffers a pronounced upfield shift when the acetyl group is removed, as shown below for the *trans* isomer.

Centre of multiplet
(Hz from TMS)

	R = COMe	R = H	Δ
1-H	311	240	71
2-H	231	216	15

(solvent D_2O, 60-MHz spectra)

The 2-methine proton of *cis*-2-methyl-4-dimethylaminomethyl-1,3-dioxolane methiodide **65**, a potent muscarinic agent, is higher field than that of the less potent *trans* derivative (**66**) the reverse being true of the 2-methyl signals (Garrison *et al.*, 1969); the stereochemistry of these

(8·49) Me   H (4·81)   (4·75) H   Me (8·67)

65   **66**

(τ values relative to DSS in D_2O)

isomers is established by unambiguous synthesis. Isomeric forms of other 2,4-disubstituted-1,3-dioxolanes show similar spectral differences which are attributed to transannular deshielding of the C-2 proton (in the *trans*) and the C-2 methyl (in the *cis*-isomer) by the 4-substituent. The results correlate with the deshielding influence of axial methyl upon a 3-axial proton in cyclohexane assessed by Booth (1966) at −0·18 ppm. Garrison *et al.* (1969) assigned configurations to some novel dioxolane isomers in this way but recommend caution in use of the method since they noted an inversion of the usual trend when the 4-substituent was a 4-tosyloxy methyl group (CH_2OTs).

The configuration of the potent cholinesterase inhibitor physostigmine (eserine) has been established as (**67**) by application of the nuclear Overhauser

(low field) Me   Me (high field)

67

effect (see p. 102). Newkome and Bhacca (1969) found that low intensity irradiation at either the 3a-Me or 8-N-Me frequency resulted in a 15% enhancement of the 8a-H signal. Hence all the substituents giving rise to these signals are *cis*. Irradiation at the high-field N-methyl (N-1) frequency did not alter the integration of the 8a-H signal. Hence the preferred conformation of 8-N-Me and 1-N-Me are *cis* and *trans* respectively to the 8a-proton.

Kato (1968) has shown that the progress of the enzymatic hydrolysis of acetylcholine by a cholinesterase isolated from horse serum may be observed by an NMR technique. The method depends on the fact that the acetyl methyl signals of Ach and its hydrolysis product, acetate, are well separated ($\Delta = 15$ Hz initially at 60 MHz). To obtain the data shown in Fig. 6.16, 0·1 M Ach and the cholinesterase preparation (20 mg/ml) are mixed in a

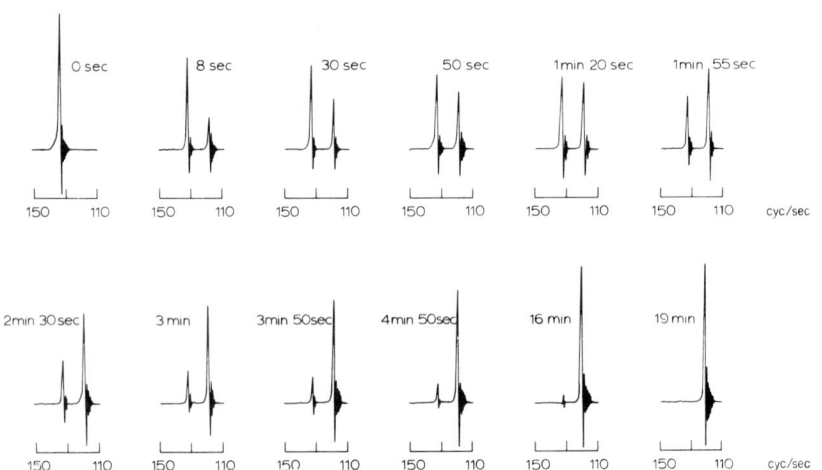

FIG. 6.16. 60 MHz PMR signals due to the $\underline{CH_3CO}$ protons of acetylcholine (lower field) and acetate (higher field) at various time intervals after the addition of cholinesterase (20 mg/ml) to 0·1 M Ach chloride in 0·2 M sodium phosphate at 39° (Kato, 1968).

D$_2$O-phosphate buffer, and the product transferred to a sample tube which is then inserted into the probe of a 60 MHz spectrometer. The first spectrum, usually recorded 8 sec after mixing, already shows that some hydrolysis has occurred (Fig. 6.16). As time progresses, the acetate signal grows at the expense of the substrate acetyl resonance and is the sole signal after 19 min. Plots of signal height against time are exponential (Fig. 6.17) and similar to those of peak areas against time; area measurements provide more accurate rate constants when high enzyme concentrations are employed because of line-broadening effects, as explained in Chapter 8 (p. 289). Kato examined the effects of varying the enzyme: substrate ratio, temperature, and pH upon the hydrolysis rate, and obtained results that agreed well with results derived in other ways. Physostigmine, edrophonium and neostigmine all acted as competitive inhibitors of the cholinesterase (Fig. 6.17) and the inhibition constants (K_i) of these antagonists, found from plots of the reciprocal of the reaction velocity against inhibitor concentration, also agreed with the corresponding literature values. This technique, which enables the direct and continuous monitoring of substrate changes under normal physiological conditions, is clearly a very attractive method of studying enzyme-substrate reactions.

The same principles may be applied to the study of enzyme reactions even when signals characteristic of substrate and product are not separated

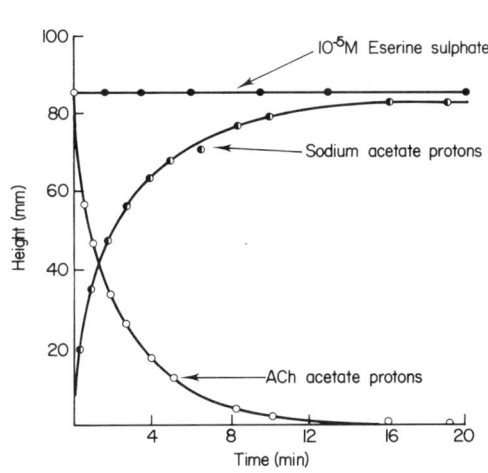

Fig. 6.17. Time course of acetylcholine hydrolysis measured by NMR (Kato, 1968).

to such a favourable degree as in the above case. For example, the N-acetyl methyl signal of N-acetyl-L-phenylalanyl-L-tyrosine and the product of its attack by pepsin (N-acetylphenylalanine) differ in chemical shift by only 4·4 Hz at 60 MHz (Sachs *et al.*, 1970). The integration step for the two signals is continuous but it may be subdivided with sufficient accuracy (as judged by the analysis of synthetic mixtures) by dropping a perpendicular onto it from the point where the two peaks intersect. The rate of hydrolysis of N-Ac-L-Phe-L-Tyr by pepsin determined by this method agrees well with previous findings and the PMR data also establish the stereospecificity of the process, the corresponding DL diastereoisomeric dipeptide being completely unaffected by the enzyme.

HISTAMINE AND ITS ANTAGONISTS

Speculations upon the role of conformational isomers in the mediation of the activity of histamine at H_1 and H_2 receptors† (Kier, 1968) has aroused interest in the conformation of histamine in the solid and solute state. Conformational studies of acetylcholine and its analogues, already described (p. 217), have been prompted by much the same reasons.

68

The bimethylene protons of histamine **68** form an AA′BB′ system typical of substituted ethanes. Four coupling constants are involved, namely $J_{AA'}$, J_{AB}, $J_{AB'}$, and $J_{BB'}$, and the spectrum is analysed by means of sum and difference parameters of these couplings (see p. 215). The fact that AA′BB′ spectra form centrosymmetric line patterns (i.e. the AA′ and BB′ signals are symmetrical about the mid-point of the entire signal) means that only one half need be used for the analysis. The two halves of the bimethylene region of the 100 MHz spectrum of histamine at pH 7 [Fig. 6.18(a)] are not equivalent however. This lack of symmetry is due to additional coupling between the higher field H_B protons (β- to the amino group of the side chain) and the H-5 proton of the imidazole ring. A symmetrical pattern results when the H-5 resonance is irradiated [Fig. 6.18(b)], and this is used for the spectral analysis. Approximate $N(J_{AB} + J_{AB'})$ and $L(J_{AB} - J_{AB'})$ values are first obtained by inspection. The N value is close to the separation between the inner member of the high field pair (line 6) and the lowest field line (1). Two other lines must have the same centre of gravity as 1 and 6 and the only possibilities are lines 3 and 4 (the latter appears as a shoulder on line 5); the four lines form one of the two quartets which make up the entire AA′ signal. The remaining lines (2, 5 and 7) form a triplet rather than a second quartet and this finding suggests that $L = 0$, i.e. $J_{AB} = J_{AB'}$. Values of N, L, ν_A and ν_B were refined by a series of computerized iterative calculations. The calculated spectrum using the values

† H_2 receptors are associated with the gastric secretion of acid induced by histamine, while H_1 receptors are involved in most of the other pharmacological properties of histamine.

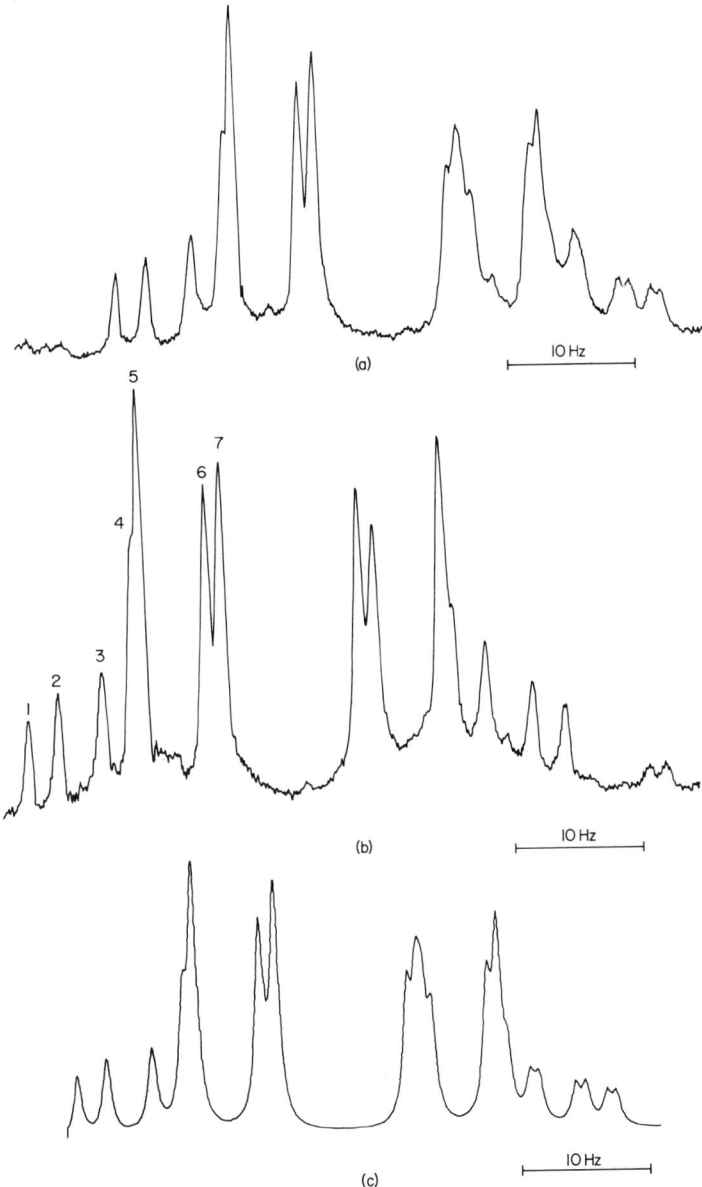

FIG. 6.18. The bimethylene 100 MHz PMR signals of histamine dihydrochloride in D₂O at pH 7. (a) normal spectrum; (b) spectrum recorded when imidazole protons irradiated; (c) calculated spectrum including coupling to the imidazole protons (Ham, 1970).

ν_A 361·08 Hz, ν_B 384·76 Hz, $N = 14·57$ Hz and $L = 0$, is shown in Fig. 6.18(c) (Ham, 1970; private communication).

Hence $J_{AB} = J_{AB'} = 7·3$ Hz. This result shows that histamine does not show a conformational preference under close to physiological conditions, the *anti* **69** and two equivalent *gauche* conformations **70** and **71** being

	69	**70**	**71**
	anti	*gauche*	*gauche*

equally populated. This result supports Kier's conclusions (1968) based on molecular orbital considerations, that the *anti* and *gauche* forms of histamine differ little in energy content. In the solid state, histamine acid phosphate monohydrate adopts the *anti* conformation **69** (Veidis and Palenik, 1969).

Some years ago a compound claimed to be the 2-positional isomer of

histamine (**72**, isohistamine) was reported (Jones, 1949), which was devoid of histaminic activity. This result was surprising because most small ring nitrogen heterocycles with a β-aminoethyl side chain are histamine agonists (Jones, 1966). The original synthesis has now been reinvestigated (Durant *et al.*, 1968). The first step is reaction between 2-chloromethylimidazole **74** and sodium cyanide giving, as Jones presumed, the 2-cyanomethyl derivative **73**. However, the PMR spectrum of the cyanide product showed that only one imidazole proton remained (one proton singlet τ 1·63) and that a methyl group was present (three proton singlet τ 7·15, diHCl salt in D_2O). Hence **74** undergoes nuclear rather than halide substitution by the cyanide ion, giving **75** and Jone's product is **76**; this derivative lacks the essential

β-aminoethyl side chain of histamine agonists and its pharmacological inactivity is explained. Authentic isohistamine **72** (PMR of diHCl salt in D_2O: 2 proton singlet τ 2·58, imidazole protons; 4 proton singlet τ 6·52, bimethylene protons) (Durant *et al.*, 1968) is a histamine agonist of a low order of potency but the maximum response of muscle to histamine or isohistamine is of the same magnitude (Kornfeld *et al.*, 1968).

Antagonists of histamine

Certain derivatives of 3-aminopropene and 4-aminobutene have anti-histaminic properties and in both series potency is dependent upon the arrangement of substituents about the carbon-carbon double bond (Adamson *et al.*, 1951; Casy and Ison, 1970). PMR spectroscopic data greatly facilitates structural and configurational assignments to these unsaturated compounds.

Triprolidine **77**, the best known example of a 3-aminopropene anti-histaminic agent, has a *cis* 2-pyridyl/H configuration and is more active than the related *trans* isomer. Models show that either the 2-pyridyl or the *p*-tolyl substituent, *but not both*, may lie in the plane of the carbon-carbon double bond; the substituent *cis* to the aminomethyl group is always forced out of this plane as a result of non-bonded interactions between the bulky *cis* groups. Hence a near planar vinylpyridine unit is only probable in the *cis* isomer **77**, while a planar *p*-methylstyrene unit should be favoured

77 *cis* py/H   **78** *trans* py/H

in the *trans* isomer **78**. The identity of the planar features is revealed by differences in both the ultra-violet and PMR spectra of the isomers. Thus the ultra-violet spectrum of triprolidine resembles that of 2-vinylpyridine while that of its isomer is similar to the spectrum of styrene. In both isomers the vinylic proton will be deshielded by the aryl group in the *cis* position (see p. 98) and data upon 2-vinylpyridine **79** and styrene **80** (Varian Associates, 1962; Tobey, 1969) shows that 2-pyridyl has the greater influence upon the proton *cis* to the aryl group. The lower field position of the vinylic signal of triprolidine (δ 6·63, oxalate in D_2O) compared with

ν_H δ 6·21

79

δ 5·61 in CDCl$_3$

80

that of its isomer (δ 6·37) thus corroborates the original assignments (Ison, 1970).

Antihistaminic 4-aminobutenes are obtained by dehydrating a 1,2-

but-1-ene

81

OH

82

but-2-ene

83

diaryl-4-aminobutan-2-ol **82** with acid. Four products may result, namely an isomeric pair of but-1-enes **81** and a pair of but-2-enes **83**. Evidence whether or not all four products form is provided by the PMR spectrum of the total product of dehydration. The spectrum of the mixture obtained from the alcohol **82** (Ar = Ph, Ar' = p-OMe.C$_6$H$_4$, R = Me), for example shows four distinct OMe and NMe signals (Fig. 6.19) and three vinylic signals, the fourth being obscured by the broad aromatic resonance. Components of the mixture are separated by fractional crystallization of the aminobutene hydrochloride salts, and progress of the purification is readily monitored by PMR spectroscopy. The but-1-enes **81** are differentiated from but-2-enes **83** by multiplicities of their vinylic signals; those of the former are broad singlets and the latter triplets (J 7 Hz). Configurational assignments rest upon predictions of the relative positions of the vinylic

trans Ar/Ar'

84

cis Ar/Ar'

85

signals of isomeric pairs. In the but-1-enes **84** and **85** the vinylic proton of the *trans* isomer should have the lower field position because it receives a deshielding contribution from both a geminal and a *cis* placed aryl group. In the but-2-enes **86** and **87** one of the vinylic protons is close (*cis*) and the

Ar'C=C H / ArCH₂ CH₂NR₂

cis Ar'/H

86

Ar'C=C CH₂NR₂ / ArCH₂ H

trans Ar'/H

87

other far removed (*trans*) from the deshielding aryl substituent; hence the vinylic resonance of the former isomer (**86**) should have the lower field position. Differences in the degrees to which aromatic and alkene planes deviate in an isomeric pair (as a result of non-bonded interactions between substituents in the planar conformation) may also contribute to differences in vinylic shifts, but this factor is likely to be of only secondary importance.

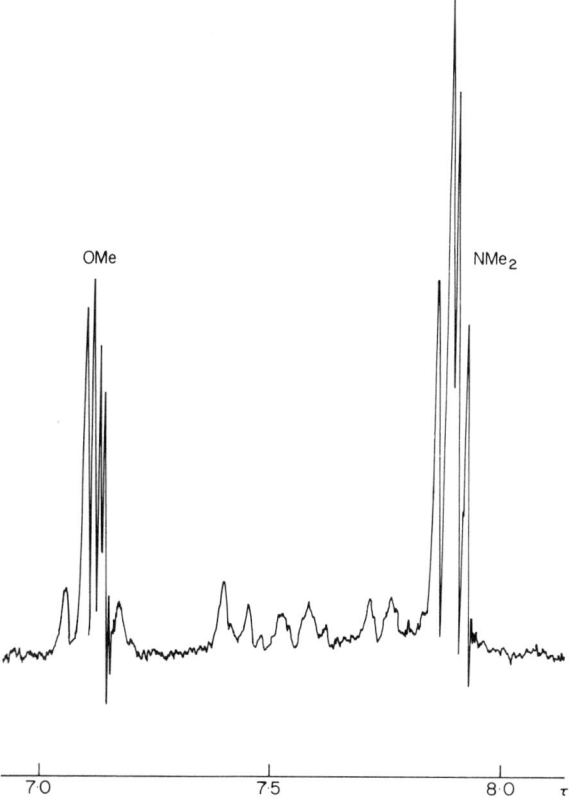

FIG. 6.19. Part of the 60 MHz PMR spectrum of the total base obtained from the dehydration of the tertiary alcohol **82** (Ar = Ph, Ar′ = p-OMe.C_6H_4, R = Me). The four lower field singlets are due to OMe and the higher field group to NMe. The solvent is $CDCl_3$.

A typical set of data and assignments are given below:

	Isomer	Vinylic chemical shift
Ph $\diagdown$ $\quad$C=CArCH$_2$CH$_2$NMe$_2$ $\diagup$ H	*trans* Ph/Ar *cis* Ph/Ar	δ 6·85 singlet δ 6·62 singlet
Ar $\diagdown$ $\quad$C=CHCH$_2$NMe$_2$ $\diagup$ $PhCH_2$	*trans* Ar/H *cis* Ar/H	δ 5·88 triplet *J* 7 Hz δ 6·17 triplet *J* 7 Hz

* Hydrochloride in CDCl$_3$, Ar=p-OMe.C$_6$H$_4$

(Casy and Parulkar, 1969a)

A more precise PMR approach to the problem of configurational assignments to isomeric alkenes is application of the principle of additivity of shielding effects (Chapter 3, p. 97). This is illustrated for the case of the isomeric but-2-enes **86** and **87** (Ar = Ar' = Ph, R = Me). Shielding values (σ) for Ph are known (Tobey, 1969) and those for CH$_2$Ph and protonated CH$_2$NMe$_2$ are found as follows. Comparison of the H$_A$·and H$_B$ chemical

$$Ph\diagdown \quad \diagup H_A-5\cdot45 \qquad\qquad Ph\diagdown \quad \diagup H_A-5\cdot62$$
$$\qquad C=C \qquad\qquad\qquad\qquad C=C$$
$$PhCH_2\diagup \quad \diagdown H_B-4\cdot49 \qquad\qquad H\diagup \quad \diagdown H_B-5\cdot14$$

$$\textbf{88} \qquad\qquad\qquad\qquad\qquad \textbf{89}$$

(chemical shifts in terms of δ in CDCl$_3$)

shifts in 1,2-diphenylprop-2-ene **88** with those of styrene **89** provide the *cis* and *trans* σ values for benzyl *positioned geminal to a phenyl group*. Thus when H-geminal to Ph in styrene is replaced by CH$_2$Ph, H$_B$ moves upfield by $-4\cdot99$ to $-5\cdot14 = +0\cdot15$ ppm (σ *cis* benzyl). Similarly *trans* benzyl $= +0\cdot17$ ppm. Chemical shift comparisons between the vinylic signals of 3-dimethylamino-1-phenylprop-1-ene hydrochloride **90** and styrene provide *geminal* and *cis* σ values for CH$_2\overset{+}{N}HMe_2$. The 100 MHz spectrum

$$Ph\diagdown \quad \diagup H_B-6\cdot45 \qquad\qquad Ph\diagdown \quad \diagup H_B-5\cdot62$$
$$\qquad C=C \qquad\qquad\qquad\qquad\qquad C=C$$
$$H_A\diagup \quad \diagdown CH_2\overset{+}{N}HMe_2 \qquad\qquad H_A\diagup \quad \diagdown H$$
$$-6\cdot81 \qquad\qquad\qquad\qquad\qquad -6\cdot65$$

$$\textbf{90} \qquad\qquad\qquad\qquad\qquad styrene$$

(σ *cis* $-0\cdot16$, σ *gem* $-0\cdot83$)

of the aminopropene (**90**) shows 7 lines in the vinylic region; the lowest field doublet is unchanged when the CH$_2$N signal is irradiated and is therefore assigned to H$_A$. The same irradiation also resolves the doublet of doublet AB pattern of the H$_A$ and H$_B$ protons which can then be analysed and leads to the chemical shifts recorded in **90**. The values σ *cis* Ph $-0\cdot39$

and σ *trans* Ph $+0.06$ are known (Tobey, 1969), and the vinylic signals of the isomers **86** and **87** (Ar $=$ Ar' $=$ Ph, R $=$ Me) may now be calculated.

$$\underset{PhCH_2}{\overset{Ph}{\diagdown}}C=C\underset{\overset{|}{H}}{\overset{H}{\diagup}}\overset{+}{CH_2NMe_2}$$

$$\delta_{C=CH} = -5.27 + \sigma \ cis \ Ph + \sigma \ trans \ CH_2Ph$$
$$+ \ \sigma \ gem \ CH_2\overset{+}{N}HMe_2$$
$$= -5.27 + (-0.39) + (+0.17) + (-0.83)$$
$$= -6.32$$

$$\underset{PhCH_2}{\overset{Ph}{\diagdown}}C=C\underset{\overset{|}{H}\ \overset{|}{H}}{\overset{\overset{+}{CH_2NMe_2}}{\diagup}}$$

$$\delta_{C=CH} = -5.27 + \sigma \ trans \ Ph + \sigma \ cis \ CH_2Ph$$
$$+ \ \sigma \ gem \ CH_2\overset{+}{N}HMe_2$$
$$= -5.27 + (+0.06) + (0.15) + (-0.83)$$
$$= -5.89$$

Observed vinylic chemical shifts are -6.30 for one isomer and -5.97 ppm for the other. Hence the former has a *cis* Ph/H and the latter a *trans* configuration, confirming previous assignments. The poorer agreement obtained for calculated and observed vinylic chemical shifts in the *trans* case is probably due to steric interactions between the adjacent phenyl and $CH_2\overset{+}{N}HMe_2$ groups invalidating to some extent the use of σ values derived from compounds in which these substituents have unhindered environments (Ison and Casy, 1971).

A series of isomeric aminobutenes have been characterized by PMR in the manner described and assayed for their ability to antagonize histamine-induced contractions of guinea-pig ileum. The most potent members all proved to be *cis* Ph/H but-2-enes and the significance of this result has been discussed in terms of uptake of antihistamine receptors (Casy and Ison, 1970).

References

Adamson, D. W., Barrett, P. A., Billinghurst, J. W., Green, A. F. and Jones, T. S. G. (1951). *Nature, Lond.* **168**, 204.

Becker, E. D. (1965). *J. Chem. Educ.* **42**, 591.

Bentley, K. W. and Hardy, D. G. (1963). *Proc. Chem. Soc.* 220.

Bible, R. H. (1965). "Interpretation of NMR Spectra". Plenum Press, New York.

Booth, H. (1965). *Tetrahedron Lett.* 411.

Booth, H. (1966). *Tetrahedron* **22**, 615.

Booth, H. and Gidley, G. C. (1965). *Tetrahedron* **21**, 3429.

Bovey, F. A. (1969). "Nuclear Magnetic Resonance Spectroscopy". Academic Press, New York.

Casy, A. F. (1966a). *Tetrahedron* **22**, 2711.

Casy, A. F. (1966b). *J. Chem. Soc.* (B), 1157.

Casy, A. F. (1968). *J. Med. Chem.* **11**, 188.

Casy, A. F. and Hassan, M. M. A. (1967). *Tetrahedron* **23**, 4075.

Casy, A. F. and Ison, R. R. (1970). *J. Pharm. Pharmacol.* **22**, 270.

Casy, A. F. and Parulkar, A. P. (1969a). *Can. J. Chem.* **47**, 423.

Casy, A. F. and Parulkar, A. P. (1969b). *Can. J. Chem.* **47**, 3623.

Casy, A. F., Chatten, L. G. and Khullar, K. K. (1969). *J. Chem. Soc.* (C), 2491.

Casy, A. F., Hassan, M. M. A. and Wu, E. C. (1971). *J. Pharm. Sci.* **60**, 67.

Chapman, O. L. and King, R. W. (1964). *J. Amer. Chem. Soc.* **86**, 1256.

Chotia, C. and Pauling, P. (1969). *Chem. Commun.* 626, 746.

Culvenor, C. C. J. and Ham, N. S. (1966). *Chem. Commun.* 537.

Culvenor, C. C. J. and Ham, N. S. (1970). *Chem. Commun.* 1242.

Cushley, R. J. and Mautner, H. G. (1970). *Tetrahedron,* **26**, 2151.

Durant, G. J., Foottit, M. E., Ganellin, C. R., Loynes, J. M., Pepper, E. S. and Roe, A. M. (1968). *Chem. Commun.* 108.

Eddy, N. B. and May, E. L. (1966). "Synthetic Analgesics Part IIB, 6,7-Benzomorphans". Pergamon, Oxford.

Eddy, N. B., May, E. L. and Mosettig, E. (1952). *J. Org. Chem.* **17**, 321.

Emsley, J. W., Feeney, J. and Sutcliffe, L. H. (1965). "High Resolution Nuclear Magnetic Resonance Spectroscopy", Vol. 1. Pergamon, Oxford.

Fedeli, W., Giacomello, G., Cerrini, S. and Vaciago, A. (1966). *Chem. Commun.* 608.

Fullerton, S. E., May, E. L. and Becker, E. D. (1962). *J. Org. Chem.* **27**, 2144.

Fulmor, W., Lancaster, J. E., Morton, G. O., Brown, J. J., Howell, C. F., Nora, C. T. and Hardy, R. A. (1967). *J. Amer. Chem. Soc.* **89**, 3322.

Garbisch, E. W. Jr. (1963). *J. Amer. Chem. Soc.* **85**, 3228.

Garbisch, E. W. Jr. (1968). *J. Chem. Educ.* **45**, 311, 402, 480.

Garrison, D. R., May, M., Ridley, H. F. and Triggle, D. J. (1969). *J. Med. Chem.* **12**, 130.

Gill, E. W. (1965). *In* "Progress in Medicinal Chemistry" (G. P. Ellis and G. B. West, eds), Vol. 4, pp. 39. Butterworths, London.

Ham, N. S. (1970). Personal communication; see also Casy, A. F., Ison, R. R. and Ham, N. S. (1970). *Chem. Commun.* 1343.

Hanson, A. W. and Ahmed, F. R. (1958). *Acta Crystallogr.* **11**, 724.

Hassan, M. M. A. (1967). Ph.D. Thesis, University of London.

Hirst, R. C. and Grant, D. M. (1964). *J. Chem. Phys.* **40**, 1909.

Inch, T. D., Chittenden, R. A. and Dean, C. (1970). *J. Pharm. Pharmacol.* **22**, 956.

Ison, R. R. (1970). Ph.D. Thesis, University of Alberta.

Ison R. R. and Casy, A. F. (1971). *J. Chem. Soc.* (B), 230.

Johnson, C. E. and Bovey, F. A. (1958). *J. Chem. Phys.* **29**, 1012.

Jones, R. G. (1949). *J. Amer. Chem. Soc.* **71**, 383.

Jones, R. G. (1966). *In* "Handbook of Experimental Pharmacology", Vol. XVIII/I, Springer-Verlag, Berlin.

Karabatsos, G. J., Sonnichsen, G. C., Hsi, N. and Fenoglio, D. J. (1967). *J. Amer. Chem. Soc.* **89**, 5067.

Kato, G. (1968). *Mol. Pharmacol.* **4**, 640.

Kawazoe, Y., Tsuda, M. and Ohnishi, M. (1967). *Chem. Pharm. Bull.* **15**, 214 and references cited therein.

Kier, L. B. (1968). *J. Med. Chem.* **11**, 441.

Kornfeld, E. C., Wolf, L., Lin, T. M. and Slater, I. H. (1968). *J. Med. Chem.* **11**, 1028.

Lister, R. E. (1964). *J. Pharm. Pharmacol.* **16**, 366.

McErlane, K. M. J. (1971). Ph.D. Thesis, University of Alberta.

Mautner, H. G. (1969). *In* "Annual Reports in Medicinal Chemistry, 1968" (C. K. Cain, ed.), pp. 230. Academic Press, New York.

Moynehan, T. M., Schofield, K., Jones, R. A. Y., and Katritzky, A. R. (1962). *J. Chem. Soc.* 2637.

Musher, J. I. and Corey, E. J. (1962). *Tetrahedron* **18**, 791.

Newkome, G. R. and Bhacca, N. S. (1969). *Chem. Commun.* 385.

Pan, Y. K. and Rogers, M. T. (1968). *Rev. Pure Appl. Chem.* **18**, 17.

Pittman, K. A. (1969). Personal communication.

Pittman, K. A., Rosi, D., Cherniak, R., Merola, A. J. and Conway, W. D. (1969). *Biochem. Pharmacol.* **18**, 1673.

Pople, J. A., Schneider, W. G. and Bernstein, H. J. (1959). "High Resolution Nuclear Magnetic Resonance". McGraw Hill, New York.

Portoghese, P. S. and Williams, D. A. (1969). *J. Hetero. Chem.* **6**, 307.

Rull, T. and Gagnaire, D. (1963). *Bull. Soc. Chim. France*, 2189.

Sachs, D. P. L., Jellum, E. and Halpern, B. (1970). *Biochem. Biophys. Acta*, **198**, 88.

Sheppard, N. and Turner, J. J. (1959). *Proc. Roy. Soc. London*, **A252**, 506.

Singh, P. and Ahmed, F. R. (1969). *Acta Crystallogr.* **B25**, 2401.

Sudmeier, J. L. (1968). *J. Phys. Chem.* **72**, 2344.

Tobey, S. W. (1969). *J. Org. Chem.* **34**, 1281.

van Gorkom, M. and Hall, G. E. (1968). *Quart. Revs.* **22**, 14.

Varian Associates (1962). High resolution NMR spectra catalog. Spectrum No. 154.

Veidis, M. V. and Palenik, G. J. (1969). *Chem. Commun.* 196.

Warren, R. J., Eisdorfer, I. B., Thompson, W. E. and Zarembo, J. E. (1966). *J. Pharm. Sci.* **55**, 144.

Whitesides, G. M., Kaplan, F., Nagarajan, K. and Roberts, J. D. (1962). *Proc. Nat. Acad. Sci. U.S.A.* **48**, 1112.

Wiberg, K. B. and Nist, B. J. (1962). "The Interpretation of NMR Spectra". Benjamin, New York.

Wu, E. C. (1970). M.Sc. Thesis, University of Alberta.

Yamaguchi, S., Okuda, S. and Nakagawa, N. (1963). *Chem. Pharm. Bull.* **11**, 1465.

Ziering, A. and Lee, J. (1947). *J. Org. Chem.* **12**, 911.

CHAPTER 7

PMR Studies of Compounds of Pharmacological Interest: Further Examples

TROPANE DERIVATIVES

There has been much interest in the PMR spectroscopy of derivatives of tropane, and an account of these studies is included here because of the relationship of this bicyclic tertiary base to the classical muscarinic antagonist, atropine.

Two studies have been directed at solving the question of whether the piperidine ring of tropine exists in a preferred chair (**1**) or boat conformation (**2**) (Chen and Le Fèvre, 1965a; Bishop *et al.*, 1966). The most valuable

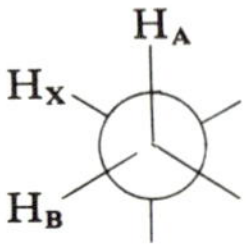

view along 4,5 bond

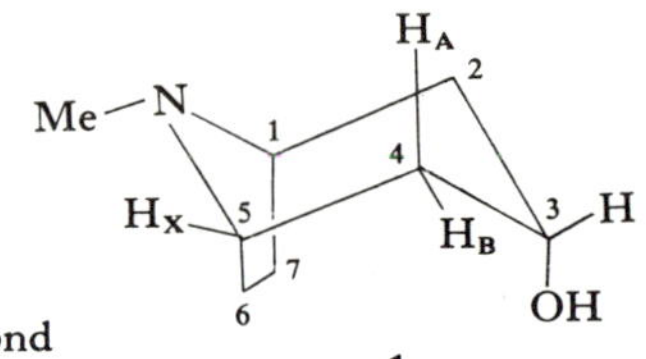

1

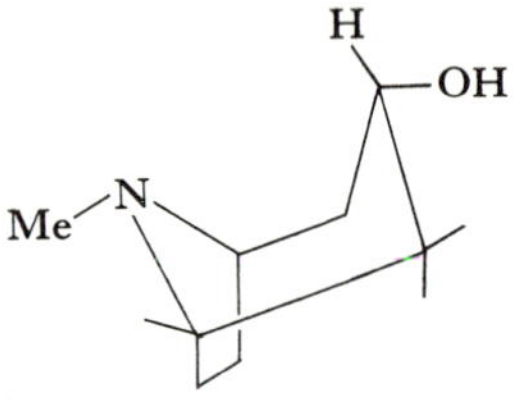

2

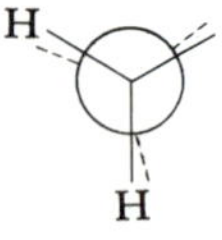

view along 4,5 bond
(for **2**)

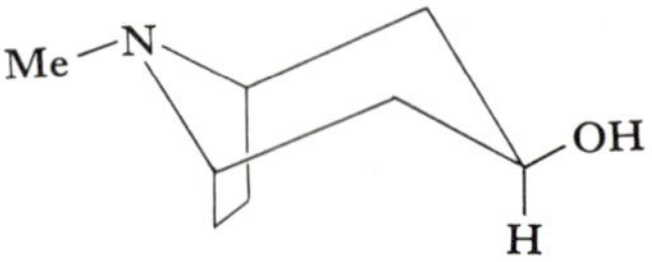

3

(β-substituents are above and α-substituents are below the piperidine ring plane)

information in this respect derives from the C-3 proton signal of tropine (Fig. 7.1). This is analysed as the X portion of two ABX systems, and the small value of its base width ($\sim$17 Hz, 60 MHz spectrum), equal to $2(J_{AX} + J_{BX})$, shows that its spin–spin coupling cannot involve a large J_{vic} value as should arise between 4H/3H *trans* protons of the boat-form **2**;

hence the 3-proton must be attached equatorially as in **1**. In contrast, the 3-methine signal of *pseudo*tropine (Fig. 7.1) is far broader (base width ~40 Hz) as expected for an axial orientation as in **3** (the actual width is evidence of ring deformation; this is discussed later). Other tropane derivatives with an axial proton at C-3 give methine signals with widths in

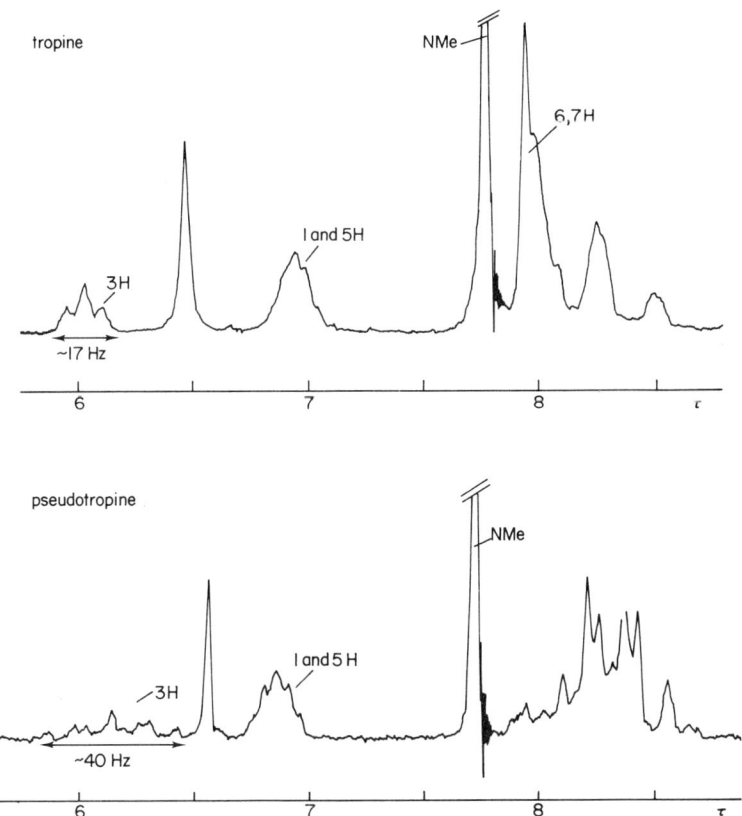

FIG. 7.1. 60 MHz PMR spectra of tropine (upper) and *pseudo*tropine (lower) in CDCl$_3$.

excess of 20 Hz since $J_{a/a}$ is usually close to 10 Hz (e.g. that of the C-3 proton of 3β-bromotropane is 34 Hz). The boat **2** is further excluded by the half width (9.5 Hz) and base width (18 Hz) of the C-1 and C-5 proton signal (Fig. 7.1). These magnitudes are consistent with the chair **1** in which four small vicinal couplings operate; in the boat, however, one "eclipsed" coupling (dihedral angle near zero) would contribute to give a signal broader than that observed. The C-1, C-5 proton signal of *pseudo*tropine has similar dimensions. The resonance signal of the bimethylene bridge protons at the 6- and 7-positions is downfield relative to the same absorption of *pseudo*tropine. This deshielding, due to the close proximity of the 3-α-hydroxy substituent (see below), is possible only in the chair **1**. The same effect is seen in related 2-β, 3-α-dibromo and 2-β-bromo, 3-α-hydroxy derivatives (Supple *et al.*, 1969).

β-4   α-4

Support for a chair tropine configuration is also provided by the near-coincidence of the tropine C-3 methine signal and the 4-proton signal of β-1-methyl-2,6-diphenylpiperidin-4-ol (**4**).† The conformations of this piperidinol and the corresponding α-isomer (*e*-4-OH) are based upon differing 4-H proton signal features (α-nonet *J* 5·1 and 11·1 Hz; β-broad triplet base width 16·4 Hz) and the lower field position (by 26 Hz) of the 2,6, proton signal in the β-isomer (Chen and Le Fèvre, 1965b). An axial 4-hydroxyl group is known to deshield axial 2,6 protons in cyclohexane derivatives (Carr and Huitric, 1964), hence the last result supports a 1,3 diaxial arrangement of these features in the β-isomer.

The nature of the C-3 methine signals of certain tropane derivatives confirms their configurations and establishes the stereochemistry of reactions employed in their syntheses (see Scheme I).

The question of the stereochemical course of tropane quaternizations has aroused much interest and controversy (Brown *et al.*, 1967; Thut and Bottini, 1968). The selectivity of such reactions is well demonstrated by some results of Fodor *et al.* (1968) on the methylation and trideuteromethyl-ation of tropane derivatives. Thus the 60 MHz PMR spectrum of tropine methiodide in DMSO-d_6 displays equi-intense singlets at δ 3·09 and 3·16 due to the two N—Me groups (see Scheme II). When tropine is quaternized with CD_3I, the lower field N—Me signal of the mixture is distinctly more intense than that at higher field whereas the reverse is true when N-CD_3-nortropine is methylated. Initially it was assumed that the higher field NMe signals were due to axial groups by analogy with simple piperidinium salts, and the data thereby held to support the preferential *axial* methylation of tropane derivatives (see Chapter 4, p. 159). Later work involving corre-lations with reference compounds whose stereochemistry was established by X-ray crystallography led to a reversal of this conclusion (Fodor *et al.*,

† The designation of **4** and its C-4 epimer is unrelated to that of tropane derivatives.

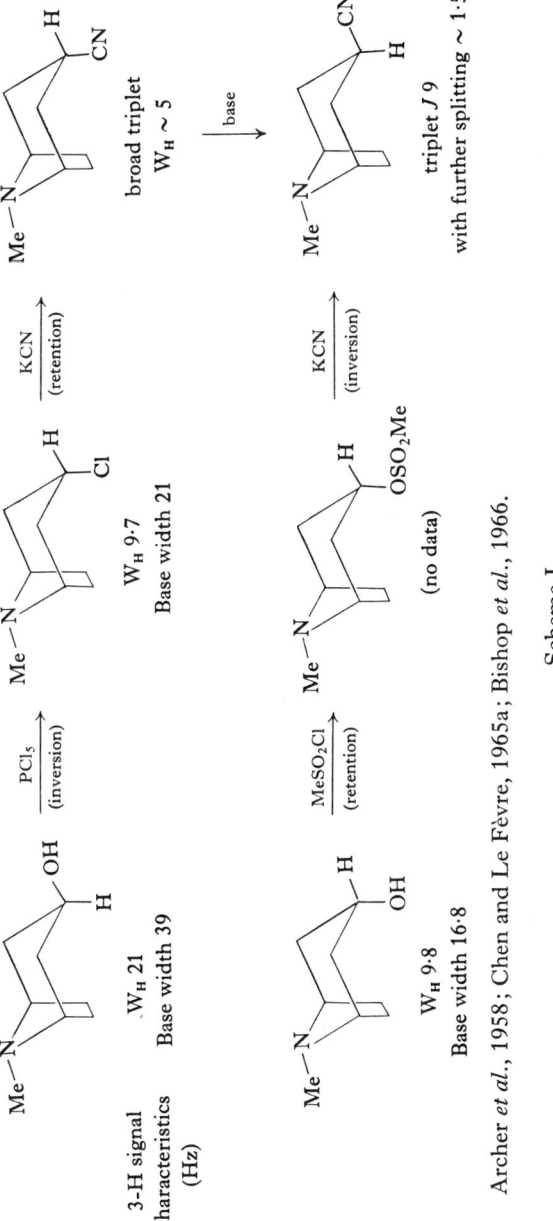

Scheme I

Archer *et al.*, 1958; Chen and Le Fèvre, 1965a; Bishop *et al.*, 1966.

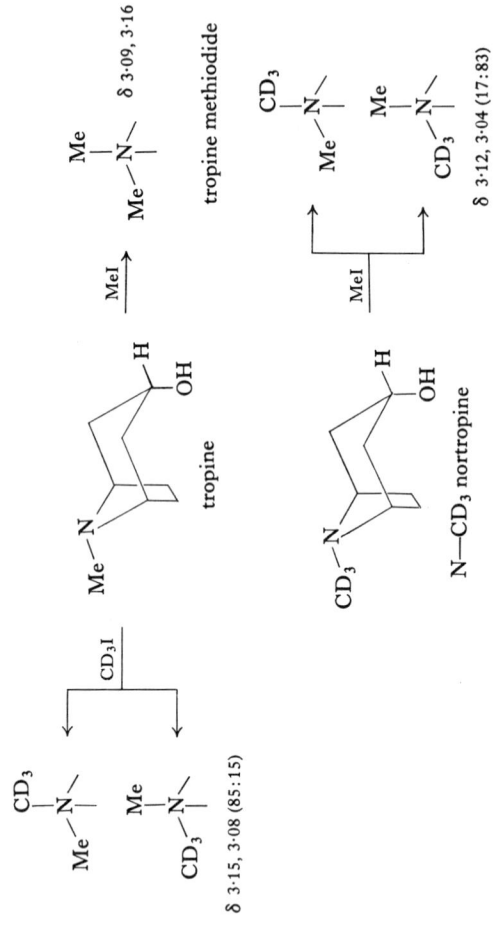

Scheme II

(Chemical shifts from 60 MHz spectra, DMSO-d₆ solvent)

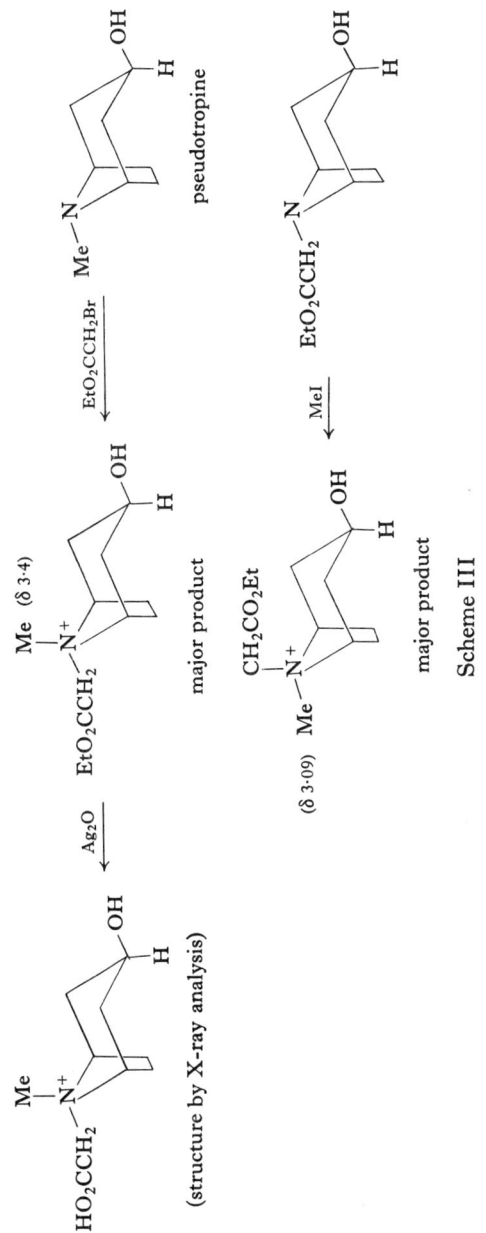

Scheme III

(Chemical shifts from 100 MHz spectra, D₂O solvent)

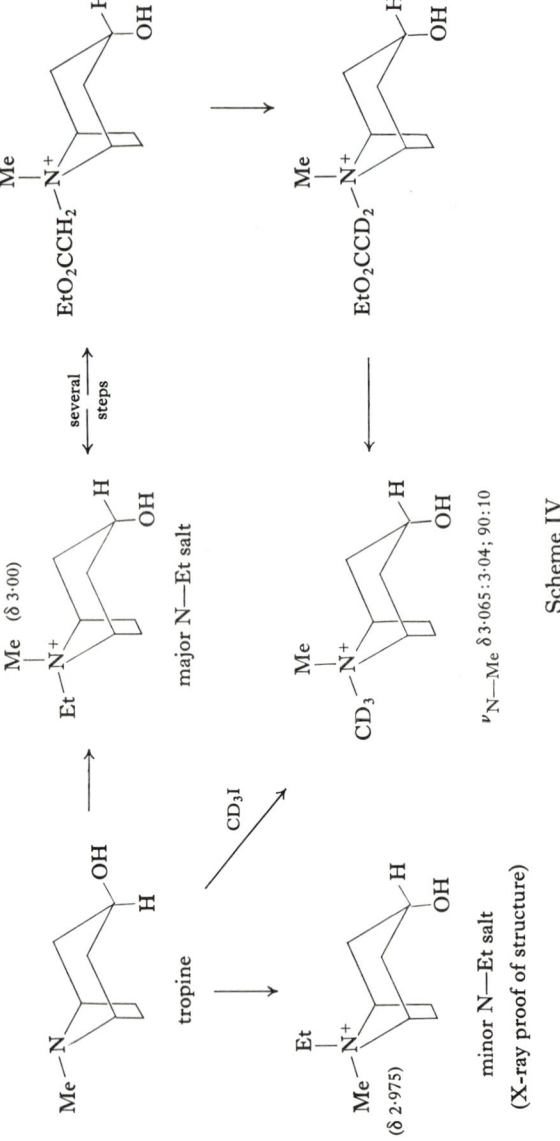

Scheme IV

(Chemical shifts from 100 MHz spectra, D₂O solvent)

1971). Thus the major product of treating *pseudo*tropine with ethoxy-carbonylmethyl bromide (EtO_2CCH_2Br) must be the product of *equatorial* attack because its hydrolysis product has been shown to have an axial N—Me substituent by X-ray analysis (see Scheme III). The N—Me resonance signal of this ethoxycarbonylmethyl derivative thus arises from an axial group and proves to be *lower* field than that of equatorial N—Me in the isomeric salt from ethoxycarbonylmethylnor*pseudo*tropine and methyl iodide (details in Scheme III). In another series of transformations (Scheme IV), the major products of ethoxycarbonylmethylation and tri-deuteromethylation of tropine have been correlated with the *a*-N-Me, *e*-N-Et quaternary salt of tropine and provide further examples of N-methyldiastereoisomeric pairs in which the *a*-N-Me resonance has the lower field position. In these examples the two N—Me signals are only just resolvable since they differ by less than 2 Hz at 100 MHz in D_2O.

It will be recalled that the axial N—Me resonance is lower field than the equatorial signal in piperidinium salts when an axial 2-methyl substituent is present (Chapter 4, p. 157). The 2,6-dimethylene bridge of tropane is analogous to a pair of α-substituents of this type and herein may lie one of the reasons for the axial N—Me signal of tropane methosalts taking the lower field position. In this connection it is of interest that the N—Me signal of the minor epimeric conjugate acid of a tropane derivative falls to

minor conjugate acid epimers

lower field of the major N—Me signal, whereas that of a simple piperidine derivative such as 1,2-dimethylpiperidine, falls to *higher* field of the principal signal (cf. Chapter 4, p. 146).

Fodor considers that the reason why equatorial quaternization is preferred in tropane derivatives is the fact that distortion of the molecule by the bimethylene bridge flattens the piperidine ring and brings the 2,4 axial hydrogen atoms closer to nitrogen than is normal in piperidine derivatives; in consequence, axial approach of an alkylating agent is opposed by un-usually large diaxial interactions with the 2,4-β-hydrogens.

N-oxide formation in atropine, tropine and related compounds is stereospecific as shown by the appearance of a single N-methyl signal in spectra of the total oxides (Mandava and Fodor, 1968). The chemical shift of the signal is close to the lower field signal of corresponding methiodides,

5   6

R = COCH(CH₂OH)Ph

now known to arise from the axial N—Me group; hence the oxides probably have the configuration **5**. Assignment of the two N—Me signals in the spectrum of scopolamine methiodide (**6**) is not readily made since the equatorial substituent is now close to the deshielding β-epoxide function (both N—Me signals of this salt are lower field than those of atropine methiodide). It is clear from these accounts of piperidine and tropane quaternizations that decisions upon the relative resonance positions of signals due to *e* and *a* N—Me groups in quaternary salts of cyclic bases are not easily made and conclusions reached must be tentative unless backed by data from reference compounds.

The bulky tropic acid residue of the pharmacologically active derivatives of tropine, atropine and scopolamine, does not radically alter the preference of the piperidine ring for a chair conformation. This is clear from the appearance of the C-3 methine resonances of these alkaloids which resemble that of tropine rather than *pseudo*tropine; thus the 3-methine signals in D_2O of atropine sulphate and tropyl acetate hydrochloride are narrow multiplets (base width about 16 Hz) while that of *pseudo*tropyl acetate is a broad multiplet (base width about 40 Hz) (Casy and Jeffery, unpublished results).

Detailed analyses of 3-axially-substituted tropane derivatives (Mandava and Fodor, 1968; Supple *et al.*, 1969) show that the extent of $H_B - H_X$ coupling (J_{BX} 0·4–1·1 Hz) is distinctly less than that of the H_A and H_X protons (J_{AX} 3·5–4·8 Hz), see **7**. These results are evidence that the piperidine

7   8

chair conformation in tropine is deformed in such a way that the H_A/H_X dihedral angle is reduced, and the H_B/H_X angle increased, as shown in **7** and **8**; examination of the ϕ/J plot of Karplus (p. 89) shows that the J_{vic}

value falls when the fully staggered dihedral angle of 60° increases as shown. There is evidence that ring deformations also occur in 3-β-substituted tropane derivatives, and this topic is discussed further below.

COCAINE AND ITS ISOMERS

A description of the PMR features of cocaine appropriately follows the last section since the tropane skeleton is common to both coca and solanaceous alkaloids.

A Dutch group (Sinnema *et al.*, 1968) have reported the PMR spectra in $CDCl_3$ of cocaine (**9**), its three geometrical isomers and other tropane derivatives. The magnitudes of spin–spin couplings involving protons at C-2, C-3 and C-4 confirm the configuration of cocaine and *pseudo*cocaine (the latter is prepared from *pseudo*ecgonine, the epimerization product of the natural alkaloid) and provide evidence of preferred conformation. The larger value of $J_{2,3}$ in *pseudo*cocaine is consistent with the configuration

cocaine **9**

$J_{2,3}$ 6·0 Hz (e/a)
$J_{3,4}$ 11·6 Hz (a/a)
        6·0 Hz (e/a)

pseudo-cocaine **10**

$J_{2,3}$ 10·4 Hz (a/a)
$J_{3,4}$ 10·4 Hz (a/a)
        6·8 Hz (a/e)

10 in which the C-2 proton is axial to an axial proton at C-3. The 3-methine signal of both isomers is similar to that of *pseudo*tropine and *pseudo*tropine benzoate in displaying the large vicinal coupling expected for the axial/axial proton orientation which obtains in all these derivatives. The magnitude of these $J_{a/a}$ couplings are, however, smaller while $J_{a/e}$ values are larger than those usually obtained in 6-membered cyclic systems. The values $J_{a/a}$ 9·8 and $J_{a/e}$ 7·0 Hz for *pseudo*tropine, for example, stand in contrast to the values $J_{a/a}$ 11·1 and $J_{a/e}$ 5·1 Hz obtained for the axial 4-methine proton of the α-epimer of **4**. These differences provide further evidence for the distortion of the piperidine ring of tropane derivatives. Models show that linkage of the 2,6 positions of piperidine by a bimethylene bridge causes the 2,6 axial bonds to move together and the 3,5 axial bonds to diverge, while nitrogen moves away from and C-4 moves nearer to the mean plane of the ring, as shown in **11**. The last two changes are substantiated by

11   12

the X-ray analysis of (−)-cocaine (Gabe and Barnes, 1963) and *pseudo*-tropine (Schenk *et al.*, 1967). Ring flattening in this manner will also cause both the H_B/H_X dihedral angle (60° in perfect chair) and H_A/H_X angle (180° ideal value) of the fragment 12 to decrease, hence J_{BX} should rise and J_{AX} fall in consequence of the Karplus $\cos^2\phi/J$ relationship.

The C-3 proton resonance of cocaine hydrochloride in deuterium oxide appears as a broad quartet while the C-2 proton signal is a doublet of doublets partially obscured by a methyl singlet (Fig. 7.2) (Casy, unpublished results). Analysis of these signals gives the values $J_{2,3}$ 7·8, and $J_{3,4}$ 10·0 and 8·8 Hz. Comparison of these coupling magnitudes with the corresponding ones for cocaine base in $CDCl_3$ shows that ring distortion is even more pronounced in cocaine (protonated) as solute in deuterium oxide and presumably also in water under physiological conditions.

Coupling constant data for allococaine and allo*pseudo*cocaine (and related free alcohols) show the 3-oxygenated functions to be axial in all derivatives as in tropine and tropine benzoate. Thus, the two $J_{3/4}$ values are both small and typical of a/e and e/e vicinal coupling (see p. 88). The complete configurational assignment of isomeric pairs in the allo-*pseudo*allo series rests

13a   13b

$J_{3,4}$ 1-2 Hz (e/e) and 5 Hz (a/e) for both

upon the interpretation of the $J_{2,3}$ values; these are 1–2 Hz for allo and 5 Hz for allo*pseudo* isomers. If the larger value is due to an a/e coupling, then allo*pseudo*cocaine must be 13b. This assumption has been made by the Dutch workers and is supported by the identical magnitude of $J_{3/4}$ a/e values and by the fact that the generally higher value of $J_{a/e}$ compared with $J_{e/e}$ in 6-membered rings (p. 91) is augmented in tropane derivatives by ring deformation effects. Electronegativity factors probably have similar influences upon coupling magnitudes since one of the coupling protons is

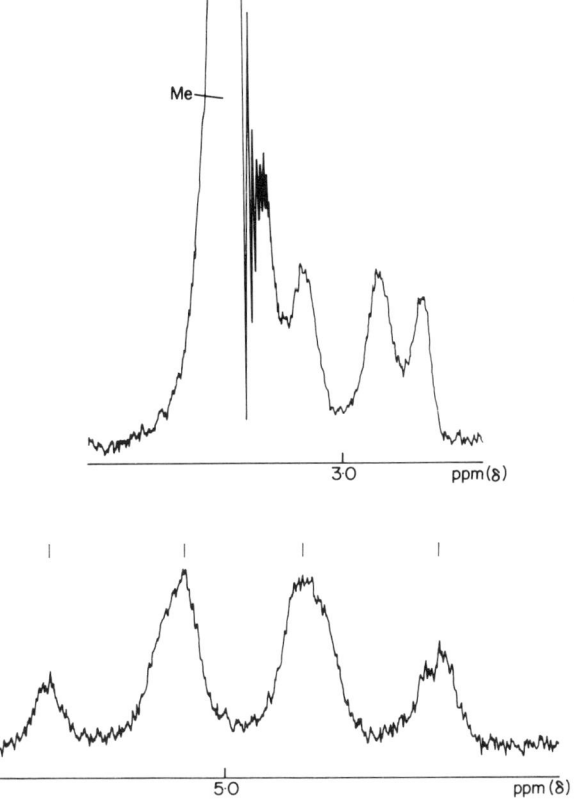

FIG. 7.2. The H-2 (upper) and H-3 (lower) 60 MHz PMR signals of cocaine hydrochloride in D₂O. Both signals were recorded at a sweep width of 100 MHz.

anticoplanar with an electronegative atom in both isomers, namely *e*-2-H/N in **13**a, and *a*-2-H/O in **13**b.

The configurations of β-eucaine (**14**) and β-isoeucaine (**15**) have similarly been assigned on the basis of PMR data (Perks and Russell, 1967); the J_{trans}/J_{gauche} ratio for **14** (~2·3) is higher than the value found in cocaine,

14 β-eucaine   **15** β-isoeucaine

C-4 signal: nonet $J \sim 5$ and 11·4 Hz; quintet $J \sim 3$ Hz hence the piperidine ring in this compound probably has a near-normal chair conformation.

EPHEDRINES AND RELATED COMPOUNDS

Several workers have reported that J_{AB}, measured from the HCOH-proton doublet, for ephedrine **16** is smaller than J_{AB} for *pseudo*ephedrine **17**

Ph	Ph
H_A——OH	H_A——OH
H_B——NHMe	MeHN——H_B
Me	Me
erythro	*threo*
16	**17**

(Hyne, 1961; Lyle and Keefer, 1966). Hyne used his data to support preferred off-staggered conformations on the basis of the Karplus relationship and hydrogen bonding data. Portoghese (1967) has extended these studies to protonated species (of greater pharmacological significance) and attempted to allow for the effect of substituent electronegativity upon J_{vic} values by the use of model compounds possessing fewer degrees of rotational freedom. Vicinal coupling constants for hydrochloride salts of *cis*

18	**19**
cis Me/Ph	*trans* Me/Ph

Solvent	J_{AB} Hz		Δ Hz
TFA	2·56	10·06	7·80
DMSO	2·47	10·02	7·55
D$_2$O	2·78	10·52	7·74
CDCl$_3$	2·67	10·36	7·69
	Ephedrine	*Pseudo*ephedrine	
	HCl	HCl	
TFA	3·74	9·6	5·86
DMSO	2·43	9·42	6·99
D$_2$O	3·63	9·21	5·58

and *trans* 3-methyl-2-phenylmorpholine were taken to represent "pure" J_{ae} and J_{aa} values since these salts are estimated to contain about 98% of conformations **18** and **19** respectively. The relative J_{AB} values of ephedrine hydrochloride and its isomer show that the rotamers **20** and **21** preponderate in the former since both involve small (*gauche*) J_{AB} couplings, and that **23** ($J_{a/a}$ involved) has the highest population in the latter. These conformers all appear to be stabilized by intramolecular hydrogen bonding of the type $^+$NH—O while **23** is also preferred on steric grounds. It is notable that the order of the J_{vic} difference between the two isomers is in complete accord with their independently assigned configurations. The higher J_{AB} values for ephedrine compared with those of the *cis* model **18** (in solvents TFA and D_2O) are probably a result of contributions of rotamer **22** which

ephedrine

H H．NH₂Me Ph OH Me **20**	H Me H Ph OH MeNH₂ **21**	H MeNH₂ Me Ph OH H **22**

pseudoephedrine

H Me NH₂Me Ph OH H **23**	H H Me Ph OH MeNH₂ **24**	H MeNH₂ H Ph OH Me **25**

involves a J_{trans} coupling; contributions from conformers **24** and **25** in the case of *pseudo*ephedrine will lower the average J_{AB} values (both involve a J_{gauche} coupling) which are, in fact, smaller than J_{AB} values for the *trans* model **19**. The ephedrine conformer **20** appears to be particularly favoured in DMSO because the J_{AB} value in this solvent is almost identical with the corresponding value of the *cis* morpholine derivative **18**; one reason advanced for this is that the bulk of the $^+$NH₂Me group is increased by hydrogen bonding to the solvent and this raises the energy of conformers **21** and **22** more than that of **20**.

The chemical shift of the benzylic proton in the *trans* morpholine hydrochloride **19** is higher field by 30–40 Hz at 60 MHz than that of the *cis* isomer; this difference is of the expected order on the basis of the chair conformations **18** and **19** because the benzylic proton will be shielded by the adjacent methyl group in the *trans* but deshielded in the *cis* isomer (Booth, 1966). A similar relationship exists for the H_A resonance of ephedrine and *pseudo*ephedrine salts (these have chemical shifts, δ 4·77 and 4·68 respectively in D_2O) and provides further support for the preferred conformations **20** and **23**. Portoghese concludes his study by speculating on the possible significance of his conformational deductions respecting the physiological activity of ephedrine and related compounds at the adrenergic receptor (Belleau, 1966). He suggests, for example, that the *gauche* OH/$\overset{+}{N}H_2Me$ orientation of the preferred ephedrine conformation may signify that ion-pair formation and hydrogen bonding to the receptor via the hydroxylic proton occur on the same side of the molecule. The lack of

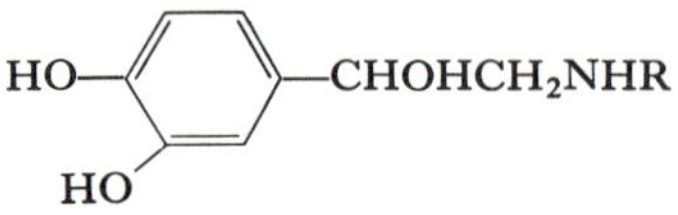

$$HO\text{—}\langle\rangle\text{—}CHOHCH_2NHR$$
$$HO$$

26

a: R = Me
b: R = *iso*Pr

conformational preference in adrenaline (**26**a) and isoprenaline (**26**b), as solutes in D_2O, provides no positive support for this view. In these catecholamines, the methylene group is adjacent to an asymmetric centre and the preferred nature of a particular conformer should be revealed by chemical shift differences between the methylene protons and the complexity of their PMR signal (see p. 208). The methylene resonances in the spectra of salts of **26**a and **26**b, in fact, appear as doublets ($J \sim 6·5$ Hz) indicative of the absence of conformational preferences (Fig. 7.3). In these spectra the methine signal is obscured by the HDO band but two lines of the anticipated triplet may be resolved.

Treatment of N,N-dimethylephedrine iodide with potassium *t*-butoxide gives *trans* 2-methyl-3-phenyloxirane while the same treatment of the corresponding *pseudo*ephedrine derivative gives the *cis* isomer (Lyle and Keefer, 1966). Both reactions proceed with inversion, as shown below for the *threo* isomer **27** (Witkop and Foltz, 1957). Vicinal coupling constants

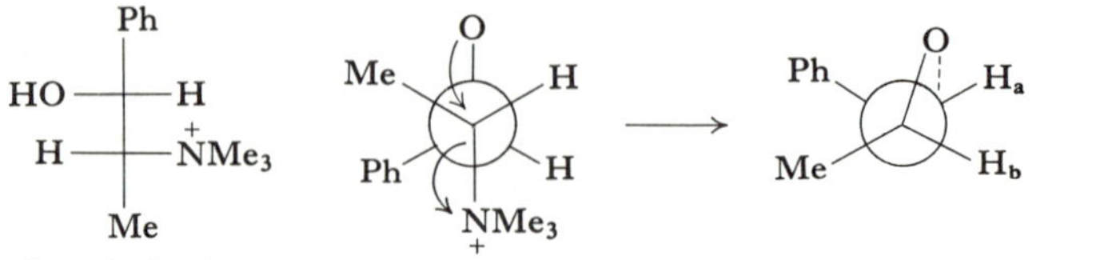

Ph		
HO ——— H		
H ——— $\overset{+}{N}Me_3$		
Me		

threo derivative

27

cis epoxide J_{ab} 4·3 Hz
trans epoxide J_{ab} 2·0 Hz

(J_{AB}) are small for both isomeric oxiranes (probably due in large part to the electronegative influence of ring oxygen) with J_{cis} the larger, and it appears generally true that the J_{AB} values of *cis* isomers are significantly greater than those of the isomeric *trans* epoxides (Lyle and Keefer, 1966). This result is claimed to be consistent with the Karplus $\cos^2 \phi / J_{vic}$ relationship but is difficult to verify from models.

Since the stereochemistry of epoxide formation from β-aminoethanols is known, knowledge of the configuration of the product epoxide, as gained from coupling constant data, allows a configurational assignment to be made to the starting material. Lyle and Keefer (1967) have applied these

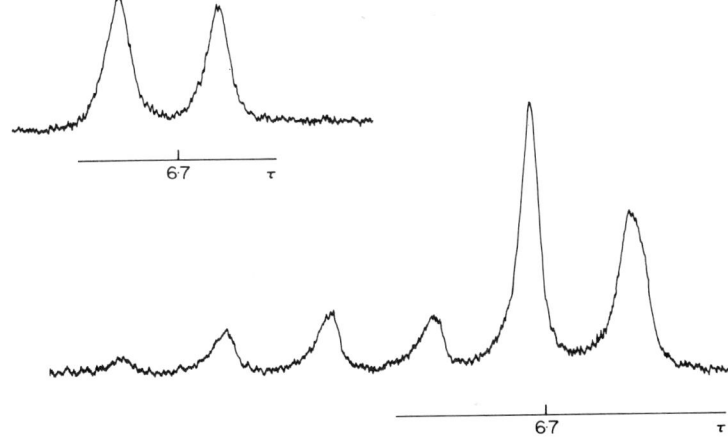

Fig. 7.3. 60 MHz PMR methylene signals of adrenaline hydrogen tartrate (upper) and *iso*prenaline hydrogen sulphate (lower) in D_2O. In the lower spectrum, the methylene doublet partially overlaps the isopropyl methine ($\underline{H}CMe_2$) signal. Both spectra were recorded at a sweep width of 100 Hz.

principles to solving the C-9 stereochemistry of cinchona alkaloids. Quinidine and epiquinidine, epimers about C-9 in **28**, were converted to the 10,11 dihydro derivatives, quaternized with benzyl chloride, and the salts heated with potassium *t*-butoxide. The PMR spectrum of the product

derived from the quinidine isomer showed a narrow one proton doublet
(J 2 Hz) at δ 4·11, low field of the methoxy singlet, which was assigned to
the epoxide proton adjacent to the quinoline ring. The spectrum of the
product from epiquinidine showed a similarly placed doublet but with a
J value of 4 Hz. The epoxide with the larger vicinal coupling constant was
given the *cis* configuration, hence the precursor aminoethanol must have
the relative configuration **30**; the quinidine derivative, which yields a

30

partial structure viewed along the C-9, C-8 bond
Q = quinoline substituent

trans epoxide must therefore have a configuration epimeric to this. When
the reaction was repeated using the benzochloride of dihydroquinidine-9d
(obtained by reducing quinidinone, **28** 9-CHOH replaced by $>$C=O,
with lithium aluminium deuteride), the spectrum of the product showed no
doublet near δ 4; this result gave proof that the oxirane proton was responsible for the δ 4·11 signal in the quinidine derived product itself.

ANTIBIOTICS

A few applications of PMR spectroscopy to investigation of the stereochemistry and other aspects of antibiotics are outlined below.

Jardetzky (1963) has analysed the PMR spectrum of chloramphenicol

$J_{AB} = 10\text{--}11$ Hz

$J_{AM} = J_{BM} = 5\text{--}6$ Hz

$J_{MX} = 2\text{--}3$ Hz (in D_2O)

threo **31**

(**31**) and its *erythro* isomer. The position of the OH resonance lines of both
diastereoisomers in acetone are at low field compared with those of the
OH of simple monohydric alcohols measured at infinite dilution in CCl_4,
and this fact indicates that the OH groups of these compounds are hydrogen
bonded (see Chapter 1). This bonding is considered to be of an intra- rather

than intermolecular nature because of the comparative concentration independence of the OH chemical shifts, while bonding to the solvent is thought unlikely on account of evidence that inter- and intramolecular hydrogen bonds are stronger than those formed between hydroxyl groups and acetone. Assuming intramolecular hydrogen bonding to be significant also in water, observed vicinal coupling constants for chloramphenicol support preferred conformations **32** for the C_3—C_2 and **33** for the C_1—C_2 fragments. A model of chloramphenicol incorporating these conformational

$\phi_{MA} < 60°$ $\phi_{XM} \sim 60°$

32   **33**

features bears a striking resemblance to a pyrimidine ribonucleotide (see below) in size, orientation of individual moieties and the distribution of electronegative groups and Jardetzky considers that this relationship may have a bearing upon the stereochemical features of structure-activity

Chloramphenicol   Uridine-5′-phosphate

relationships in the chloramphenicol series. Any alteration of the geometry of the propanol side chain, as in the *erythro* isomer in which conformational preferences about the C-1, C-2 bond are less as seen by the larger J_{MX} value (6 Hz), destroys the similarity to the ribose moiety.

Finishing touches to the complete stereochemical definition of streptomycin and its congeners (**34**) follows from some coupling constant data upon protons attached to carbon atoms forming part of the C—O—C linkage of the monosaccharide units of the antibiotics (these are the anomeric protons) (McGilveray and Rinehart, 1965; Claes *et al.*, 1969). The PMR spectra of streptomycin **34** (R=CHO) and dihydrostreptomycin **34** (R=CH$_2$OH) contain two signals in the anomeric proton region. The signal

streptose
fragment ⟶

anomeric protons

34 $R' = -C\begin{smallmatrix} =NH \\ \diagdown NH_2 \end{smallmatrix}$

from that of the N-methyl-L-glucosamine fragment is differentiated by its downfield shift (about 20 Hz at 60 MHz) in corresponding sulphates, the deshielding influence of the methylamino substituent increasing on protonation; chemical shift values of the streptose anomeric proton are almost the same in free base and sulphate forms. The coupling constant of the glucosamine proton ($J \sim 3$ Hz) corresponds clearly to an axial-equatorial H-1, H-2 relationship and thus to the α-L-configuration. The anomeric proton signal of streptose occurs as a broad singlet ($J \sim 1$ Hz), the small extent of coupling showing that the C-1 and C-2 protons of the 5-membered furanoside ring are *trans* to one another (see p. 355), hence streptose also has an α-L-configuration.

Griseofulvin (**35**), like chloramphenicol, is a member of a stereochemical quartet of isomers only one of which has antifungal properties (Grove, 1963) and knowledge of its molecular shape may advance our understanding of structure-activity relationships in this field. Levine and Hicks (1968) have

35   **36**

(partial structure showing ring C)

deduced the C-4 configuration of griseofulvol **36**, the exclusive product of reduction of griseofulvin by sodium borohydride, and the preferred ring C conformation of this secondary alcohol, from PMR data. Methylation of griseofulvol by methyl iodide and sodium methoxide in methanol was expected to occur with retention of configuration but the PMR spectrum of the ether and precursor alcohol differed markedly in the δ 1·5–3·06 region, suggesting an inversion process. This suspicion was supported by the isolation of an isomeric ether, following use of dimethylformamide, NaH and MeI, which had a similar PMR spectrum to griseofulvol. Spectral analysis of the "inverted" ether was possible by 1st order approach. The triplet of doublets at δ 2·32 ppm (1 proton) (see Fig. 7.4) must be identified with one of the C-5 methylene protons, all other assignments being ruled out by chemical shift or multiplicity considerations. Since two large (14 Hz) *J* values are required for this splitting pattern, it follows that a 1,2 *trans* diaxial coupling is present in addition to geminal coupling. The value 14 Hz is unusually large even for 1,2 *trans* diaxial coupling and *J* values of this

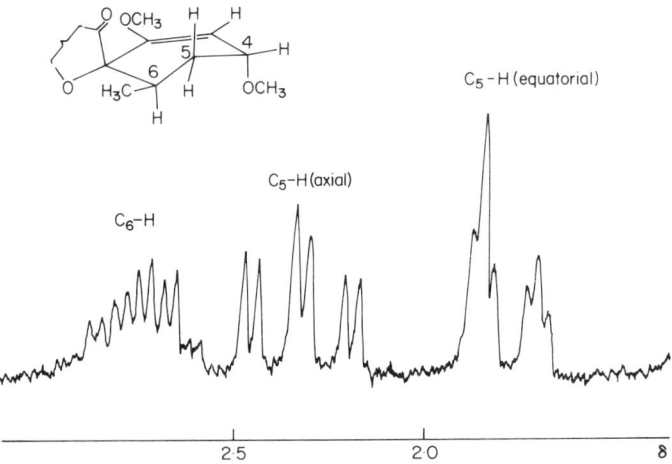

FIG. 7.4. Part of the 100 MHz spectrum of 4 α-griseofulvol methyl ether in CDCl₃. The upper drawing represents the ring C conformation of this compound (Levine and Hicks, 1968).

magnitude appear to be characteristic of spiranes of this type (De Jongh and Wynberg, 1965). The small vicinal coupling (*J* 3·5 Hz) must involve a C-4 or a C-6 equatorial proton, the former being identified as correct on account of the 5·5 Hz splitting of the vinylic proton (see below). It follows that the 14 Hz vicinal coupling is to an axial proton at C-6, a condition which requires the ring C conformation to be represented as shown in Fig. 7.4.

Dreiding models of substituted cyclohexenes such as **37** show that the 3*e*-H/2-H dihedral angle is smaller than the 3*a*-H/2-H angle, and the values $J_{2/3a}$ 2·8 and $J_{2/3e}$ 5·0 Hz are estimated on the basis of the cos²φ/*J* relationship of Karplus. Actual values obtained in this and related cases deviate somewhat from these values due to π-bond contributions (Garbisch,

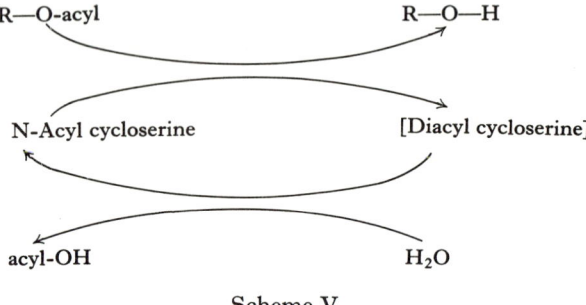

37   View along C-2, C-3 bond

1964) but it appears safe to make configurational assignments in cyclohexenes on values of this order.

Returning to the griseofulvol studies, since the alcohol is epimeric with the "inverted" methyl ether, its ring C conformation is as represented in Fig. 7.4 with the *e* C-4 H replaced by OH and the *a* C-4 OMe by H. The axial conformation of the C-4 proton in griseofulvol is confirmed by the narrower (2 Hz) splitting of the vinylic signal.

There is reason to suppose that acetyl derivatives of cycloserine **38** might assist acyl transfer reactions similar to those catalysed by imidazole (Mandell *et al.*, 1963) (see Scheme V), and that this catalytic role may have a bearing

R—O-acyl   R—O—H

N-Acyl cycloserine   [Diacyl cycloserine]

acyl-OH   H₂O

Scheme V

upon the antibiotic properties of cycloserine. A diacetyl derivative of cycloserine can, in fact, be prepared which is rapidly hydrolysed in water at pH 7·0

38   39   40   41

to the mono acetyl form **39** and this compound thus fulfills one of the requirements of Scheme V. Two structures (**40** and **41**) may be written for diacetyl
cycloserine, and the validity of **40** in which ring nitrogen rather than carbonyl
oxygen is acetylated has been established by PMR spectroscopy as follows
(Milne and Cohen, 1967).

The ring protons signal of the spectrum of diacetylcycloserine in pyridine-
d_5 may be analysed as an ABX system. Details are provided in the original
paper and the analysis provides a good example of the algebraic procedure
as opposed to the geometric method described for the comparable analysis
of spectrum of β-methylacetylcholine (p. 221). The AB portion due to
the H_A and H_B (see **40**) forms two quartets which overlap to give a 7 line
signal (Fig. 7.5). The quantities J_{AB} (repeated four times), D_+ (half the

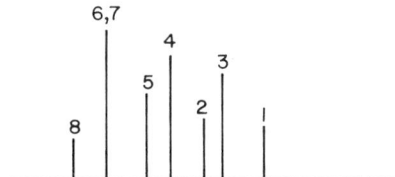

Fig. 7.5. Calculated AB portion of the ABX spectrum of diacetycycloserine.
Lines 2, 4, 6 and 8 are A lines and lines 1, 3, 5 and 7 are lines from the B proton.

separation of lines 8 and 4, or 6 and 2), and D_- (half the separation of lines
7 and 3, or 5 and 1) are measured. These yield the values

$$|\nu_A - \nu_B| = 21\cdot6 \text{ Hz}$$

and

$$|(J_{AX} - J_{BX})| = 3\cdot2 \text{ Hz} \qquad (1)$$

The separation between the centres of the A quartet and the B quartet is
$\frac{1}{2}|(J_{AX} + J_{BX})|$, thus

$$|(J_{AX} + J_{BX})| = 19\cdot5 \text{ Hz} \qquad (2)$$

Hence, from (1) and (2),

$$J_{AX} = 8\cdot1 \text{ Hz and } J_{BX} = 11\cdot4 \text{ Hz}$$

To complete the analysis, the quantity $\frac{1}{2}(\nu_A + \nu_B)$ is required. This is the
mean of the centres of the quartet sub-spectra (Bovey, 1969), and is obtained
by taking the sum of the frequencies of lines 4 and 6 (centre lines of one
quartet) and 3 and 5 (centre lines of the second quartet) and dividing by four.
It follows that

$$(\nu_A + \nu_B) = 556\cdot6 \text{ Hz}$$

but

$$(\nu_A - \nu_B) = 21 \cdot 6 \text{ Hz}$$

hence

$$\nu_A = 289 \cdot 1 \text{ Hz and } \nu_B = 267 \cdot 5 \text{ Hz.}$$

The X portion of the spectrum must be symmetrical about ν_X which is therefore 325·6 Hz. Recombining the PMR parameters by means of a digital computer gave a theoretical spectrum which corresponded with that observed.

The spectra of N-acetylcycloserine (Table 7.1, No. 1) and N-acetyl-2-methylcycloserine (No. 3) (models for the N,N-diacetate **40**), and N-

TABLE 7.1

Coupling constants and chemical shifts derived from the NMR spectra of cycloserine derivatives

(in Hz from TMS, at 60 MHz)

	Compound	J_{AB}	J_{AX}	J_{BX}	J_{XY}	ν_A	ν_B	ν_X
1.	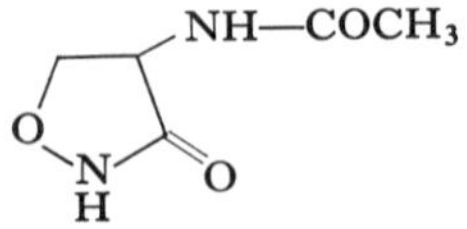 (ring structure) NH—COCH$_3$	8·6	8·5	9·9	6·1	293·1	259·8	328·7
2.	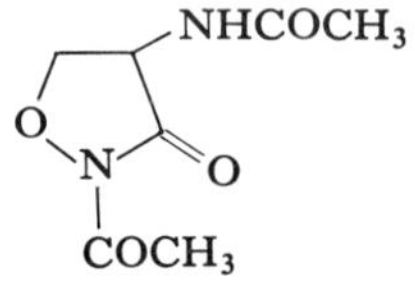 (ring structure) NHCOCH$_3$; N—COCH$_3$	8·1	8·1	11·3	7·3	289·1	267·5	325·6
3.	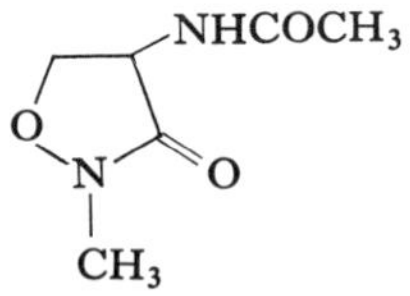 (ring structure) NHCOCH$_3$; N—CH$_3$	8·5	8·4	9·7	7·0	283·1	249·6	316·0
4.	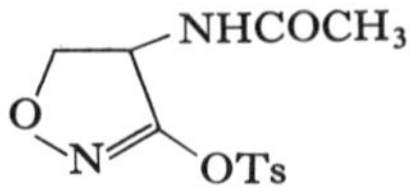 (ring structure) NHCOCH$_3$; OTs	9·8	10·1	5·9	8·4	283·8	268·3	359·2
5.	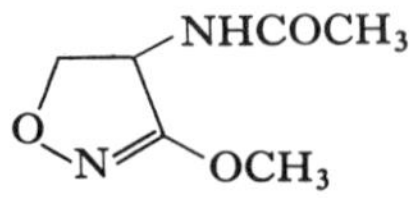 (ring structure) NHCOCH$_3$; OCH$_3$	9·4	9·3	6·4	7·8	280·2	260·1	341·2

acetyl-O-methylcycloserine and N-acetyl-O-tosylcycloserine (Nos 5 and 4) (models for the N,O-diacetate **41**) were recorded and analysed in the same way. Inspection of the seven parameters derived for each compound (Table 7.1) shows that a clear distinction can be made between Nos 1 and 3 on the one hand, and Nos 4 and 5 on the other and that parameters for diacetyl cycloserine place it in the former group. Most parameter differences between the two groups, although of value for empirical correlations, cannot be readily interpreted. The lower field position of the H_X signal in Nos 4 and 5, however, must be due to the fact that the H_X proton is allylic when the double bond is in the ring, and the higher field position of this proton in the diacetylcycloserine spectrum clearly supports structure **40** rather than **41**.

PENICILLINS

Most penicillins have a simple PMR spectrum. That of the potassium salt of benzyl penicillin (**42a**) in D_2O, for example, consists of two closely

a: R = CH₂Ph

b: R =

c: R = CH(NH₂)Ph

42

placed singlets near δ 1·5 due to the non-equivalent 2-methyl groups, and singlets near δ 3·6 and 4·3 due to benzylic CH_2 and 3-H protons respectively. The most informative signal is a doublet of doublets near δ 5·5 due to the β-lactam protons at C-5 and C-6 (Fig. 7.6). This resonance is of the AB type because of the small $\Delta\nu_{AB}$ value and most penicillins display a 4 line signal of this nature (Green *et al.*, 1965). Its appearance varies due to changes in $\Delta\nu_{AB}$, thus that of cloxacillin (**42b**) forms a well separated doublet of doublets with a $\Delta\nu_{AB}$ value of about 0·2 ppm, while that of ampicillin (**42c**) is virtually a singlet (Fig. 7.6). In all cases, the vicinal coupling constant between the H-5 and H-6 protons falls in the range 4–5 Hz. Barrow and Spotsweed (1965) examined a series of simple β-lactam derivatives and consistently found $^3J_{cis}$ (4·9–5·9 Hz) to be greater than $^3J_{trans}$ (2·2–2·5 Hz). Thus the β-lactam protons of the penicillins are almost certainly *cis* orientated unless fusion of the lactam to the thiazolidine ring radically alters the shape of the 4-membered ring. Several penicillin derivatives that have been epimerized at C-6 have now been reported (Wolfe and Lee, 1968; Johnson *et al.*, 1968) and in all the vicinal coupling magnitude between H-5 and H-6 is close to 2 Hz, as required for *trans* geometry about these positions.

The chemical shift separations of the β-lactam protons of cephalosporin derivatives generally exceed those of the penicillins and this fact can be

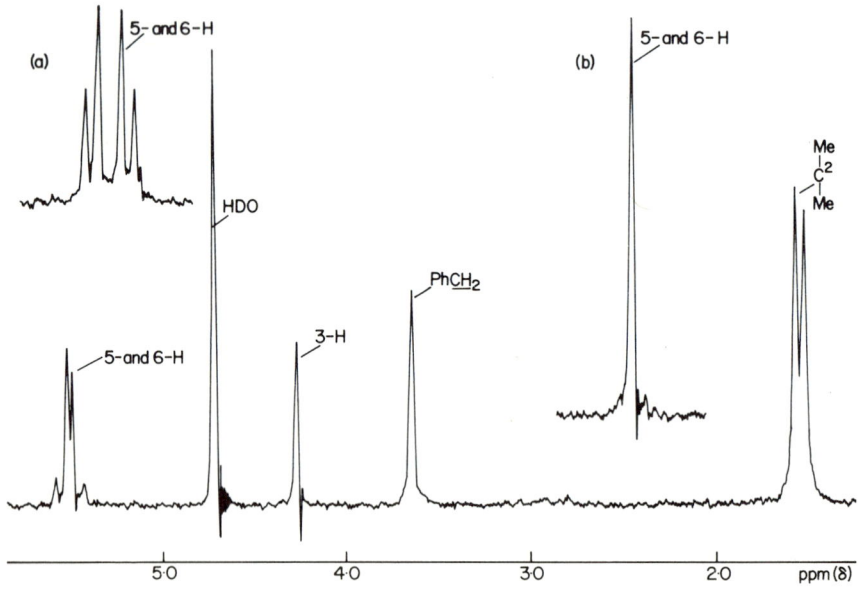

FIG. 7.6. 60 MHz PMR spectrum of benzylpenicillin (**42a**) in D$_2$O. Insert (a) shows the β-lactam protons signal of cloxacillin (**42b**) and insert (b), that of ampicillin (**42c**). The last two signals are centred near δ 5·7 and 5·5 respectively.

used to distinguish the two classes; an example is shown in **43** (Green *et al.*, 1965).

43

H-6   τ 4·73 doublet (*J* 5)

H-7   τ 3·67 quartet (*J* 5, 8·2), doublet (*J* 5) when NH
         coupling removed by deuteration.

Solvent: pyridine

The stereochemistry of the thiazolidine ring of penicillins and penicillin sulphoxides has been established by an elegant combination of nuclear Overhauser effects (NOE) and chemical shift data (Cooper *et al.*, 1969). In this study, the first task was to assign the two *t*-methyl and all the ring proton signals of the spectra of derivatives of phenoxymethylpenicillin

44

(**44**, penicillin V). This was done for the ring protons from their splitting patterns; thus H-3 is a singlet, H-5 a doublet, and H-6 a quartet (in solvent CCl_4 or C_6D_6, H-6 is coupled with H-5 and N—H; in D_2O the N—H coupling is absent and the H-5 and H-6 protons form an AB signal as explained above).

The α- and β-methyl signals were assigned from NOE data. The two possible conformations of the thiazolidine ring of the penicillin nucleus are shown in **45** and **46**. In **45** the 2α- and 2β-methyl groups lie close to H-3

45

46

whereas in **46** only the 2β-methyl protons are proximal to H-3. Thus the β-methyl protons in either conformation should contribute to the intramolecular relaxation of H-3, while the α-methyl protons should relax H-3 in conformation **45** only. In addition, proton H-5 should be relaxed by α-methyl in conformation **46** but not in **45** (all these conclusions are made from examination of Dreiding models). Results of the NOE experiment upon the methyl esters of phenoxymethylpenicillin and the corresponding sulphoxide and sulphone are given in Table 7.2. It follows that penicillins with a sulphide atom prefer conformation **45** (H-3 signal relaxed by both α- and β-methyl, H-5 influenced by neither) while sulphoxides and sulphones have the conformation **46** (H-3 relaxed by only one of the methyl groups, H_5 relaxed by the methyl group which is *without effect* on H-3). A further

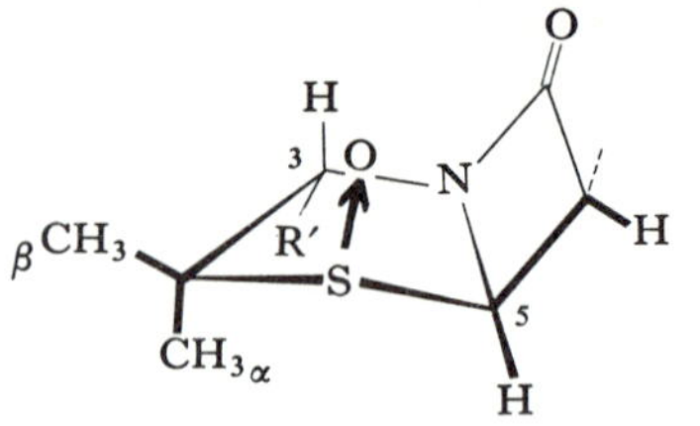

47

conclusion is that the low field methyl signal is due to the β-group which is *cis* to H-3, while the high field signal is due to the group *trans* to H-3 and *cis* to H-5.

TABLE 7.2

Percentage increase in integrated intensities

Methyl signal irradiated	Sulphide —S—	Sulphoxide —SO—	Sulphone —SO$_2$—
High field	H-3: 7–13% increase	H-3: negligible effect	H-3: negligible effect
	H-5: negligible effect	H-5: 14% increase	H-5: 11% increase
Low field	H-3: 21–22% increase	H-3: 26–27% increase	H-3: 15–22% increase
	H-5: negligible effect	H-5: negligible effect	H-5: negligible effect

The next problem was to determine whether the S—O bond of the sulphoxides was *cis* or *trans* to H-3 [(S) or (R) respectively in terms of the Cahn–Ingold–Prelog nomenclature (1956)]. This was done by comparing predicted and observed chemical shift changes consequent upon conversion of a sulphide to a sulphoxide in the conformation **47** and the alternative in which the S—O bond is *trans* to H-3. The S—O function is an anisotropic group (see p. 93) and its shielding effects are believed to be similar to those of a carbon–carbon triple bond (these are shown in **48**). In the latter case the magnitude of nuclear screening (σ) upon a given proton is given by the expression

$$\sigma = \Delta_x \frac{(1 - 3\cos^2\theta)}{3R^3}$$

where R = distance between the proton under study and the electrical centre of gravity of the anisotropic bond, θ = the angle between the direction

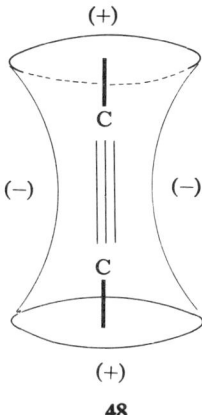

48

R and the symmetry axis of the anisotropic bond, and $\varDelta_x = $ a constant (anisotropy) characteristic of the bond under consideration (McConnell, 1957). Calculations based upon this expression and a Dreiding model of the (S)-form (**47**) showed that the α-methyl and H-5 protons should be shielded and the 3-H proton deshielded relative to their resonance positions in the sulphide; opposite conclusions were drawn when calculations were based on the (R)-isomer. The experimental results agreed qualitatively with those calculated in the first instance, and this evidence supports an (S)-configuration for the sulphoxide.

Aromatic solvent-induced shifts (ASIS) corroborate this assignment as follows. Ledaal (1968) has proposed that the geometry of benzene-solute collision complexes involving solutes containing any polar functional group can be rationalized in terms of one common model. The model presumes that the dipole axis of the polar functional group in the solute molecule is located along the six-fold axis of symmetry of the benzene system with the positive end of the polar function nearest and the negative end farthest away from it. This is illustrated in **49** for the dimethylsulphoxide-benzene complex and accounts for the upfield shift of the methyl protons. Assuming

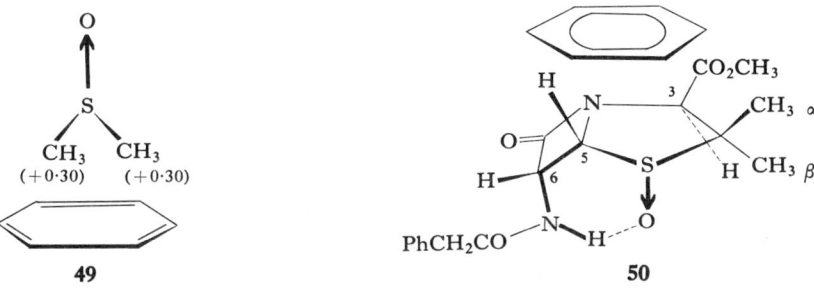

49   **50**

the same principles to govern the geometry of the penicillin sulphoxide-benzene collision complexes, a model such as **50** is anticipated (Barton *et al.*, 1969); this requires that the 2α-methyl and H-5 protons be strongly shielded while the remaining protons be only marginally affected. If the sulphoxide has the R-configuration the solvent will approach the molecule from the β-face and the 2β-methyl protons should be shielded. The experimental results (Table 7.3) support the S configuration, but it is to be noted that the β-methyl group is also shielded, although to a lesser degree. Barton *et al.* (1969) obtained the isomeric sulphoxide and found that in this case, the β-methyl signal was the more shielded as predicted for an R sulphoxide configuration.

TABLE 7.3

Group	Normal sulphoxide	Isomeric sulphoxide
α-Me	+0·64	+0·09ˑ
β-Me	+0·41	+0·24
H-3	−0·07	−0·05
H-5	+0·99	+0·34
H-6	+0·04	+0·88

Positive values indicate upfield shifts, negative values downfield shifts, i.e. values as $\tau C_6 D_6 - \tau CDCl_3$.

The final piece of evidence concerns the N—H resonance (Cooper *et al.*, 1969). In the sulphide (**44**, R=Me) this signal moves downfield by 1·21 ppm when $CDCl_3$ is replaced by DMSO-d_6 as solvent because of the deshielding which follows the formation of a hydrogen bond (N—H $\cdots$ $\bar{O}$—$\overset{+}{S}Me_2$) (see p. 27). In the S-sulphoxide, a strong intramolecular bond already exists as in **50**. This is clear from the fact that its NH resonance is downfield of the corresponding signal due to the sulphide when both compounds are examined in $CDCl_3$. This result supports the configurational assignment as does also the fact that the NH signal of the sulphoxide fails to shift downfield when $CDCl_3$ is replaced by DMSO-d_6 as solvent.

STEROIDS

Much information upon the PMR features of steroids is presented in a text by Bhacca and Williams (1964) and a review by Page (1970). The spectra of most steroids are characterized by a broad, ill-resolved, band in the region upfield of δ 2·5 due to ring protons from which emerge narrow singlets due to the angular methyl groups at C-18 and C-19. These features are clear in the spectrum of cortisone acetate, a typical example (Fig. 7.7); in this, singlets due to the C-19 and C-18 methyl groups and to methyl of the acetate function are clearly apparent as are also the vinylic signal and the C-21 methylene quartet. The last signal displays unusually large geminal coupling ($J \sim 18$ Hz) and the significance of this fact will be discussed later. The chemical shifts of the angular methyl groups are influenced by ring substituents and by the geometry of the A/B ring junction. Thus a β-bromo substituent shifts the 19-methyl signal downfield by 14 Hz (at 60 MHz)

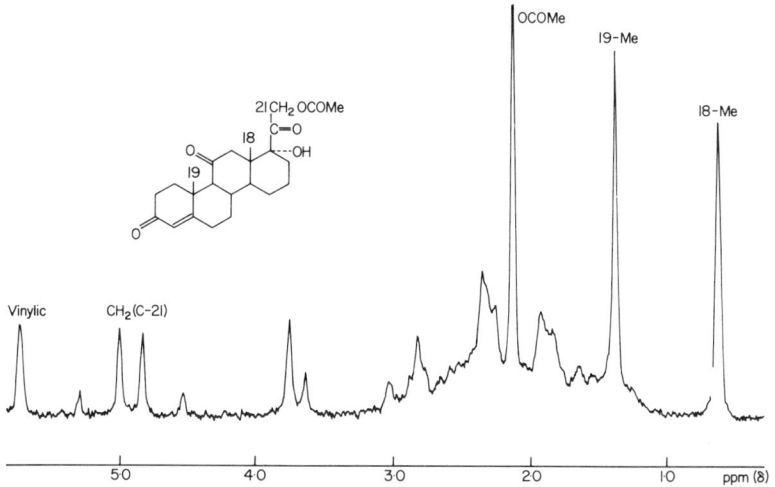

FIG. 7.7. 60 MHz PMR spectrum of cortisone acetate in $CDCl_3$.

compared with its chemical shift in 5α14α-androstane whereas an α-bromine causes only a 4·5 Hz low field shift. The further removed C-18 methyl signal is unaltered by these substituents. The effects of more than one substituent are generally additive and the large amount of substituent data

	5α,14α-androstane	2β-Br	2α-Br
19−Me	47·5	61·5	52·0
18−Me	41·5	41·5	41·5

(chemical shifts in Hz from TMS in $CDCl_3$ recorded at 60 MHz)

available thus allows structural and configurational assignments to be made to uncharacterized derivatives. Much of the data is due to Zürcher (1963) and its application is fully discussed in Bhacca and Williams' text (1964). A few examples of recent work are given below.

The conformation of the functional groups of pharmacologically active steroid molecules is a topic of current interest because such knowledge may lead to a better understanding of the way in which these agents act (Bush,

51

1962). PMR studies of the C-17 side chain of cortisone (**51**) and related compounds are relevant in this connection (Cole and Williams, 1968). The evidence depends on an analysis of effects of the C-20 carbonyl function upon the geminal coupling constant between the non-equivalent protons at C-21; a brief outline of the principles involved is given first. The numerical value of J_{gem} is raised (equivalent to an algebraic decrease because J_{gem} constants are usually negative) by about 1·9 Hz for each adjacent π-bond (associated for example, with a C=O or C=C function) in a freely rotating situation (Barfield and Grant, 1963). In more rigid systems, the π-electron contribution depends on the dihedral angle between the methylene group and the p-orbital of the adjacent sp^2 carbon atom. This is low in situations where the methylene protons lie on *opposite sides* of the p-orbital as in cyclohexanone derivatives (**52**). In these J_{gem} values are about -12 to

52

(the C—H bond which eclipses C=O is shown "bent" for clarity)

-13 Hz, close to the value for methane ($-12\cdot4$ Hz). Values in cyclopentanone systems (**53**), where both methylene protons must lie to the *same side* of the π-orbital, are significantly higher (-16 to -17 Hz). Cookson *et al.* (1966),

53

in a useful review on geminal coupling constants, illustrate the effects of ϕ on J by data upon the 12-methylene group in 11-ketosteroids. In normal

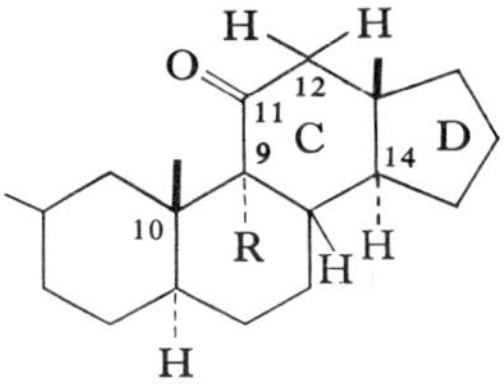

54 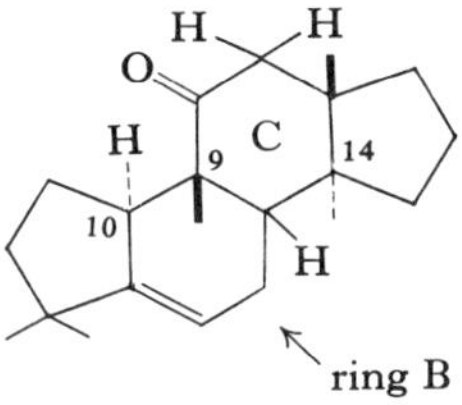   **55**

steroids (**54**, R = H), where ring C is only slightly distorted from an ideal chair cyclohexanone by the *trans* fusion to ring D, $J = -12$ to -12.5 Hz. The small decrease in J to -13 to -14 Hz produced by substitution at C-9 (**54**, R = Me or Br) can be explained by a small downward rotation of the C=O group (slightly decreasing ϕ) that moves the substituent R away from the 12α- and 14α-H atoms. In 9β,14α-dimethyl compounds with a *cis*-junction of rings B and C (**55**) J falls to -16 Hz. Here a chair conformation for ring C suffers more angle-strain than a boat and brings the 14α-methyl group very near to the 10α-H atom (these interactions cannot be represented satisfactorily in two-dimensional drawings and are best appreciated by the examination of Dreiding models). The boat conformation brings the two β-methyl groups very close but a balance can be struck, when the C=O group more nearly bisects the CH$_2$ angle (as in **53**). This must lead to a numerical increase in J_{gem} and high values are characteristic of boat

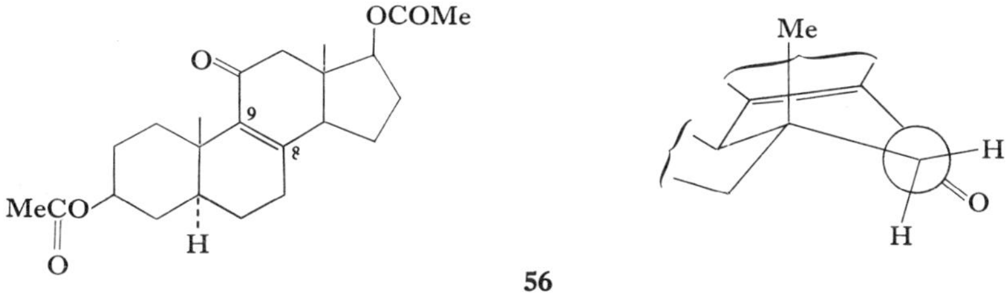

56

conformations of cyclohexanones. Finally, in the $\Delta^{8,9}$-11-ketone (**56**) where ring C is held almost in the conformation most favourable for π-contribution, $J = -16$ Hz.

Geminal 21-H, 21-H' coupling constants of the 17-side chain of several pregn-4-ene derivatives (e.g. **57** a–d) fall in the range −19·7 to −16·7 Hz

	a	b	c	d
57 J_{gem} −16·7		−17·5	−16·4	−18·2 Hz
(in CDCl$_3$)			(in C$_6$D$_6$)	

	e	f	g
	−19·7	−17·5	−17·8
			(in CDCl$_3$)

indicative of high π-contributions (cf. Fig. 7.7). Hence the C-21 protons must be approximately symmetrically disposed with respect to the π-orbital of the 2-carbonyl group and, furthermore, lie on the same side of this orbital. Two conformations satisfy this requirement, namely **58** and **59** shown for a 21-CH$_2$OH group; infra-red data ($\nu_{C=O}$ of the 21-methyl analogue increases when replaced by CH$_2$OH or CH$_2$OMe, indicative of coplanarity of the OR and C=O functions) supports the former.

(C—OH bond "bent" for clarity)

PMR and other data (to be discussed later) support a conformation (**60**) for the pregn-20-one skeleton in which the carbonyl group is directed over, and the 20-21 bond away from ring D whereby interactions of the side chain with the 15-angular methyl and 16β-H are minimized. By analogy the two conformations **61** and **62** are proposed in both of which maximum π-orbital contributions to J_{gem} should occur. Conformer **61** is preferred on the basis of absence of infra-red evidence for hydrogen bonding between the 17- and 21-OH groups (feasible in **62** because the two groups are close) and evidence that stable conformations of a tetrahedral carbon atom bonded

60   61   62

to a carbonyl involved the eclipsing of C=O by a single bond of the sp^3 carbon (as in **58**). The numerical increase in J_{gem} which follows replacement of 17α-H by 17α-OH (cf. **57** a and b; c and d) is consistent with the decrease in stability of conformations other than **61** that results from the electrostatic repulsion between the two hydroxyl dipoles (furthest removed in **61**), while the increase in the 17α,21-diol **57e** relative to the 17α-OH,21-acetate f and 21-methoxyl compound g is explained by the stabilization of **61** by the weak C=O/21-OH hydrogen bond and reduction in size of the 21-substituent which makes the eclipsing of the carbonyl group more favourable. Kier (1968), has proposed a favoured cortisone conformation from molecular orbital calculations which has features similar to **61** and **62** and has drawn comparisons with conformers of histamine and serotonin. An X-ray study (Cooper and Duax, 1969) has now shown that the solid state conformation of the cortisone side chain is very similar to that deduced from PMR evidence.

PMR evidence for the pregnan-20-one conformation derives from solvent studies of angular methyl chemical shifts in some steroidal ketones, e.g. **63** and **64** (Williams and Bhacca, 1965). In many derivatives of this type the C-18 and C-19 protons come to resonance at higher field when CDCl$_3$ is

63   64

Resonance $\varDelta$		Resonance $\varDelta$	
H-19	0·30	H-19	0·01
H-18	0·00	H-18	0·25

$(\varDelta = \nu_{C_6H_6} - \nu_{CDCl_3}$ in ppm$)$

replaced by benzene as solvent. It is concluded that ketones in benzene solution form a solute-solvent collision complex in which the aromatic π-electrons interact with the partial positive charge on the carbonyl carbon atom in such a manner that they are as far as possible from the partial negative charge on oxygen, as in **65**.† In such complexes, axial methyl groups should be shielded, whereas α-equatorial groups should either be

$$\varDelta = \nu_{C_6H_6} - \nu_{CDCl_3}$$

deshielded or barely influenced, as a result of the diamagnetic anisotropy of the benzene ring. The data obtained showed that benzene causes shielding (relative to CDCl$_3$) of methyl resonances lying behind a plane drawn through the carbonyl carbon atom and perpendicular to the direction of the C—O bond (see **66**). The signals of α-axial and γ-axial groups were most affected while those due to α-equatorial groups (these groups lie on the boundary of the described plane) showed little or no solvent change. The fact that the C-18 protons of 5α-pregnan-20-one (**67**, R = H) showed only a minor shift ($\varDelta = 0.03$ ppm) in spite of their proximity to the 20-keto function implies that the C-18 methyl group must lie near a plane drawn through carbonyl and at right angles to the C—O bond. Conformations **68** and **69**

† A model in which the plane of the benzene ring is steeply inclined to the overall plane of the steroid molecule has also been proposed (Williams and Wilson, 1966).

fit this requirement, while **70** does not. (In these drawings the conformation of the C-17 side chain is depicted as viewed along the C-20–C-17 bond, cf. illustration alongside **68**. The spatial relationships are more readily appreciated by the examination of Dreiding models). A decision between **68** and **69** may be made from solvent shift data upon the 16α-methyl analogue (**67**, R = Me). In **68** both the 16α- and 13β-methyl groups lie close to the reference plane, while in **69**, the 16α-group lies well behind the plane and should therefore be shielded when CDCl₃ is replaced by benzene as solvent. Since very small $\varDelta$ values are obtained for both 13-β and 16α-methyl resonances, **68** must be the preferred conformation.

In recent years much interest has been directed towards the synthesis of azasteroids (Martin-Smith and Sugrue, 1964), and PMR spectroscopy has played an important part in this work. Ring fusions in derivatives with nitrogen at position 8 of the steroid nucleus (**71**) may either be *trans* or *cis*

71

trans

72

cis

73

and are analogous to those of the quinolizidines **72** and **73** respectively. In *trans* quinolizidines the nitrogen lone-pair and adjacent bridgehead hydrogen have a *trans* diaxial relationship, while in each of the *cis* conformers, their dispositions are *gauche*. Uskokovic *et al.* (1964) have found that the chemical shift of the bridgehead hydrogen in *trans* quinolizidines is higher field (above τ 6·2) than that of *cis* isomers (below τ 6·2). In the compounds examined, the *trans* signals could not be resolved (they fell amongst a broad ring proton multiplet) but the *cis* were downfield of this band and their splitting patterns allowed a differentiation between the two alternative forms to be made. Thus the PMR spectrum of the *cis*-2-(*p*-chlorophenyl)-1,2,3,4,6,7-hexahydro-11bH-benzo[a]quinolizidine (**74**) showed a

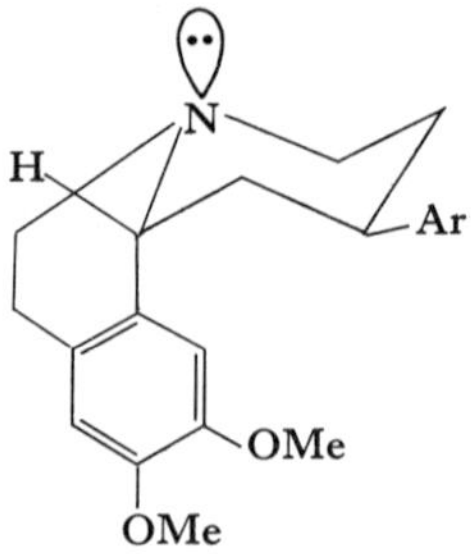

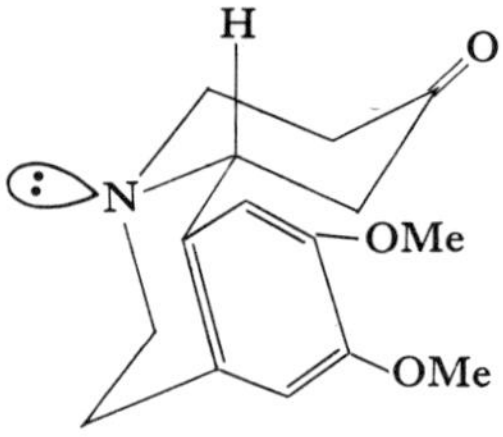

$$Ar = p\,ClC_6H_4$$

74   **75**

one proton triplet near τ 6·0 with couplings (J 5 Hz) typical of a/e and e/e
vicinal protons. The ketone (**75**) is an example of the alternative *cis* form
(**73**); its bridgehead proton gave rise to a doublet of doublets at τ 5·85 with
splittings $J_{ae} = 5$ Hz and $J_{aa} = 11$ Hz.

These and other principles have greatly aided stereochemical assignments
to 8-azasteroids. Brown *et al.* (1966) obtained azasteroids of type **76** in

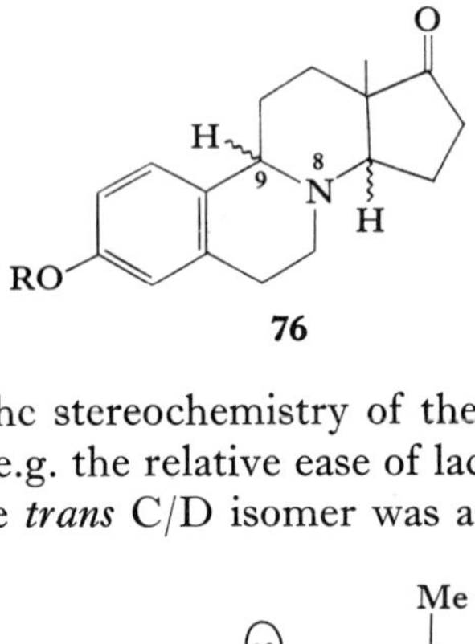

76

three isomeric forms. The stereochemistry of the C/D ring junction was
known from other data (e.g. the relative ease of lactam formation in one of
the synthetic steps). The *trans* C/D isomer was assigned the structure **77**

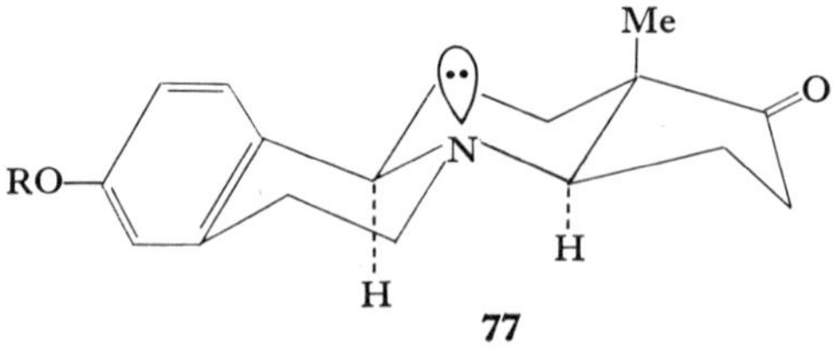

77

on the basis of (i) strong Bohlmann bands in its infra-red spectrum (this
is evidence that the lone-pair on nitrogen is *trans*-diaxial to at least two
hydrogens on adjacent carbon atoms), and (ii) the absence of a C-9 (bridgehead) proton signal below τ 6·8. The other two isomers both had a *cis*
C/D ring junction. One showed Bohlmann bands in its infra-red spectrum
and hence must have the *trans* quinolizidine structure **78** or **79**. These differ

78　　　　　　　　79

in the axial (**78**) and equatorial (**79**) orientation of the angular methyl group, and the work of Bhacca and Williams, earlier discussed (p. 273), which correlates changes in the chemical shift of a methyl group proximate to a carbonyl on passing from CDCl$_3$ to benzene as solvent allows a distinction between the two to be made. The upfield shift which occurs in this case (τ_{Me}: 9·0 in CDCl$_3$; 9·24 in C$_6$H$_6$) is consistent with the axial arrangement as in **78**. The second *cis* C/D isomer is related to *cis* quinolizidine **73** because, (i) its infra-red spectrum showed no Bohlmann bands, and (ii) its PMR spectrum displayed a C-9 proton signal at τ 6·2 downfield of the main proton

80　　　　　　　　81

band. The two structures **80** and **81** are possible, and the absence of a significant angular methyl solvent shift (τ_{Me}: 8·83 in CDCl$_3$; 8·79 in C$_6$H$_6$) shows the former to be preferred. The splitting pattern of the angular hydrogen atom at C-9 could not be used to differentiate between the two *cis* quinolizidine conformations **80** and **81** because this signal was masked by the methoxy resonance. It was resolvable in the corresponding 3-acetate, however, and appeared as a broad doublet rather than the quartet expected for the *gauche* and *trans* diaxial relationship between the C-9 proton and the two C-11 protons in conformation **80**. This result, together with the fact that the base was slowly oxidized by mercuric acetate (a reaction requiring the C-9 hydrogen to be *trans* diaxial to the lone-pair) is evidence that the isomer in question exists as a mixture of **80** and **79**.

A related stereochemical problem has been reported by Bhacca *et al.* (1968) who found that reduction of the 12-ketoiminium salt **82** by hydride

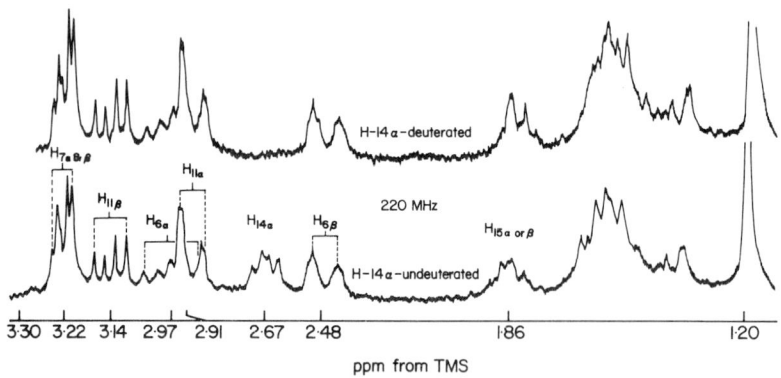

reagents or catalytic means produced a mixture of two components. One of these was identified as **83** by direct comparison with the known 12-desoxy derivative (Brown *et al.*, 1966, see preceding text); it had a *trans* quinolizidine structure as confirmed by the nature of its C-9 PMR signal. This fell near τ 6·2 and displayed typical $J_{a/a}$ (11 Hz) and $J_{a/e}$ (4·25 Hz) coupling. The C-14 proton signal provided information about the C/D ring junction since it showed small couplings ($J \sim 5$ Hz) typical of an equatorially orientated atom. In both isomers derived from **82**, the C-14 proton signal was unresolvable in either 60 or 100 MHz spectra because the C-6, C-7 and C-11 protons resonated in the same region. Examination of the 220 MHz spectra of these isomers, however, allowed a first order assignment. Part of the spectrum of the second isomer is shown in Fig. 7.8. The quartet near

FIG. 7.8. Part of the 220 MHz spectrum of the azasteroid **84** (the second isomer of the text) in CDCl$_3$. In the upper spectrum the proton at C-14 is replaced by deuterium.

τ 2·67 is attributed to the C-14 proton because the signal is absent in the spectrum of the deuterated analogue (obtained by treating **82** with LiAlD$_4$). The second isomer showed no Bohlmann bands in its infra-red spectrum and displayed a C-9 proton signal below τ 6·2. Both facts show the isomer to be based on *cis* quinolizidine. The angular methyl group at C-13 must be axial because its PMR signal moved upfield when CDCl$_3$ was replaced by benzene as solvent. The 14-H signal (identified in the 220 MHz spectrum as described) was a doublet of doublets with coupling constants (11 and 7 Hz) of a magnitude requiring the proton to be axially placed. These data are accommodated by the two structures **84** and **85**. The fact that the

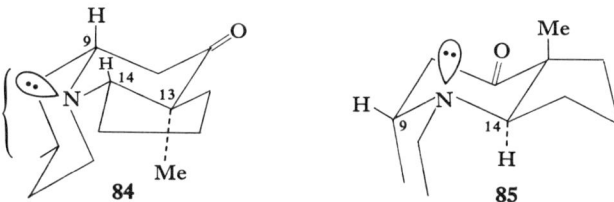

14-H chemical shifts for the two isomers are almost identical suggests that their protons have similar magnetic environments. Since 14-H(β) is *cis* to the lone pair in **83** it is reasonable to assume that it is also *cis* to this feature in the second isomer. Hence structure **84** is assigned to the second isomer since a *trans* diaxial relationship obtains in **85**.

References

Archer, S., Bell, M. R., Lewis, T. R., Schulenberg, J. W. and Unser, M. J. (1958). *J. Amer. Chem. Soc.* **80**, 4677.

Barfield, M. and Grant, D. M. (1963). *J. Amer. Chem. Soc.* **85**, 1899.

Barrow, K. D. and Spotsweed, T. M. (1965). *Tetrahedron Lett.* 3325.

Barton, D. H. R., Comer, F. and Sammers, P. G. (1969). *J. Amer. Chem. Soc.* **91**, 1529.

Belleau, B. (1966). *Pharm. Rev.* **18**, 131.

Bhacca, N. S. and Williams, D. H. (1964). "Applications of NMR Spectroscopy in Organic Chemistry". Holden-Day, San Francisco.

Bhacca, N. S., Meyers, A. I. and Reine, A. H. (1968). *Tetrahedron Lett.* 2293.

Bishop, R. J., Fodor, G., Katritzky, A. R., Soti, F., Sutton, L. E. and Swinbourne, F. J. (1966). *J. Chem. Soc.* (C), 74.

Booth, H. (1966). *Tetrahedron* **22**, 615.

Bovey, F. A. (1969). "Nuclear Magnetic Resonance Spectroscopy." Academic Press, New York and London.

Brown, D. R., Lygo, R., McKenna, J. and McKenna, J. M. (1967). *J. Chem. Soc.* (B), 1184.

Brown, R. E., Lustgarten, D. M., Stanaback, R. J. and Meltzer, R. I. (1966). *J. Org. Chem.* **31**, 1489.

Bush, I. E. (1962). *Pharm. Rev.* **14**, 317.

Cahn, R. S., Ingold, C. K. and Prelog, V. (1956). *Experientia* **12**, 81.

Carr, J. B. and Huitric, A. C. (1964). *J. Org. Chem.* **29**, 2506.

Chapman, O. L. and King, R. W. (1964). *J. Amer. Chem. Soc.* **86**, 1256.

Chen, C. Y. and Le Fèvre, R. J. W. (1965a). *J. Chem. Soc.* 3473.

Chen, C. Y. and Le Fèvre, R. J. W. (1965b). *J. Chem. Soc.* 3467.

Claes, P., Vanderhaeghe, H., Totté, J. and Slincke, G. (1969). *Bull. Soc. Chim. Belg.* **78**, 151.

Cole, W. G. and Williams, D. H. (1968). *J. Chem. Soc.* (C), 1849.

Cookson, R. C., Crabb, T. A., Frankel, J. J. and Hudec, J. (1966). *Tetrahedron* Supplt **7**, 355.

Cooper, A. and Duax, W. L. (1969). *J. Pharm. Sci.* **58**, 1159.

Cooper, R. D. G., Demarko, P. V., Cheng, J. C. and Jones, N. D. (1969). *J. Amer. Chem. Soc.* **91**, 1408.

De Jongh, H. A. P. and Wynberg, H. (1965). *Tetrahedron* **21**, 515.

Fodor, G., Medina, J. D. and Mandava, N. (1968). *Chem. Commun.* 581.

Fodor, G., Chastain, R. V., Frehel, D., Cooper, M. J., Mandava, N. and Gooden, E. L. (1971). *J. Amer. Chem. Soc.* **93**, 403.

Gabe, E. J. and Barnes, W. H. (1963). *Acta Cryst.* **16**, 796.

Garbisch, E. W. (1964). *J. Amer. Chem. Soc.* **86**, 5561.

Green, G. F. H., Page, J. E. and Staniforth, S. E. (1965). *J. Chem. Soc.* 1595.

Grove, J. F. (1963). *Quart. Rev.* **17**, 1.

Hyne, J. B. (1961). *Can. J. Chem.* **39**, 2536.

Jardetzky, O. (1963). *J. Biol. Chem.* **238**, 2498.

Johnson, D. A., Mania, D., Panetta, C. A. and Silvestri, H. H. (1968). *Tetrahedron Lett.* 1903.

Kier, L. B. (1968). *J. Med. Chem.* **11**, 915.

Ledaal, T. (1968). *Tetrahedron Lett.* 1683.

Levine, S. G. and Hicks, R. E. (1968). *Tetrahedron Lett.* 5409.

Lyle, G. G. and Keefer, L. K. (1966). *J. Org. Chem.* **31**, 3921.

Lyle, G. G. and Keefer, L. K. (1967). *Tetrahedron* **23**, 3253.

Mandava, N. and Fodor, G. (1968). *Can. J. Chem.* **46**, 2761.

Mandell, L., Moncrieff, J. W. and Goldstein, J. H. (1963). *Tetrahedron* **19**, 2025.

Martin-Smith, M. and Sugrue, M. F. (1964). *J. Pharm. Pharmacol.* **16**, 569.

McConnell, H. M. (1957). *J. Chem. Phys.* **37**, 226.

McGilveray, I. J. and Rinehart, K. L. (1965). *J. Amer. Chem. Soc.* **87**, 4003.

Milne, G. W. A. and Cohen, L. A. (1967). *Tetrahedron* **23**, 65.

Page, J. E. (1970). *In* "Annual Reports on NMR Spectroscopy" (E. F. Mooney, ed.), Vol. 3, p. 149. Academic Press, London and New York.

Perks, F. and Russell, P. J. (1967). *J. Pharm. Pharmacol.* **19**, 318.

Portoghese, P. S. (1967). *J. Med. Chem.* **10**, 1057.

Schenk, H., MacGillavry, C. H., Skolnik, S. and Laan, J. (1967). *Acta Cryst.* **23**, 423.

Sinnema, A., Maat, L., van der Gugten, A. J. and Beyerman, H. C. (1968). *Rec. Trav. Chim.* **87**, 1027.

Supple, J. H., Pridgen, L. N. and Kaminski, J. J. (1969). *Tetrahedron Lett.* 1829.

Page, J. E. (1970). *In* "Annual Review of NMR Spectroscopy" (E. F. Mooney, ed.), Vol. 3. Academic Press, London.

Thut, C. C. and Bottini, A. T. (1968). *J. Amer. Chem. Soc.* **90**, 4752.

Uskoković, M., Bruderer, H., von Planta, C., Williams, T. and Brossi, A. (1964). *J. Amer. Chem. Soc.* **86**, 3364.

Williams, D. H. and Bhacca, N. S. (1965). *Tetrahedron* **21**, 2021.

Williams, D. H. and Wilson, D. A. (1966). *J. Chem. Soc.* B, 144.

Wolfe, S. and Lee, W. S. (1968). *Chem. Commun.* 242.

Witkop, B. and Foltz, C. M. (1957). *J. Amer. Chem. Soc.* **79**, 197.

Zürcher, R. F. (1963). *Helv. Chim. Acta* **46**, 2054.

CHAPTER 8

Biochemical Aspects—1

In this and the following chapter some applications of NMR spectroscopy to biochemistry are described. Since the review of Kowalsky and Cohn (1964), an increasingly large number of biochemical NMR studies have been reported and the more recent work has been surveyed in a recent issue of Chemical Reviews (Rowe *et al.*, 1970). Biochemical examples receive full attention in various periodical accounts of NMR spectroscopy such as those published in Analytical Chemistry (e.g., Heeschen, 1968) and by the Chemical Society (e.g., Feeney, 1969). In this chapter, the main topics discussed are proteins and their simpler components. Relaxation time studies are also included; these are relevant to all classes of biopolymer but most of the reported work involves proteins. Relaxation data are of particular value to enzymologists concerned with enzyme-substrate interactions, and pharmacologists and medicinal chemists who are trying to gain insight into drug-receptor interactions. PMR studies of carbohydrates, nucleosides, nucleotides and nucleic acids form the content of Chapter 9.

AMINO ACIDS AND PEPTIDES

The PMR spectra of most naturally occurring amino acids have been reported (Bystrov *et al.*, 1969; Rowe *et al.*, 1970), and Bovey's collection of NMR data (1967) is a good source book in this regard. Details of the 220 MHz spectra of 21 amino acids are also available (Bak *et al.*, 1968). These data clearly have potential value in the analysis of mixtures derived, for example, from protein hydrolysates. The presence of aromatic amino acids such as phenylalanine and tryptophan in mixtures (or in proteins themselves, see later) may readily be detected by resonance signals near δ 7·5 while the multiplicity of signals in the methyl region gives information about the aliphatic chain, e.g. alanine, doublet (MeCH); methionine, singlet (MeS);

valine, doublet of doublets ($Me_2CH\overset{*}{C}$). Spectra of some simple mixtures of amino acids which illustrate these analytical aids have been reported by

Sen and Wu (1969). The PMR characteristics of the CO_2Me resonances of N-trityl methyl esters of amino acids may also be applied to identification problems (Webb *et al.*, 1969). In methyl esters of α-amino acids, the carbomethoxy peak moves upfield by 0.2–0.9 ppm after N-tritylation (see example below for derivatives of glycine). These comparisons involve a solvent

$$H_2NCH_2CO_2Me \cdot HCl \xrightarrow[\text{2. ClCPh}_3]{\text{1. NEt}_3 \text{ (to free base)}} Ph_3CNHCH_2CO_2Me$$

ν_{Me} δ 3.87 in D_2O δ 3.60 in $CDCl_3$,
$\Delta\nu = 0.27$ ppm

change and the removal of a proton from charged nitrogen but it was shown that neither factor had much influence on the CO_2Me resonance; hence the diamagnetic shift observed is attributed to the trityl group. The 24 α-amino esters examined could be divided into two groups: those showing shifts of 0.25–0.27 ppm (glycine, sarcosine and proline), and those having shifts of 0.59–0.97 ppm. Aromatic α-amino acids had the largest $\Delta\nu$ values. These differences are explained in terms of anticipated preferred conformations. The ester methyl protons will be most influenced by the trityl function in the *gauche* conformation **1**, and least in the *anti* form **2**. As R gets larger, **1**

is more favoured, hence the upfield shift of the CO_2Me resonance should increase; thus a small effect is obtained in the case of glycine ($\Delta\nu = 0.27$ ppm) where R = H, and a larger effect in that of valine ($\Delta\nu = 0.78$ ppm) where R = $CHMe_2$. In the aromatic derivatives this steric factor is probably augmented by an additional shielding effect due to the R group itself; in the phenylalanine case for example (see **3**) $\Delta\nu$ is 0.85 ppm. When the trityl amino group is further removed from the ester function, negligible shifts are found, e.g. β-alanine $\Delta\nu$ 0.1 ppm and γ-aminobutyric acid $\Delta\nu$ -0.03 ppm.

Recently a method for the analysis of mixtures of amino acids and of peptides based on ^{19}F NMR has been proposed (Sievers *et al.*, 1969). The mixture to be analysed is treated with trifluoroacetic anhydride and the fluorine NMR spectrum of the resultant N-trifluoroacetylated (N-TFA) product recorded with acetone as solvent. The fluorine chemical shift is very sensitive to small differences in shielding hence most N-TFA derivatives of amino acids have distinctive ^{19}F chemical shifts, as illustrated in Fig. 8.1; relative proportions of the components may be judged from the peak heights of each $COCF_3$ signal. Dipeptides may also be differentiated

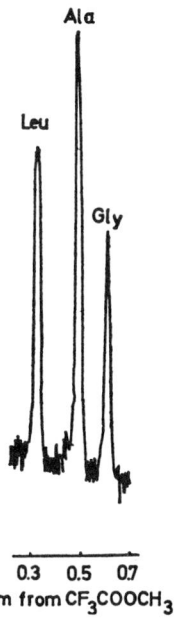

Ala

Leu

Gly

0.3 0.5 0.7
ppm from CF_3COOCH_3

Fig. 8.1. ^{19}F 84·6 MHz NMR spectrum of a mixture of N-TFA-L-leucine (11 mg), N-TFA-L-alanine (14 mg) and N-TFA-glycine (6 mg) in acetone (0·6 ml) (Sievers *et al.*, 1969).

in this manner even when two peptides differ only in their second amino acid units, e.g. the ^{19}F resonances of methyl esters of N-TFA-glycyl-L-leucine and N-TFA-glycyl-L-alanine are readily resolved at 84·6 MHz. This procedure is analogous to the analytical use of trifluoroacetates of alcohols (Chapter 1, p. 32).

A PMR procedure for determining the optical purity and absolute configuration of amino acids has already been described in Chapter 5.

The synthesis of peptides presents many problems of structure, optical purity and configuration which are capable of solution by PMR spectroscopy, as several studies of peptides have demonstrated. The amino acid sequence in simple peptides may be determined by making use of the fact that the resonance line of a proton group next to charged nitrogen ($\overset{+}{H}NCX$) moves upfield when the proton is removed, while that of a group adjacent to an ionized carboxylic acid ($XC\overset{-}{C}O_2$) moves downfield when the acid receives a proton (see Chapter 2, p. 70). Thus, in the case of a dipeptide

$$\underset{\underset{R \qquad R'}{\mid \qquad \mid}}{\overset{+}{H_3N}CHC\overset{\overset{\displaystyle O}{\displaystyle \|}}{}NHCHCO_2^-}$$

(which exists as a zwitterion at neutral pH), addition of base shifts the R signal to *higher* field while addition of acid will shift R' to *lower* field (Sheinblatt, 1966, 1967). If the peptide has the alternative structure,

$$\underset{\underset{R' \qquad R}{\mid \qquad \mid}}{\overset{+}{H_3N}CHC\overset{\overset{\displaystyle O}{\displaystyle \|}}{}NHCHC\overset{-}{O}_2,}$$

these pH changes will likewise move the R and R' signals but in the *opposite* sense. Hence groups adjacent to the nitrogen and carboxylate functions may be identified by the direction of their shift, whence follows the amino acid sequence. Spectra of glycylalanine (Gly-Ala) and alanylglycine (Ala-Gly) illustrate the method (Fig. 8.2). In the Gly-Ala spectra, the glycyl CH₂ singlet suffers an upfield shift (+33 Hz) after adding base, and the alanyl CH a downfield shift after acid (−14 Hz). In the Ala-Gly spectra, it is the alanyl CH signal which moves *upfield* after base, while the glycyl CH₂ resonance moves *downfield* after acid. The central member of a tripeptide may be identified by the failure of its resonance lines to change significantly with variation of pH. Care must be taken in the case of peptides containing more than two ionizable features, e.g. thiol derivatives.

Just as the chemical shifts of comparable proton groups of diastereo-isomers may differ (Chapter 5, p. 170) so too may those of isomeric peptides of the type L-X-L-Y and L-X-D-Y (Halpern *et al.*, 1967a). Differences in the N-terminal resonances (assigned by observing shifts in spectral lines with varying pH, as just described) are seen even in the spectra of L-Ala-L-Ala and L-Ala-D-Ala which, although small, permit identification of the two components in a mixture (Fig. 8.3). The presence of an aromatic amino acid in the peptide greatly enhances the methyl shift difference. Given the reference spectra, diastereoisomeric optical purity, relative configuration, and the location of the alanyl position can all be derived from the spectra of peptides obtained from optically active alanine and

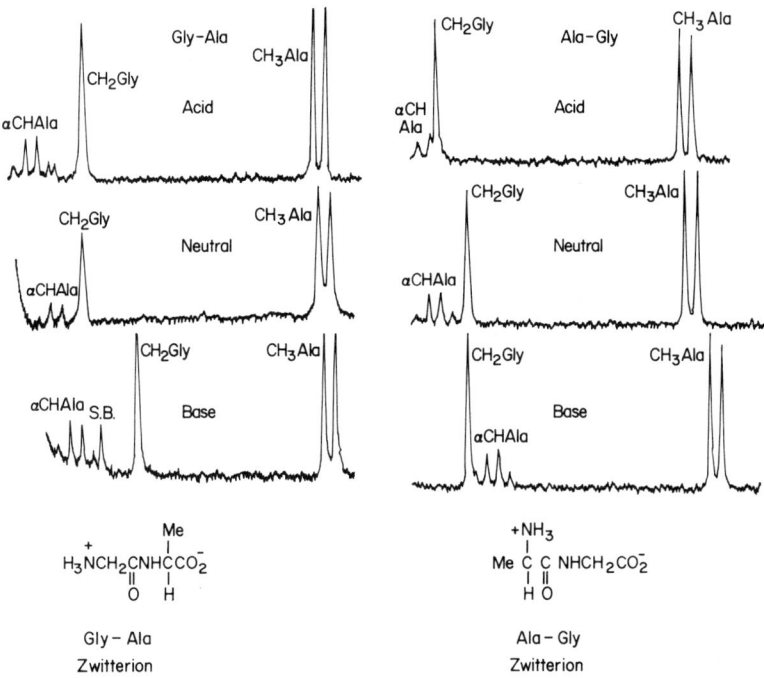

FIG. 8.2. 60 MHz PMR spectra in D₂O of glycylalanine (left hand side) and alanylglycine (right hand side) at different pH values (Sheinblatt, 1966).

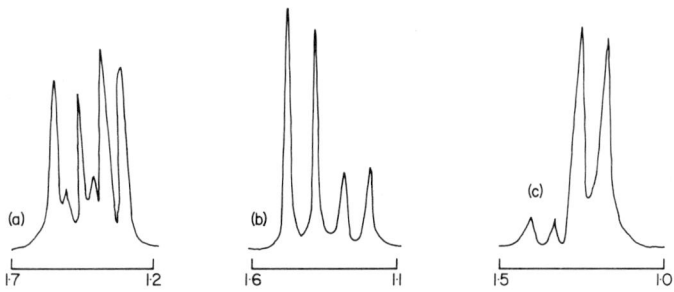

FIG. 8.3. 60 MHz PMR spectra in the region of the methyl resonance of alanyl peptides. (a) L-alanyl-D-alanine and D-alanyl-D-alanine (1:9, pH 5 in D₂O); (b) L-alanyl-L-phenylalanine and L-alanyl-D-phenylalanine (3:1, pH 3 in D₂O); (c) L-phenylalanyl-L-alanine and L-phenylalanine-D-alanine (1:9, pH <1 in D₂O) (Halpern *et al.*, 1967a).

phenylalanine. Diastereoisomeric N-acyl peptide derivatives, as encountered in peptide syntheses, also differ in their spectra; this makes PMR spectroscopy a convenient tool for assessing the relative degrees of racemization associated with various synthetic procedures, and may be

$$PhCH_2CCO_2H \quad + \quad H_2NCH(Me)CO_2Me \quad \longrightarrow \quad PhCH_2CCONHCH(Me)CO_2Me$$

NHCOMe		NHCOMe
4	**5**	**6**

Me resonance (Hz from TMS at 60 MHz)

L-L	81·0
L-D (or D-L)	73·5

quantitative if diastereoisomeric signals do not overlap (Halpern *et al.*, 1967b). In this way it was shown that the two L-amino acid derivatives **4** and **5** yield the L-L peptide **6** containing 10% of the D-L (or L-D) isomer when carbonyldiimidazole is used as the coupling agent, and 50% when dicyclohexylcarbodiimide is so employed. The influence of other factors e.g. structure of the acyl function and the amino acids upon racemization can also be judged accurately by this method.

Dipeptides may be converted to corresponding cyclo-forms (diketopiperazines) without change of configuration (see **7**) (Nitecki *et al.*, 1968),

LL (*cis*)

7

and the cyclo-diastereoisomers also differ in their PMR characteristics especially if aromatic groups are present (Westley *et al.*, 1968). Thus the methyl resonances of DL (*trans*) and LL (*cis*) cyclo-alanylphenylalanine differ by 34 Hz. The higher field position of the *cis* signal is attributed

8

trans

ν_{Me} 86·5

9

cis

ν_{Me} 52·5 Hz from TMS

60 MHz in TFA

to aromatic shielding which arises as a result of a preferred conformation in which the benzene ring faces the piperazine ring (see **9**); equatorial methyl as in **8** is too far removed to be affected in this way (Kopple and Marr, 1967). These PMR differences serve as a further check of the optical purity of dipeptides and also are of potential value in establishing the configuration of diketopiperazines isolated from natural sources.

Evidence for the preferred rotational states of the $N_{(6)}H—C_{(5)}H$ fragment in the alanine dipeptide derivatives (**10**) has been obtained from the determination of the $^3J_{NH-CH}$ coupling constants and consideration of their magnitudes in relation to the dihedral angle of the system in question (Bystrov *et al.*, 1969). In spectra of these peptides, it is first necessary to differentiate $N_{(3)}H$ and $N_{(6)}H$ signals which both appear as broad doublets in a variety of solvents. In benzyloxycarbonyl derivatives (**10a**) the doublets are well separated, and that at higher field (δ 5·87 in $CDCl_3$) is assigned to $N_{(6)}H$ on the basis of comparisons with the spectra of ethyl N-methyl-carbamate (ν_{NH} δ 5·16) and physostigmine (δ 5·33), compounds which both

$$\overset{O}{\overset{\|}{}}$$

contain the HNCO feature of urethanes. The less shielded acyl $N_{(3)}H$ proton comes to resonance at much lower field (δ 6·96). Assignment of NH signals in N-acetyl derivatives (**10b**) is more difficult because the NH chemical shifts are very close. Analysis is achieved via identification of the 2- and 5-methyl resonances. These form two overlapping doublets in D_2O; that at higher field is unchanged when the ester group is partially hydrolysed

$$R—\overset{\overset{O}{\|}}{C}—NH—\underset{5}{\overset{\overset{Me}{|}}{CH}}—\overset{\overset{O}{\|}}{C}—NH—\underset{2}{\overset{\overset{Me}{|}}{CH}}—CO_2Me$$

10   (a) R = PhCH$_2$O; (b) R = Me

(by addition of NaOD) whereas the lower field doublet is reduced in intensity and a new doublet appears higher field than the original group (Fig. 8.4). The low field doublet must then be due to 2-methyl because only this group will be influenced (shielded) by the change $CO_2Me \rightarrow \bar{C}O_2$ (cf. Sheinblatt, 1966). Double resonance now makes possible the rest of the assignments (Fig. 8.4). Irradiation at the C-2 methine frequency simultaneously decouples the 2-methyl and $N_{(3)}$-H doublets, while irradiation at the C-5 methine frequency will decouple the C-5 and $N_{(6)}$-H signals. Observation of mutually collapsing methyl and NH signals thus allows unambiguous assignments to be made.

The $^3J_{NH-CH}$ coupling constants of the $N_{(6)}H—C_{(5)}$ fragment fall in the range 8·3–7·8 Hz. These values, on the basis of a coupling constant-dihedral

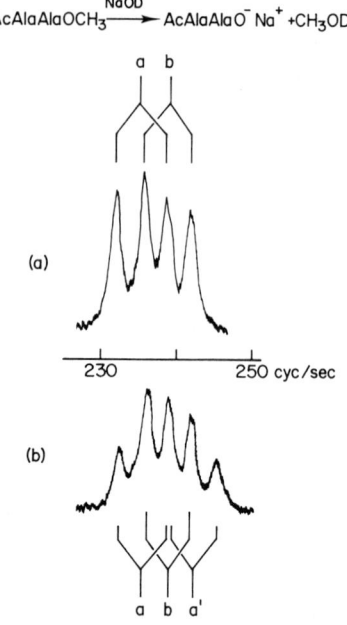

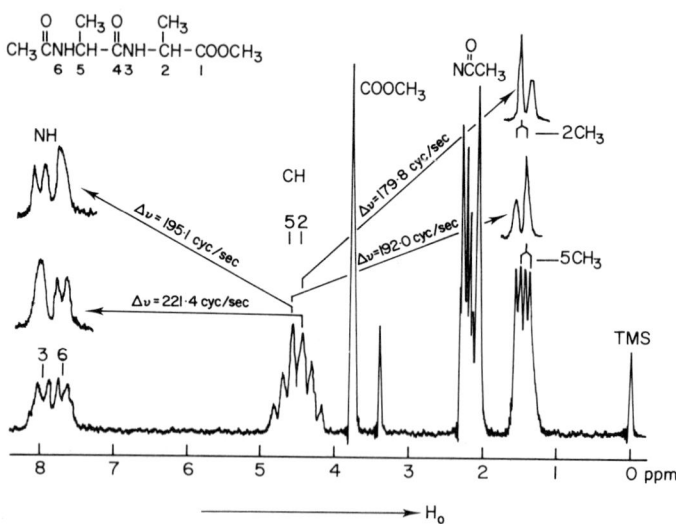

Fig. 8.4. (a) 100 MHz PMR spectral region of the Me signals of N-acetyl-D-alanyl-D-alanine methyl ester in D_2O. (b) The same with addition of NaOD. The frequency scale is in Hz with reference to dioxan as internal standard. (c)Double resonance 60 MHz PMR spectrum in N-acetyl-D-alanyl-D-alanine methyl ester in $(CD_3)_2CO$ (Bystrov *et al.*, 1969).

angle relationship developed in the same paper ($\phi = 0$, $J \sim 8$ Hz; $\phi = 90°$, $J < 1$ Hz; $\phi = 180°$, $J \sim 9.5$ Hz), suggest a marked preference for *cis* rotamers (**11**) and this conclusion is corroborated by infra-red evidence of intra-molecular hydrogen bonding (only the *cis* rotamer brings the $N_{(6)}$ carbonyl function in close proximity to the $N_{(3)}$ hydrogen atom).

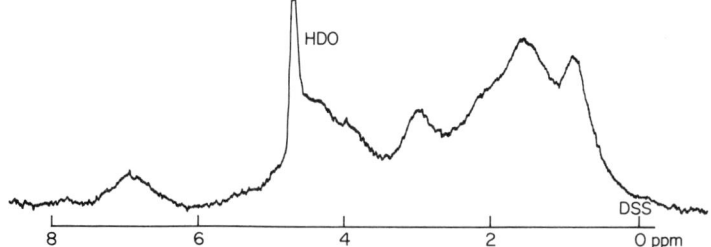

11

(front atom is nitrogen)

POLYPEPTIDES AND PROTEINS

A good account of the use of PMR spectroscopy in polypeptide and protein structural studies has been given by Bradbury and Crane-Robinson (1968) and the choice of topics discussed in the following part of this chapter

HDO

DSS

8   6   4   2   0 ppm

FIG. 8.5. 60 MHz PMR spectrum of ribonuclease, 11% in D_2O, pH 7·5, at 38° (Ferguson and Phillips, 1967).

has been guided by this review. In general, proteins give highly complex PMR spectra as a result of the overlapping of resonance peaks from similar chemical groups in slightly differing magnetic environments. The spectrum of ribonuclease is typical of the class (Fig. 8.5); this enzyme has the distinction of being the first protein to be examined by PMR spectroscopy, its 40 MHz spectrum being reported by Saunders *et al.* in 1957. A further factor adding to the complexity of protein spectra are the broad resonance widths normally exhibited by biopolymers, especially those of very high molecular weight. This broadening is due to anisotropic magnetic dipolar interactions between magnetic nuclei. In the case of small molecules in solution, broadening interactions of this type are largely averaged out by molecular tumbling so that resonance half-widths of 1 Hz or less are common. For larger molecules, rotation is slower (the rates of intra and inter molecular motions in the system are termed correlation times) and anisotropic dipolar broadening becomes significant. Local motions of segments of a large molecule such as the flexing of the polymer chain of a protein in the random-coil configuration, may reduce dipolar broadening relative to that for a rigid conformation of the same molecule, as mentioned later (McDonald and Phillips, 1969). Further discussion of resonance line shape is given when the applications of relaxation time measurements are described (p. 304).

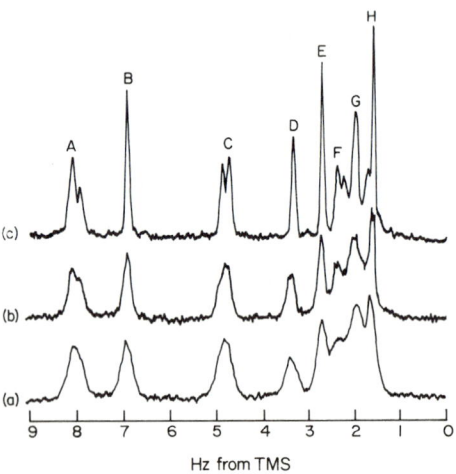

Fig. 8.6. PMR spectra of a solution of copoly (Glu 42%, Lys 28%, Ala 30%) in TFA showing the differences in resolution obtainable at different frequencies. (a) 60 MHz; (b) 100 MHz; (c) 220 MHz. Peak assignments: (A) amide NH, (B) lysine $NH_3{}^+$, (C) αCH, (D) lysine ϵCH_2, (E) glutamic γCH_2, (F) glutamic βCH_2, (G) lysine, β, γ, δCH_2, (H) alanine CH_3 (Bradbury and Crane-Robinson, 1968).

Spectral simplification may be achieved by increasing the strength of the applied field as is well illustrated in the spectra of a simple random co-polymer recorded at 60, 100 and 220 MHz (Fig. 8.6). However, even the most powerful spectrometer available, operating at 220 MHz (magnetic field 51,700 gauss), does not give sufficient resolution to allow the spectrum to be completely deciphered (Markley *et al.*, 1968b). In spite of this, important results have been obtained from study of the aromatic resonances which not only fall in a clear spectral range but are present in comparatively small numbers. Two studies in this regard are now discussed, namely those of insulin and ribonuclease.

Insulin, a comparatively small biopolymer (molecular weight ~5000) and of established structure, is cleaved by performic acid into two polypeptide chains (A and B) as S-sulphonates. Spectra of these fragments, of intact insulins, and of twenty-one commonly occurring amino acids were recorded at 220 MHz in TFA or CF_3CO_2D as solvent (Bak *et al.*, 1968); the results allow certain resonances, especially those in the aromatic region, to be assigned to specific protons in various amino acid residues of the A and B chains of insulin. In general, it appears that only moderate changes in the proton resonances of free amino acids follow their incorporation into a polypeptide chain, hence data upon the monomers may be used to make assignments to the spectra of polymers.† All spectra were single recordings of 5% (amino acids) and 10% (protein) solutions, and direct observation of one protein resonance per molecule was possible. In protein-PMR studies, cumulative spectra averaged with a computer of average transients (CAT) are commonly required because the net concentration of a particular amino acid residue may be low even in strong solutions of high molecular weight proteins. Some of the assignments made are now illustrated. The 1500–1600 Hz region of the spectrum of natural bovine A chain S-sulphonate displays three bands of intensities equivalent to 3,4 and 4 protons (Fig. 8.7β); intensities were obtained by integrating (by planimeter) small sections of the polypeptide spectra, chosen so as to represent regions with a known total number of protons. The lowest field band is assigned to the $^+NH_3$ protons of the terminal glycine residue since this band disappears after deuteration. Chain A is known to contain two tyrosine residues and has no other aromatic components. Tyrosine itself displays two doublets (A_2X_2 spectrum) at 1600 (*ortho* CH) and 1547 Hz (*meta* CH) which coincide fairly closely with the central and high field bands of Fig. 8.7β. The three-peak appearance of the central band shows that the magnetic environments of the *ortho* protons differ slightly in the two residues. The same spectral region of the B chain S-sulphonate (Fig. 8.7δ) is more complex. Of the integrated total of 31 protons, 6 are due to ^+NH (3 from terminal phenyl-alanine and 3 from the lysine side chain). Aromatic residues known to be present are 2 tyrosine, 3 phenylalanine and 2 histidine. Monomer charac-teristics of phenylalanine are two bands centred near 1640 and 1612 Hz of intensity ratio 3:2, and of histidine, a low field band at 1921 (H-2) and a high band at 1685 Hz (H-4) (see **12**). Peak a (Fig. 8.7γ) is assigned to the H(4) protons (one from each of the 2 histidine residues) while e, f, g, h and i

† A procedure for computing spectra of proteins in the random-coil configuration has been proposed, based upon the PMR characteristics of amino acids and small peptides (McDonald and Phillips, 1969). Allowance is made for spin–spin coupling and for the broader half-width of protein resonances (an average of 10 Hz is assumed for a signal not split by spin–spin interactions). Good correspondence between computed and actual spectra was found for lysozyme, ribonuclease, pepsin, trypsin and other examples.

are assigned to the tyrosine aromatic protons (cf. the appearance of the same signals of chain A, Fig. 8.7α). Signals b (3 protons), c–d (8) and part of e, f, g (8 − 4 = 4 protons) must be due to the 3 phenylalanine residues (15 protons in all). Two of the residues have almost coincident signals (close to those of the monomer) and represent 6 protons of the c–d band (integral 8) and the unassigned 4 of the e, f, g band. Signals of the third

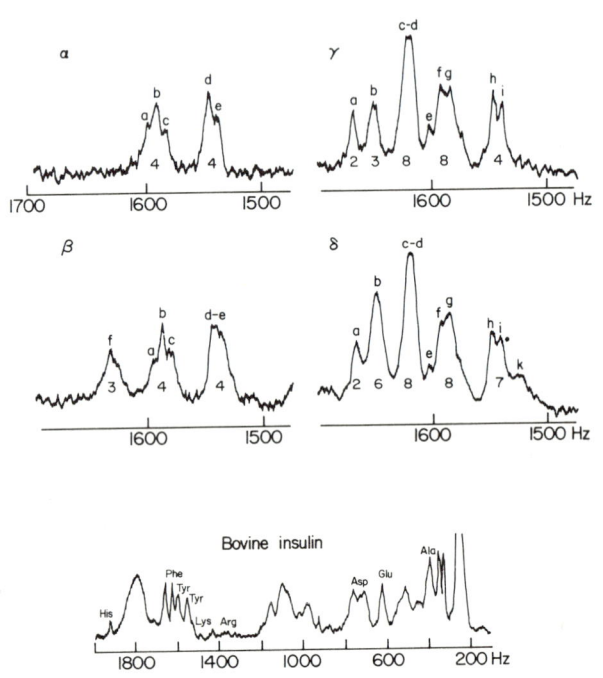

Fig. 8.7. 220 MHz spectra of deuterated bovine A chain S-sulphonate (α); natural bovine A chain S-sulphonate (β); deuterated bovine B chain S-sulphonate (γ); natural bovine B chain S-sulphonate (δ). Region 1500–1680 Hz relative to TMS (internal). The number of assigned proton resonances is indicated by 2, 3, 4, 6, 7, 8. The bottom spectrum is that of bovine insulin (Bak *et al.*, 1968). The solvent is TFA.

residue are moved to lower field forming a 3 proton band b and a 2-proton band which overlaps the lower field aromatic band of the other two residues (c–d, total integral 6 + 2 = 8). This analysis thus accounts for all signals in the aromatic region in terms of the known amino acid content of the A and B chains; furthermore, it enables environmental differences between identical residues to be detected. The spectrum of bovine insulin is less well resolved than those of its constituent chains (Fig. 8.7); nevertheless, signals due to many of its amino acid residues in the 1600 and high field (200–800

Hz) regions can be assigned. The H-2 signal of histidine near 1900 Hz is clearly defined and the unusually low field position of this signal has been particularly well exploited in the next example, ribonuclease.

Ribonuclease (RNase), a single chain enzyme of molecular weight near 17,000, is composed of 124 amino acids; four of these are histidine residues which occur at positions 12, 48, 105 and 119 in the chain. Jardetzky and his

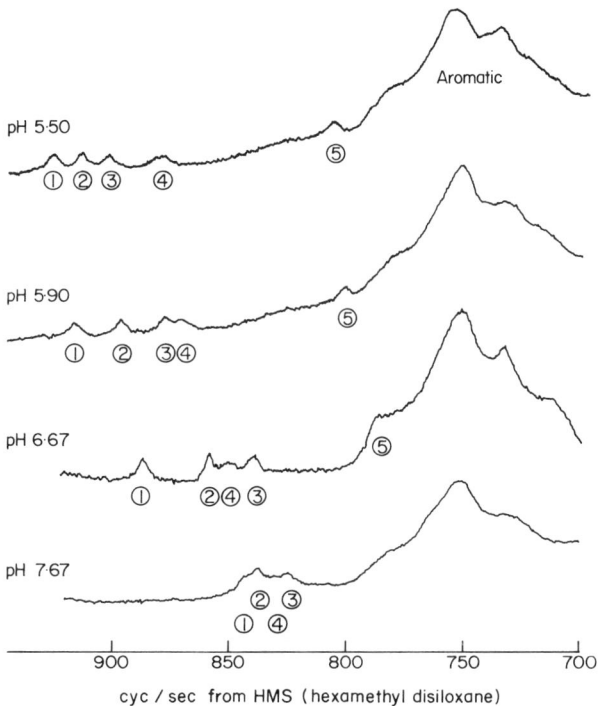

Fig. 8.8. 100 MHz PMR spectra of the aromatic region of ribonuclease (6.5×10^{-3} M in 0.2 M acetate, D_2O buffer) at different pH values. Peaks 1 to 4 are the 4 C-2 His peaks, and peak 5 is one of the C-4 His peaks (Cohen, 1969).

colleagues (Meadows *et al.*, 1967) recorded the spectrum of the enzyme at 100 MHz in a D_2O-acetate buffer system. RNase concentrations employed were about 0.01 molar ($\sim$10%) and spectra were the average of 20-180 CAT sweeps. Under these conditions all 4 C-2 protons of the histidine (His)† residues are clearly resolved (Fig. 8.8). Peak 5 is one of the C-4 protons of the His residues, the other 3 being buried under the large aromatic envelope

† The abbreviations His (histidine), Phe (Phenylalanine) and Tyr (tyrosine) are used henceforth in this account.

of Phe and Tyr absorptions. The chemical shifts of the signals vary with pH and reflect the degree of proton uptake at the basic centre of the imidazole nucleus (**12**), the C-2 (and C-4) resonance moving downfield as the propor-

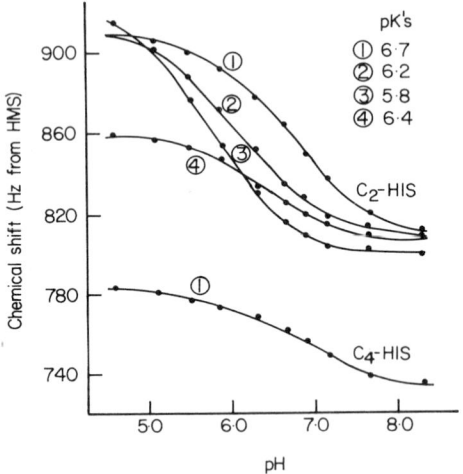

$$R = CH_2CHNH(\overset{O}{\underset{\|}{C}}\diagdown)$$

12

tion of protonated species increases. If the C-2 (or C-4) chemical shift is plotted against pH, a typical S-shaped titration curve is obtained and the

FIG. 8.9. Titration curves of histidine C_2-H peaks and one histidine C_4-H peak of RNase A. Reported pK's are for 32°C in 0·2 M deuteroacetate buffer (Meadows *et al.*, 1968).

pH at the midpoint corresponds with the pK_a of the imidazole residue (see Chapter 2). In this way the pK's of the 4 histidine residues were found to be roughly 5·8 (peak 3), 6·2(2), 6·4(4) and 6·7(1), (uncorrected for the presence of D_2O) (Fig. 8.9). The titration curve of the single C-4 resonance gave a pK_a of 6·7 and was thus identified with the same residue as peak 1.

The next problem was to assign peaks 1–4 to specific histidine residues. The proton of peak 4, which is in a quite different magnetic environment from those of the others, especially in the protonated state, is tentatively assigned to His-48 because this is known to be chemically inactive and X-ray data points to it being partially buried in the peptide chain. Under these conditions the His-48 imidazole ring might be shielded by its neighbours. Peaks 1 and 4 are unchanged in the presence of 3'-CMP and 5'-CMP (cytidine monophosphate), competitive inhibitors of the enzymatic activity of RNase. However peak 3 is broadened by 3'-CMP, while both peaks 2 and 3 shift downfield in the presence of 5'-CMP (Fig. 8.10). It is well

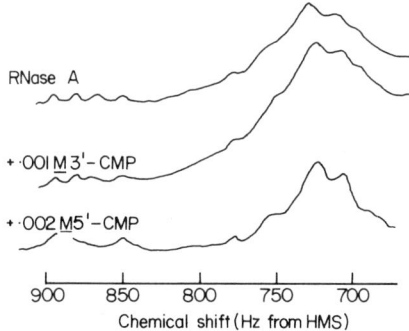

Fig. 8.10. Spectra of 0·1 M RNase A at pH 5·37 in the presence of inhibitors (RNase A alone, 14 CAT scans; RNase A + 0·001 M 3'-CMP, 50 CAT scans; RNase A + 0·002 M 5'-CMP, 25 CAT scans) (Meadows *et al.*, 1967).

established that His-12 and His-119 are involved in enzymatic activity and that the 2 residues are located near each other at the active site. The C-2 imidazole protons of these residues are therefore the ones most likely to suffer a change in environment upon uptake of an inhibitor; hence peaks 2 and 3 are assigned to residues 12 and 119. By elimination, peak 1 is probably due to His 105.

The next piece of evidence comes from the examination of 1-carboxy-methyl-His-119 RNase **13** and 3-carboxymethyl-His-12 RNase **14**, i.e. RNase derivatives known to be substituted at specific residues (Crestfield

13   **14**

R = rest of enzyme molecule

et al., 1963). The titration curves of peaks 1 and 4 are unaffected by the substituents at His-12 or His-119, but the histidine residues of peaks 2 and 3 *both* increase in pK_a to values of 6·7 and 6·9 respectively in the carboxymethyl derivatives. This result supports the view of the active site histidine residues being close to each other in the folded structure of the protein, since substitution on either of the two histidines changes the magnetic environment of both. The pK_a increase seen is unlikely to be due to a widespread conformational change because neither the pK_a values nor the chemical shifts of the other 2 histidine residues are affected. The

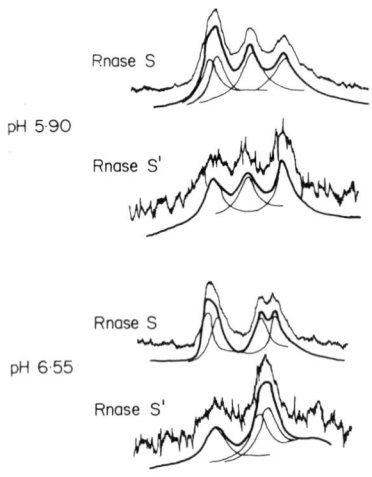

FIG. 8.11. Histidine C_2-H peaks of RNase S and RNase S′ (with His-12 C_2-H deuterated) at 2 pH's. Fitted curves, obtained with a Dupont 310 curve resolver assuming equal areas for all peaks, are shown under each spectrum (Meadows *et al.*, 1968). RNaseS is identical with RNaseA except for cleavage of the 20-21 peptide bond.

problem now is to differentiate the His-12 from the His-119 signal. To do this, use is made of the fact that the imidazole C-2 protons of histidine and its derivatives exchange for deuterium when the compounds are stored in D_2O (the C-4 protons do not exchange); in histidine and histidylhistidine, exchange is complete after 2 days at 37° and pH 9. It proved possible to remove selectively the C-2 proton of the His-12 residue as follows. When RNase is treated with a proteolytic enzyme (subtilism) obtained from *b. subtilis* it is cleaved at the 20–21 position to give the 20-residue peptide component (S-peptide) and the shortened protein (S-protein) (Richards and Vithayathil, 1959). These are separated, the S-peptide maintained at 40°C for 5 days at pH 7 (when no C-2 proton is detectable by PMR) and then combined to give the reconstituted enzyme (RNase S′). The C-2 resonance region of RNase S′ shows 3 peaks instead of 4 (Fig. 8.11) and it can be seen that the missing His-12 proton is in the region where peaks 1 and 2 overlap at nearly all pH values in the spectrum of RNase itself. Peak 1 had already been assigned to His-105, therefore peak 2 must arise from the C-2 proton of His-12. The complete assignment is therefore: (1) 105, (2) 12, (3) 119, (4) 48.

The influence of various inhibitors, anions, solvent and temperature upon the histidine residues of RNase have been studied by following changes in the C-2 proton resonances with the aim of learning more about the mechanism of action of the enzyme in solution (Meadows and Jardetzky, 1968; Roberts *et al.*, 1968). It is deduced, for example, that only the protonated forms of the histidine residues 12 and 119 play an active role in the interaction of RNase with cytidine-3′-phosphate (3′-CMP **15**), because the pH for maximum binding of this inhibitor (pH 5·5) corresponds with that at which the His-12 and His-119 residues are fully protonated, as is evident from the titration curves (Fig. 8.12). The data for these curves were recorded

15

from spectra of RNase containing the saturation level of 3′-CMP, and demonstrate that only His-12 and His-119 are affected by the inhibitor, as mentioned earlier (under these conditions, C-2 of His-48 is only visible at low pH values).

Another approach to the problem of simplifying the spectra of proteins is that of selective deuteration. A deuterated staphylococcal nuclease has been obtained by growing a strain of *staphylococcus aureus* in D_2O upon a mixture of deuterated and protonated amino acids; this medium was derived by the hydrolysis of protein isolated from algae grown in 99% D_2O (Markley *et al.*, 1968a, b). The isotopic composition of the individual amino acids was determined from the PMR spectra of the separated components. Resonance lines were undetectable except for cases in which D/H exchange occurred during hydrolysis, e.g., at the 2 and 6 positions of the tyrosine ring and the 2-position of histidine. The only fully protonated amino acids added to the growth medium were tryptophan (Try) and methionine (Met); one residue of Try and 4 of Met are present in staphylococcal nuclease. The aromatic region of normal and selectively deuterated enzyme are shown in Fig. 8.13. The C_2—H protons of the 4-His residues are clear in both spectra and the identical chemical shifts of corresponding signals suggests that the 2 enzymes do not differ in conformation. Signals due to the single Try residue are assigned on the basis of its unique splitting pattern. The 7 Tyr residues give rise to singlets, because only the 2- and 6-positions are

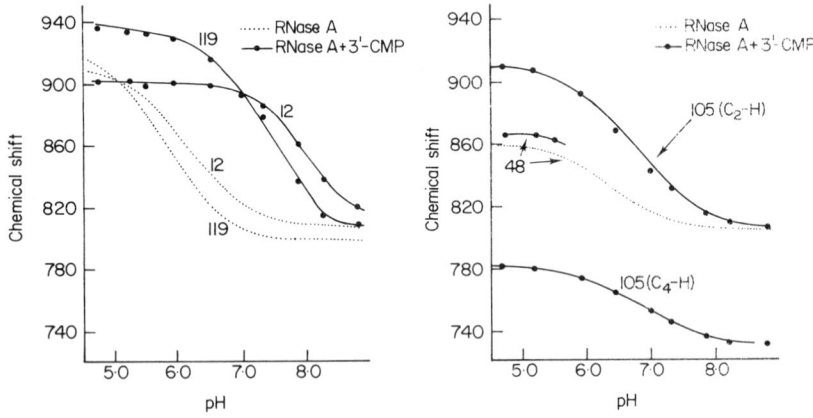

Fig. 8.12. Titration curves of His peaks in the RNase-3'-CMP complex. Concentrations are RNase, 0·0065 M; 3'-CMP, 0·03 M. Above pH 7 the enzyme is no longer saturated with inhibitor. Dashed lines show titration curves of His peaks in RNase alone. In the case of the 105 His curves (right hand plot) the lines for the complex and RNase alone almost coincide and only a single line is shown (Meadows and Jardetzky, 1968).

protonated, and these are assigned as shown. In the aliphatic spectral region, it proved possible to assign singlets due to the 4 methionine S-methyl groups; two were close to the resonance position of the monomer (247 Hz) while the other pair were much higher field (169 and 162 Hz). Other assignments also deviated in the same direction from corresponding values in monomers and simple peptides, e.g. the proton signals of the single Try residue lie 53 Hz upfield from that of the dipeptide tryptophanylalanine. Upfield shifts of these magnitudes could be due to aromatic diamagnetic shielding as occurs upon stacking of aromatic rings, and may therefore provide evidence about the tertiary structure of the protein molecule. The synthesis of proteins fully deuterated except for leucine side chains has also been reported (Crespi *et al.*, 1968; Crespi and Katz, 1969).

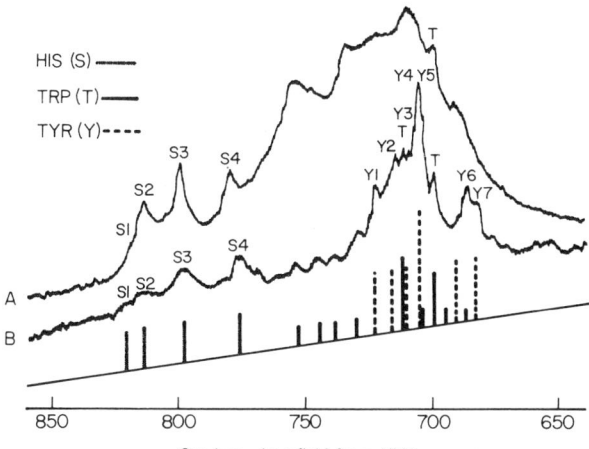

FIG. 8.13. 100 MHz PMR spectrum of (a) 20% solution of protonated staphylococcal nuclease in D_2O, 60 scans; (b) 6% solution of selectively deuterated analogue of nuclease, 129 scans. The bar graph indicates the predicted intensities of the spectral lines. The nuclease from the Foggi strain contains 4 histidine (S), 7 tyrosine (Y), 1 tryptophan (T) and 3 phenylalanine residues (Markley *et al.*, 1968a).

PMR spectroscopy provides a good means of monitoring helix-coil transitions of polypeptides provided resonance lines characteristic of each form can be resolved (Ferretti, 1967). A good example is the case of poly-γ-benzyl-L-glutamates (Bradbury *et al.*, 1968). The spectrum of the polypeptide with a degree of polymerization (dp) of 92 shows duplicate signals for both the amide NH and αCH resonances (Fig. 8.14). Changes in the

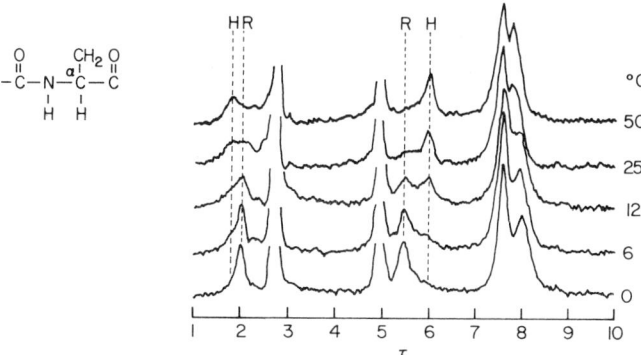

FIG. 8.14. 100 MHz PMR spectra of poly-γ-benzyl-L-glutamate dp 92 showing the temperature induced helix-coil transitions. N—H signals (H = helix and R = random coil) fall near τ 2·5 and αCH near τ 5 (Bradbury *et al.*, 1968). Solvent, 8% TFA plus 92% $CDCl_3$.

intensities of these signals occur as the temperature is varied. It is known from optical rotatory dispersion (ORD) data that the polymer is almost random at $0°$ and largely helical at $50°$ in the solvent mixture 8% TFA-92% $CDCl_3$. On these grounds, the higher field αCH and the lower field NH are assigned to the helical form and their partners to the random coil. The αCH peaks are particularly well separated (50 Hz at 60 MHz) and changes in their relative areas provide a means of analysing the random coil-helix content of a particular solution. Area measurements are more difficult when polymers of higher dp are studied because the two components of the αCH resonance peak broaden markedly with increasing size (Fig. 8.15) due to increasing correlation times. For the largest polymer (dp 640), 2 peaks are no longer seen and, instead, there is a broad single band at an inter-

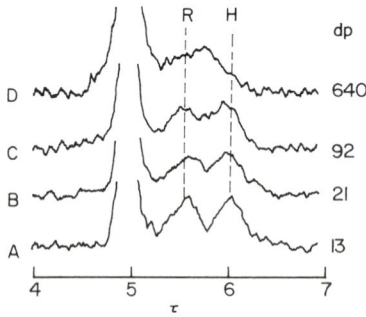

FIG. 8.15. 60 MHz PMR spectra of poly-γ-benzyl-L-glutamate polymers showing the effect of molecular weight on the components of the αCH peak. Solvent compositions were chosen so as to give approximately 50% helix (from ORD measurements) for all samples (Bradbury *et al.*, 1968).

mediate position. Greater accuracy may be achieved by recording spectra at 100 MHz, as illustrated in Fig. 8.16 for the polymer of dp 92. The overlapping curves may be analysed by using the shape of the non-overlapping edges of the benzyl CH_2 resonance and the αCH resonance peak of the helical form to obtain the complete shape of the respective peaks. In this way the scatter of results obtained was less than 5% in the determination of the ratio

$$\frac{\text{area of the } \alpha CH \text{ helix peak}}{\text{total area of } \alpha CH \text{ peak}} \text{ for the different solutions.}$$

This technique is considered to permit an unambiguous determination of the helix-content which can be used to calibrate analytical methods based on ORD and optical density (McTague *et al.*, 1964). Spectra of partially helical states of the largest polymer (dp 642) display a single broad band even at 100 MHz and the analysis described above cannot be applied in such cases.

Signal widths are broader in the more rigid helical conformations of polypeptides and differences in the line widths of helical and random coil signals provide information about their relative degrees of rotational freedom. Similarly, but on a different scale, proton resonance signals due to proteins in folded conformations are broader than those in the unfolded state because the side chains attached to the backbone of the folded molecule are restricted in their movements and are less exposed to the solvent. Pronounced sharpening of various proton bands have been observed upon

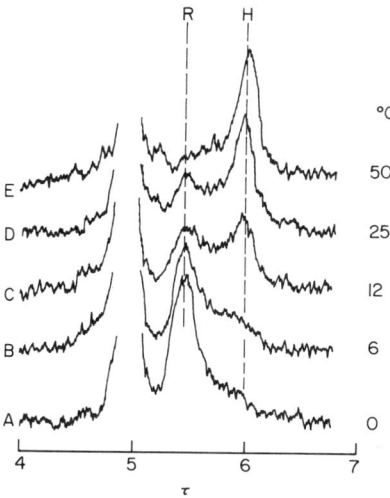

Fig. 8.16. 100 MHz PMR spectra of poly-γ-benzyl-L-glutamate dp 92 showing the αCH peaks as the polymer passes through the helix-coil transition (Bradbury *et al.*, 1968).

conformational unfolding (denaturation), as is illustrated in the case of lysozyme (Fig. 8.17).

The thermal unfolding of α-chymotrypsin at acid pH has been followed quantitatively by a study of changes with temperature in the high field portion of the PMR spectrum of the protein (Hollis *et al.*, 1967). A pronounced change in the region of 0·6 ppm (corresponding to absorptions of the methyl protons of valine, leucine and isoleucine) occurs as the temperature is raised, the peak intensity increasing and the overall signal width narrowing (Fig. 8.18). Denaturation of the protein may also be followed by changes in optical density at 293 mμ, and plots of relative intensity changes of spectra (ultra-violet or PMR) against temperature are similar and enable the fraction of molecules in the denatured state at any temperature to be

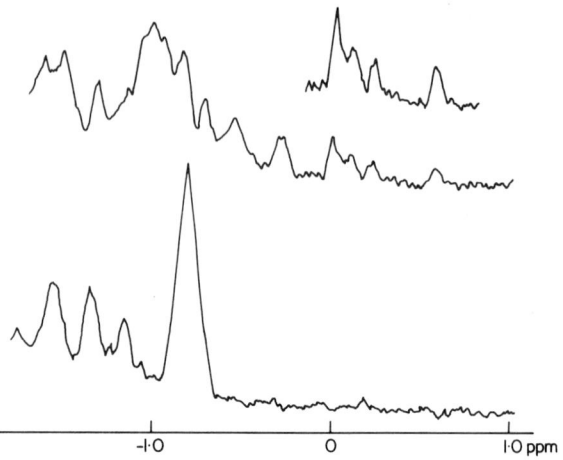

FIG. 8.17. 220 MHz PMR spectra in the high field region of lysozyme at 64°, native (top), and at 78°, denatured (bottom) in D_2O, pD 5·5 (McDonald and Phillips, 1967).

calculated. The PMR spectrum remains sensitive to temperature even after denaturation is judged complete (from the levelling off of the ultra-violet intensity versus temperature plot), and allowance is made for this factor in the calculations. On this basis, dimethionine sulphoxide chymotrypsin, a derivative in which the 2 methionine residues of α-chymotrypsin are oxidized to sulphoxide groups, is judged to be about 40% unfolded at pD 4 and 20°C.

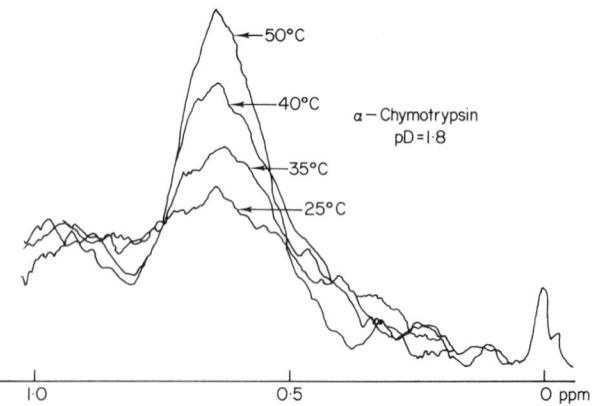

FIG. 8.18. Part of the 100 MHz PMR spectra of α-chymotrypsin at pD 1·8 and various temperatures. Spectra are the time average of 25 scans of approximately 0·8% protein solutions (Hollis *et al.*, 1967).

The 220 MHz spectrum of lysozyme, an enzyme obtained from egg-white, exhibits the usual increase in resolution and sharpening of peaks as the temperature is raised (in this case from 30 to 80°) that is attributed to denaturation of protein molecules (McDonald and Phillips, 1967). Striking changes in the high field region of the spectrum occur when the temperature is lowered below 74° which reflect conformational changes taking place as the protein refolds to assume its native form. The sharp intense high field resonance at δ 0·85, assigned to the methyl groups of leucine, isoleucine and valine, decreases in intensity while at the same time resonances emerge in the δ 0·7 to −0·7 range (Fig. 8.17). The unusual positions of these signals are considered due to high-field ring current shifts induced in the C—H protons of amino acid residues such as leucine, etc. whose side chains are close to the faces of the aromatic groups of histidine, phenylalanine, tyrosine and tryptophan residues in the protein *folded* conformation. The minimum approach of the protons, for example, of a methyl group to a benzene nucleus along the six-fold axis of the ring would be about 3·0 Å and in this approach the protons should be subject to a diamagnetic ring current field equivalent to a shift of about 1·8 ppm (see p. 93). On this basis shifts experienced by certain methyl and methylene groups of lysozyme (at least 1·4 ppm) are of the anticipated order.

Abnormally high field resonances are also seen in the spectrum of the native form of cytochrome C, a member of a class of proteins that contain a polypeptide chain linked to a porphyrin ring with a single iron atom at its centre. There is no PMR absorption in the region immediately upfield from the δ 0·85 resonance line of the spectrum of the denatured form of either the oxidized (ferri) or reduced (ferro) form of the proteins recorded above 60°. As the temperature is reduced, 2 groups of satellite bands appear, one in the region δ +0·5 to −1·0, and the other even higher field between δ −2·0 and −3·8. The former is concluded to have a similar origin as that observed in lysozyme, i.e. C—H protons shielded by the aromatic rings of phenylalanine, etc. The shielding magnitudes experienced by protons of the second band are too great to be accounted for by the ring current effects of small aromatic ring systems (the separation between methyl protons and a benzene ring would need to be 2·1 Å if they are to experience a shift of 4·2 ppm and this distance is less than the permitted minimum). A far more likely shielding source in regard to the second band is the highly conjugated porphyrin unit of the protein molecule since large ring current shifts of 10 ppm or more have been associated with this extensively de-localized ring system (Caughley *et al.*, 1967). In the spectrum of the methyl ester of coproporphyrin 1 (**16**), for example, the signal of the 4 equivalent methine protons (α, β, γ, δ) has a chemical shift of δ 9·96 which is about 2·6 ppm lower field than the benzene proton signal, while the ring methyl

M = Me

P = $CH_2CH_2CO_2Me$

π-electron system
extends over the closed conjugated
path indicated by heavy lines

16

signal (δ 3·55) also has an unusually low field position (cf. $\underline{Me}C_6H_5$ in toluene, δ 2·32) (Becker *et al.*, 1961). On the other hand, the $\overline{NH}$ protons, positioned at the centre of the ring current system, come to resonance at the abnormally high field position of δ −3·89; this chemical shift is about 13 ppm higher field than that of the NH resonance of pyrrole. The absence of high field signals in the spectra of metal complexes of porphyrins confirms their assignment to NH.

The spectrum of the oxidized form of cytochrome C shows severely broadened signals at about δ +25 (low field) and δ −33 (high field) in addition to the high field resonances already mentioned (Kowalsky, 1965). These bands are absent in reduced cytochrome C and are thus attributed to interactions between the delocalized unpaired electron of the paramagnetic haeme iron atom and protons on the ligands bound directly to Fe^{3+}. Shifts so induced may either be to high or low field and are termed contact shifts (Eaton and Phillips, 1965), and the resonance lines are broadened as a result of the shortening of relaxation times caused by the paramagnetic iron atom (see p. 305). Information about ligands coordinated to Fe^{3+} in haeme proteins is potentially available by examination of contact resonance bands since they have the advantage of being far removed from overlapping resonances in the normal chemical shift range; work of this nature has been reported for haemoglobin and related compounds (Wüthrich *et al.*, 1968; Davis *et al.*, 1969). The recent use of shift reagents containing a paramagnetic ion for spectral simplification is based on similar principles (Sanders and Williams, 1971).

THE STUDY OF SPECIFIC MOLECULAR INTERACTIONS BY NUCLEAR MAGNETIC RELAXATION MEASUREMENTS

The potential utility of NMR as a method for the study of molecular interactions, particularly of the type in which small molecules are bound by macromolecules, is now well recognized. A full account of the principles involved has been given by Jardetzky (1964) and only a brief outline of some of the salient features is given below.

In a PMR experiment protons in the lower energy spin state (α) absorb energy and move to the higher energy level (β). A key factor which determines the shape of the resultant PMR signal is the rate at which the nucleus returns to the lower energy level since the width of a spectral line is proportional to the reciprocal of the average time the system spends in the excited state (width $\propto$ 1/lifetime in excited state). The rate at which nuclei in the β state revert to the α state is termed the "relaxation time". Sharp signals are obtained from slow rates (long residence time in β-state) and broad signals for fast rates (short residence time in β-state). Very fast rates, as occur in solids and macromolecules, yield signals which are so broad that they often escape detection. There are two mechanisms of relaxation. In one, there is a redistribution of absorbed energy within the spin system in which every downward transition of a nucleus ($\beta \rightarrow \alpha$) is accompanied by an upward transition of another nucleus ($\alpha \rightarrow \beta$); this is termed "spin–spin" or "transverse relaxation" and is characterized by the time constant T_2 and relaxation rate $1/T_2$. In the other mechanism, the energy released by a $\beta \rightarrow \alpha$ transition is lost as thermal energy to the lattice, a term used to describe the entire molecular system irrespective of the physical state of the material; this process is called "spin–lattice" or "longitudinal" relaxation and is characterized by the time constant T_1 and relaxation rate $1/T_1$. Spin–spin relaxation rates ($1/T_2$) are obtained from measurements of the width of a PMR signal at half its maximum height (Fig. 8.19). Details of the procedure are given later together with a method for measuring $1/T_1$. Jardetzky (1964) has made an extensive analysis of all the factors which influence the $1/T_1$ and $1/T_2$ values. In both cases these include contributions due to intermolecular interactions, to paramagnetic species in solution, and to any inhomogeneity of the applied magnetic field. The most significant factor from the point of view of the study of molecular interactions, however, is the "intramolecular" contribution. This depends on the distance between a nucleus and its nearest neighbour(s) and also upon a constant called the "correlation time" which can be taken as a crude measure of the time during which 2 spins maintain a given orientation relative to one another.

The relationship for the intramolecular term, written in a simplified form is

$$\frac{1}{T_{1(\text{or}2)}} \cong C\tau_c \sum \frac{1}{\langle r_{ij} \rangle^6} \tag{1}$$

where r_{ij} is the average distance between nuclei i and j, C is a constant, and τ_c the correlation term.

It follows that if r remains constant, the relaxation rate increases as the correlation time becomes longer; this increase will be reflected in a progressive broadening of the magnetic resonance signal. Factors which increase τ_c need to be considered next. Correlation times for small molecules freely rotating and tumbling in solution are very short. However, when a small molecule is bound to a macromolecule its rates of molecular motion, particularly rotational motion, are generally greatly diminished, and restrictions of this nature will be revealed by increases in the relaxation

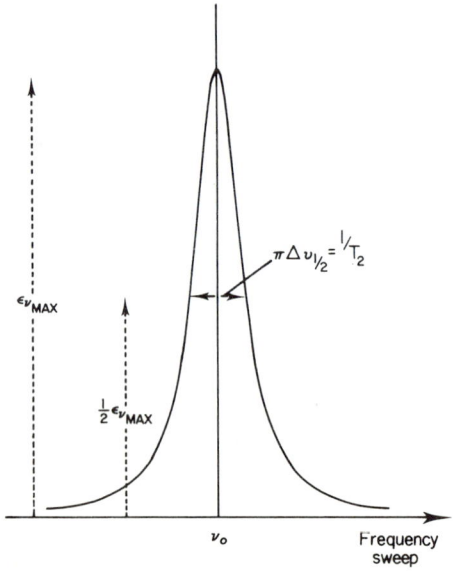

Fig. 8.19. Line shape of an NMR absorption signal showing the relationship of the line width at half height to the relaxation time (Cohen, 1969).

rates of the protons in the bound molecules. Herein lies the fundamental basis for the application of relaxation measurements to molecular interactions. It is probable that the rotational freedoms of the various parts of a bound molecule will not be affected to the same extent, the more tightly attached units being more restricted (often termed "stabilized" or "frozen") than features not so directly involved at the site of binding. In such cases a *selective* change in correlation times will occur and be detected by a *selective* broadening of the PMR signals in the spectrum of the bound molecule.

Meaningful interpretations of relaxation rate measurements require making allowance for contributions due to nonspecific effects and the elimination, as far as possible, of factors other than that of the intramolecular term. A correction for intermolecular contributions may be made by extrapolating the observed $1/T_1$ or $1/T_2$ values to infinite dilution in a deuterated solvent, thus eliminating terms arising from solute-solute and solute-solvent interactions respectively. The field inhomogeneity term can be evaluated by carrying out relaxation measurements at 2 different field strengths (e.g. 60 and 100 MHz). The system should be free of paramagnetic impurities, including dissolved oxygen, so that the paramagnetic term is a minimum; any residual contributions in this respect can be gauged by a change in the relaxation rate of the solvent signal. It is fortunate that the dominant contribution to the relaxation rate of any given proton arises from its nearest neighbour in consequence of the $1/r^6$ relationship [see Equation (1)]. This fact simplifies the interpretation of changes in $1/T_{1(2)}$ in terms of a change of correlation time or internuclear distances. The distance factor needs to be considered in the case of groups like $R_2CH-CHR_2'$ when the relaxation rate can be varied by a factor of 7 by rotation about the C—C bond from a *trans* to an eclipsed H/H orientation, but not

for groups such as CH_3, CH_2 and in which the distances between adjacent protons are fixed.

Although $(1/T_2)$ rates involve more variables than spin–lattice relaxation rates, changes in their values may be detected readily from a normal PMR spectrum and gauged quantitatively from line width measurements. Most relaxation studies discussed here, therefore, are based on $1/T_2$ measurements. When signals are narrow, factors other than the intramolecular contribution are difficult to assess, and in these cases $1/T_1$ values provide more accurate data. In most organic liquids and solutions T_1 and T_2 have similar or identical values and data are usually expressed in terms of one or the other even when specific methods for measuring T_1 and T_2 are employed. Most of the examples which follow are studies of the influence of a macromolecule upon the PMR spectrum of a small molecule. Resonance bands due to the former are not usually apparent in the composite spectrum because their widths are so broad that the bands are lost in the noise level of the base line.

The first application of these principles to be reported in full was a study of the binding of benzylpenicillin (penicillin G) to bovine serum albumin (Fischer and Jardetzky, 1965); this work will be described in some detail because it illustrates well the techniques and safeguards necessary in relaxation experiments and also shows how the results may be interpreted.

The spectrum of penicillin G in D_2O displays a series of narrow signals. When albumin is present all the lines are broadened but the extent of broadening varies considerably, as shown in Fig. 8.20. Values of $1/T_2$ for

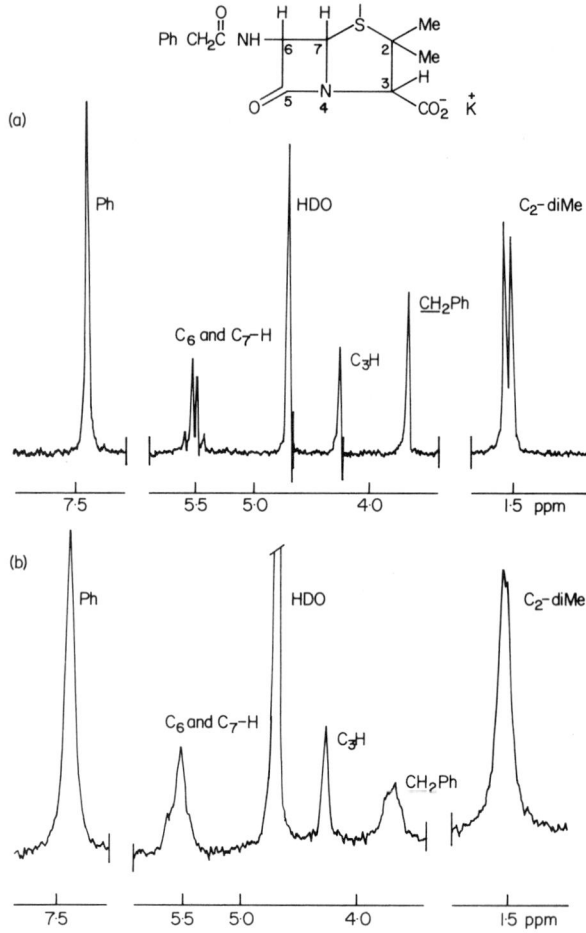

Fig. 8.20. 60 MHz PMR spectrum of (a) penicillin G (0·33 M) in D$_2$O; (b) penicillin G (0·33 M) plus bovine serum albumin (12%) in D$_2$O.

the various PMR signals of penicillin G are obtained from the spectral line widths using the formula

$$1/T_2 \ (\text{sec}^{-1}) = \pi \Delta\nu_{1/2}$$

where $\Delta\nu_{1/2}$ is the line width at one-half the maximum peak height. Line width measurements require a high degree of accuracy and it is essential that spectra be obtained under non-saturation conditions. Peaks are recorded employing sweep widths lower than the normal value of 1 Hz per mm but at normal rates (~2 Hz/sec). Line widths are measured on peaks bracketed between 2 side bands of fixed frequency included in the spectrum by means of an audio oscillator; this procedure is more reliable than that of using the calibration of the chart paper itself. It is also important to record the spectrum sufficiently on either side of a peak so that the base line position may be located correctly.

In the absence of binding agents, the PMR signals of small molecules are narrow ($\Delta v_{1/2}$ <1 Hz) and T_2 values cannot be obtained accurately from their widths at half height. In such cases the longitudinal relaxation time (T_1) is measured using the direct saturation method. In this procedure a resonance line is saturated by application of a strong RF field. When the RF field is suddenly reduced to a non-saturation value, the signal recovers exponentially with a time constant which is essentially T_1 (Fig. 8.21). In most applications of relaxation time measurements to the study of the interactions between large and small molecules, it is assumed that T_1 equals T_2, as already stated.

A number of control experiments are necessary to establish the extent to which changes in relaxation times may be attributed to binding factors. Marked changes in both the chemical shifts and $1/T_2$ values of resonance lines of penicillin G occur as the concentration of the antibiotic in D_2O is raised from 0·1 to 2·0 M (moles per litre); data for the aromatic signal, for example, are as follows: v 450 Hz from standard, $\Delta v_{1/2}$ 5·84 Hz (0·5 M); v 440 Hz, $\Delta v_{1/2}$ 19·4 (2 M). Allowance for the contribution of drug–drug interactions to the observed line broadening must therefore be made. It is found, however, that changes comparable to those seen on addition of albumin to dilute penicillin solutions (e.g. 0·2 M) are not observed in the absence of protein until a concentration of 1–1·5 M is reached. In simple solutions, changes in relaxation rates are paralleled by changes in chemical shift as the antibiotic concentration is increased. This is not the case when line broadening follows addition of albumin, e.g. aromatic signal for 0·5 M penicillin v 448 Hz (0% protein); 447·7 Hz (5% protein); 447·9 Hz (10% protein). It is assumed, therefore, that the peak broadening observed in penicillin-albumin solutions is not due to interactions between drug molecules.

The line widths of penicillin PMR signals broaden progressively as the concentration of albumin is increased, and the possibility of these changes being non-specific effects due to increases in the viscosity of the solution must be examined. The fact that the viscosity of a 5% solution of albumin in D_2O is only 1·25 times that of pure D_2O, an increase too small to be the sole cause of the observed effects, proves, however, that viscosity changes are not the prime cause of line broadening. In contrast, the influence of γ-globulin, a protein which is known not to bind penicillin, may be accounted for by viscosity changes, and the small influence of this protein upon $1/T_2$

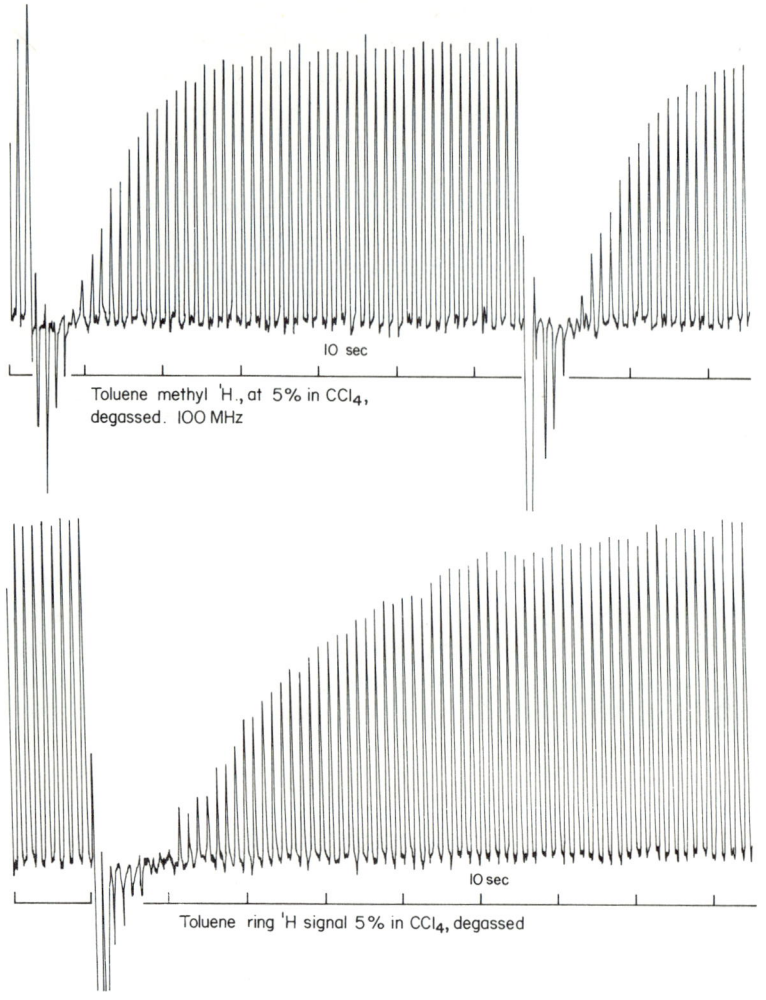

Toluene methyl ^{1}H., at 5% in CCl$_4$, degassed. 100 MHz

10 sec

Toluene ring ^{1}H signal 5% in CCl$_4$, degassed

10 sec

FIG. 8.21. Recording showing recovery of the toluene aromatic and methyl PMR signals following saturation (Martin, 1970).

T_1 is obtained using the equation

$$A_t = A_\infty(1 - e^{-t/T_1})$$

where A_∞ = equilibrium intensity and A_t = intensity at time t.

Rearranging, $A_\infty - A_t = A_\infty e^{-t/T}$

$$\ln(A_\infty - A_t) = \ln A_\infty \frac{-t}{T_1}$$

or

$$\log(A_\infty - A_t) = \frac{\text{constant}}{2\cdot3} - \frac{t}{2\cdot3T_1}$$

The slope of a plot of $\log(A_\infty - A_t)$ against t in seconds gives $-1/2\cdot3T_1$, whence T_1 may be calculated.

compared with that of albumin (Fig. 8.22) provides further evidence that line broadening due to the latter protein is a specific effect.

The results of a series of measurements for a fixed penicillin and varying albumin concentrations are shown in Fig. 8.23. Although the CH$_2$Ph peak is the broadest, the relaxation rate of the aromatic (Ph) peak is changed by a larger factor ($\times$18 approx.) than the rates of either the C$_3$H peak ($\times$6 approx.) or the CH$_2$Ph peak ($\times$11 approx.). Consideration of relative increments rather than absolute extents of broadening, leads to the conclusion that it is the aromatic protons of the penicillin molecule that are primarily involved

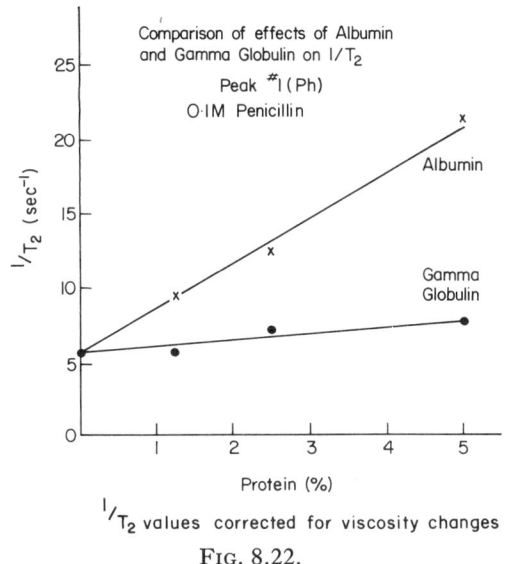

FIG. 8.22.

in the binding of the drug to the protein. If a non-specific mechanism were operating, addition of more penicillin to a given penicillin-albumin solution should *increase* the width of the lines as a result of increases in viscosity and drug-drug interactions. In fact, *decreases* are observed. This must mean that broadening is due to a specific interaction between drug and protein which is saturable, i.e., a given albumin molecule can interact with only a limited number of penicillin molecules at one time. Once saturation conditions are obtained, addition of more penicillin increases the population of unbound molecules and the line widths of the PMR signals decrease accordingly. It follows, and is confirmed by experiment, that the relaxation rate depends on the penicillin-albumin ratio rather than the albumin concentration itself.

A better understanding of the reason why albumin has a pronounced effect on line widths but a negligible one on the chemical shifts of PMR signals, follows from estimates of the fraction of penicillin molecules bound to the protein. This fraction is expected to be small because of (i) the large molar ratio of drug:protein (about 200:1) and (ii) the number of binding sites per albumin molecule is believed to be less than 5 on the basis of studies using very dilute penicillin solutions. Dialysis experiments give a rough estimate

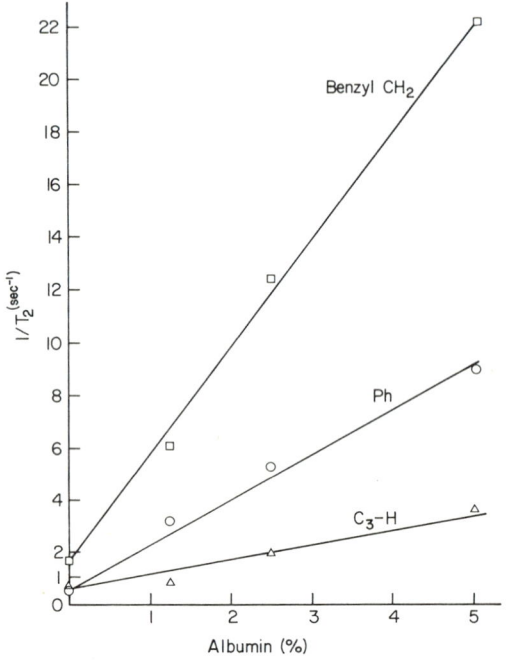

Fig. 8.23. Effect of albumin on the $1/T_2$ relaxation values of three PMR signals of penicillin G (Fischer and Jardetzky, 1965).

of the extent of binding and show that there are never more than 2 or 3 molecules of penicillin bound per molecule of albumin. This result explains the lack of effect of albumin on chemical shifts. Even if the chemical shift of a proton group is radically changed upon drug binding (a maximum of about 3 ppm caused by placing a proton into the diamagnetic region of an aromatic ring structure can be estimated), the contribution of the bound species to the time-averaged signal of the free and bound molecules would be insignificant because only 1% or so of the molecules are in the bound form (this argument assumes rapid exchange, see below). The fact that the PMR signal—which is due largely to free species—displays significant broadening means, therefore, that the relaxation times of the free and bound forms must differ by much larger factors than those of the chemical shifts of the two species. Signals due to the bound form would not be detected in isolation because of their very low intensities and broad nature, but are revealed by their ability to modify the free signal, a phenomenon which is made possible by the rapid exchange rate between free and bound forms. An apposite way of expressing this situation is to say that "the exchange process acts as an amplifier for the detection of the relatively small fraction of hapten (small molecule) bound by its effect on the spectrum of the excess free hapten" (Metcalfe *et al.*, 1968).

To explain the results of the penicillin-albumin experiments quantitatively, a kinetic model is proposed which takes account of the exchange rate between free and bound molecules (T_{exchange}) as well as the relaxation times. If the 3 rates are related as

$$\frac{1}{T_{\text{exchange}}} < \left(\frac{1}{T_2}\right)_{\text{free}} < \left(\frac{1}{T_2}\right)_{\text{bound}}$$

Rate: very slow   slow   fast

2 superimposed lines, 1 broad (and probably unresolvable) and 1 narrow, would be seen. This result does not occur. If the relationship is

$$\left(\frac{1}{T_2}\right)_{\text{free}} < \frac{1}{T_{\text{exchange}}} < \left(\frac{1}{T_2}\right)_{\text{bound}}$$

Rate: very slow   slow   fast

relaxation of the bound molecules will be complete before exchange takes place hence relaxations of the free molecules cannot be modified by the bound forms. The relaxation rates of the free molecules, however, will be determined by the exchange rate and all peaks observed (essentially representing the free molecules) will have approximately equal widths. Again, this result is not obtained. Finally, if the relationship is

$$\left(\frac{1}{T_2}\right)_{\text{free}} < \left(\frac{1}{T_2}\right)_{\text{bound}} < \frac{1}{T_{\text{exchange}}}$$

Rate: slow   fast   very fast

each relaxation process will have contributions from both free and bound species, and the time averaged relaxation rate of the two forms is given by

$$\frac{1}{T_2} = B\left(\frac{1}{T_2}\right)_{\text{bound}} + (1 - B)\left(\frac{1}{T_2}\right)_{\text{free}}$$

$$= \left(\frac{1}{T_2}\right)_{\text{free}} + B\left[\left(\frac{1}{T_2}\right)_{\text{bound}} - \left(\frac{1}{T_2}\right)_{\text{free}}\right] \qquad (2)$$

where B is the fraction of the total penicillin bound. This model correlates best with the experimental data.

The problem is how to calculate $(1/T_2)_{bound}$. Let P and A be the total penicillin and albumin concentrations respectively, and assume that there are n non-interacting binding sites on each protein molecule.

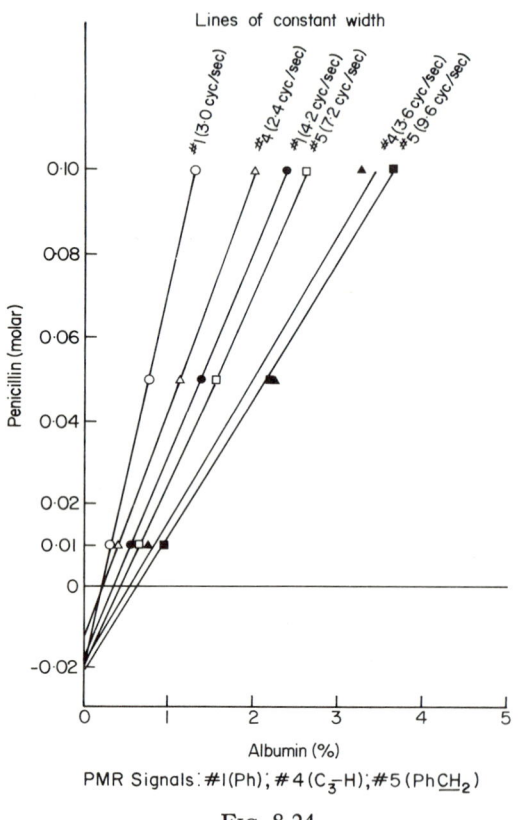

Lines of constant width

Penicillin (molar)

Albumin (%)

PMR Signals: #1(Ph); #4(C$_3$-H); #5(Ph$\underline{CH}_2$)

Fig. 8.24.

Then, on the basis of the law of mass action,

$$K = \frac{[\text{unbound penicillin}][\text{unbound sites}]}{[\text{bound penicillin}]}$$

$$= \frac{P(1-B)(nA-PB)}{PB}$$

or, on rearranging

$$P = \frac{K}{B-1} + \frac{nA}{B} \tag{3}$$

To evaluate B [and hence $(1/T_2)_{\text{bound}}$ from Equation (2)] and K, P and A must be plotted under conditions where B is a constant. P and A values which satisfy this requirement are established by finding various combinations of P and A concentrations that produce some arbitrarily chosen line width since the $\Delta \nu_{1/2}$ value will reflect a specific bound:free ratio. A plot of P and A values corresponding to 2 different widths and presented for each of 3 different peaks is shown in Fig. 8.24. From Equation (3) it follows that plots of P against A, with B constant, will be straight lines with slope n/B and will intersect the $A = 0$ axis at $-K$ (for $B \ll 1$). The straightness of the lines in Fig. 8.24 and the fact that they all extrapolate to roughly the same point are support for the model chosen. To evaluate $(1/T_2)_{\text{bound}}$ from Equation (2) it is necessary to know B. Fig. 8.24 gives the value n/B. If calculations are carried out using $n = 1$ the results can be fitted to any value of n by

$$\left(\frac{1}{T_2}\right)_{\text{bound}, \, n} = \left(\frac{1}{T_2}\right)_{\text{bound}, \, n=1}, \quad \text{divided by } n$$

Since interest lies chiefly in the relative values of $(1/T_2)_{\text{bound}}$ for various proton groups, knowledge of the true value of n is not essential. Results of these calculations, shown in Table 8.1, illustrate the radical differences in

TABLE 8.1

Calculation of relaxation rates for bound penicillin

Peak	Width Hz	$1/T_2$ sec^{-1}	$1/T_{2\,\text{free}}$	n/B	$1/T_{2B}$	Av. $1/T_{2B}$	$T_{2\,\text{free}}/T_{2B}$
Ph	3·0	9·4	0·555	621	5550	4990	9000
Ph	4·2	13·2	0·555	350	4430		
C_3—H	2·4	7·55	0·667	372	2560	2380	3560
C_3—H	3·6	11·3	0·667	207	2200		
CH_2Ph	7·2	22·6	2·0	332	6850	6725	3360
CH_2Ph	9·6	30·2	2·0	234	6600		

free and bound $1/T_2$ values (assuming, as is likely, low values of n) and also make obvious the fact that the phenyl group is much more affected by binding than are the other groups. The selective effect of binding on phenyl indicates that this group is intimately involved in the binding process, while other portions of the molecule retain freedom of motion to a much greater extent.

Changes in pH over the range 6·1 to 8·3 have little effect on the $1/T_2$ values of lines in penicillin G-albumin spectra; this result suggests that the binding mechanism does not depend strongly on an ionic interaction between the carboxyl group of penicillin G and protonated basic centres of histidine present in the protein molecule. Further to this point, if an ionic bond were involved, it should be weakened, by an increase in the ionic strength of the solution. However, addition of salt to an 0·1 M penicillin G 2·5% albumin solution caused an enhancement of line broadening. This supports the view that the penicillin-albumin interaction is hydrophobic in nature (phenyl is a very hydrophobic structure) as it is well known that salt tends to increase interactions between water-repellant groups. The extent of binding would be expected to decrease as the temperature is raised and it is, in fact, found that line widths decrease as the temperature increases, e.g. $1/T_2$ of phenyl signal of 0·2 M penicillin–2·5% albumin: 10°, ~18 sec^{-1}, 60° ~5 sec^{-1}.

Phenoxymethylpenicillin (penicillin V), another antibiotic which is known to bind to albumin, also exhibits line broadening of its PMR signals in the presence of protein. The methylene signals of penicillin G and V do not overlap and can be distinguished in the spectra of mixtures. The spectrum of a solution of the two penicillins and albumin shows slightly less broadening of the V(CH$_2$) peak and much less broadening of the G peak compared with controls. Thus there is a competition for binding sites on the albumin molecule. The spectrum of a penicillin-albumin mixture is not affected by 6-aminopenicillanic acid (**17**, R = H) or by its acetyl derivative (**17**, R = COMe). In contrast, the line width of the methylene signal of a penicillin-albumin spectrum is reduced in the presence of phenoxy-

17   **18**

acetic acid (**18**, R = OH) and phenoxyacetamide (**18**, R = NH$_2$). All these results further stress the dominant role played by the aromatic group of penicillin in its binding to albumin.

An extensive study of the binding of antibacterial sulphonamide drugs to bovine serum albumin has also been made (Jardetzky and Wade–Jardetzky, 1965). Longitudinal and transverse relaxation times of the various PMR signals of the sulphonamides were measured as in the penicillin investigations and comparable control experiments were carried out by which non-specific contributions to relaxation rates could be assessed. Three representative examples were chosen, namely, sulphacetamide **19** (aliphatic side chain), 5-methyl-3-*p*-aminobenzenesulphonamidoisoxazole

19   20   sulphaphenazole   21

20 (single-ring side chain), and 1-phenyl-5-*p*-aminobenzenesulphonamido-pyrazole 21 (two-ring side chain). The aromatic signal of the *p*-amino-benzenesulphonamide (PABS) moiety forms a typical AA'BB' line pattern; the 4 strongest lines form a symmetrical pair of doublets each of which tend to be fused by broadening as illustrated in the case of spectra of another sulphonamide, sulphathiazole (Fig. 8.25). In cases like this where the resonance is not a singlet, the $1/T_2$ values are calculated from the formula

$$1/T_2 = \pi(\Delta v'_{1/2} - J)$$

where $\Delta v'_{1/2}$ is the width at half height of the fused peak and J is the peak separation in the absence of broadening, i.e. when $\Delta v_{1/2} \ll J$.

The effect of albumin upon the relaxation rates of the PABS and methyl protons signals of 0·1 M sulphacetamide in D_2O is shown in Fig. 8.26. The relaxation rate of the PABS signal increases by a factor of 44 whereas that of the methyl increases 14-fold as the protein concentration is increased from 0 to 10%. This difference shows that the PABS moiety is preferentially stabilized by the interaction of the sulphacetamide molecule with albumin, whereas the acetyl group retains much greater freedom of motion, and is evidence that the aromatic ring is the primary binding site. As judged from line width changes, maximum binding occurs at about pH 6 while binding increases when salt is added (supporting a hydrophobic rather than ionic interaction).

In most sulpha-drugs in clinical use, the sulphonamide group is linked to a heterocyclic substituent, hence it was felt that the binding of sulph-acetamide might not be typical of the group. However, relaxation studies of the isoxazole derivative 20, in which line widths of the PABS and isoxazole proton and methyl signals were recorded, again demonstrated that the PABS moiety is the primary site of binding.

The last example 21 contains two potential binding sites of an aromatic nature. In this case overlap of signals complicates line width measurements. The width of the phenyl peak of the pyrazole nucleus is estimated after subtracting the lower field part of the PABS signal and the pyrazole peak from the observed envelope, the widths of these signals being assumed identical to those of their higher field counterparts. Line width measurements demonstrate pronounced $1/T_2$ increases for both the PABS and phenyl protons upon addition of albumin, while $1/T_2$ of the pyrazole

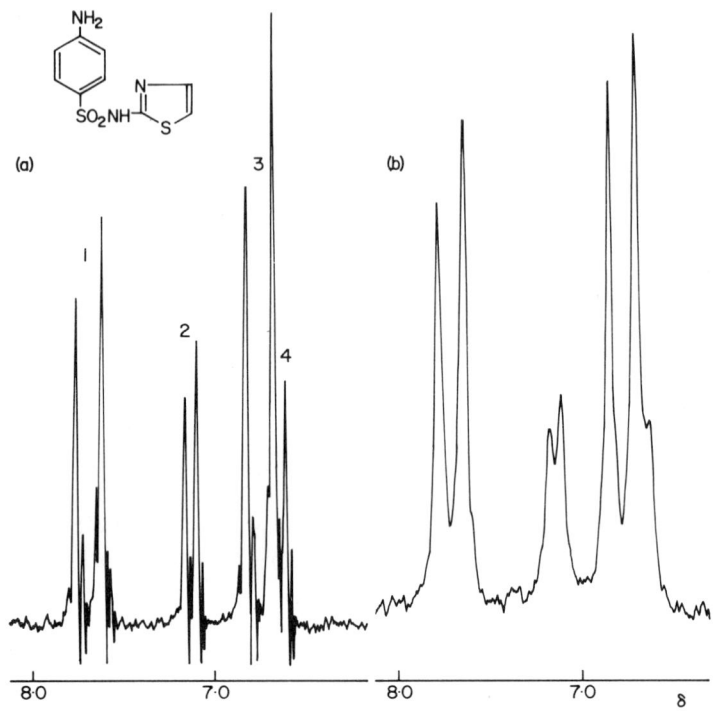

Fig. 8.25. Part of the 60 MHz PMR spectrum of (a) 10% sulphathiazole sodium in D$_2$O; (b) 10% sulphathiazole sodium plus 10% bovine serum albumin in D$_2$O. Groups 1 and 3 are the two halves of the *p*-aminobenzenesulphonamido AA′BB′ signal, while the doublet of doublets 2 and 4 is the signal of the thiazole protons.

protons is much less affected. These results are inconsistent with phenyl of the pyrazole nucleus being the primary binding site since the

$$\left(\frac{1}{T_1}\right)_{Ph} > \left(\frac{1}{T_1}\right)_{pyrazole} > \left(\frac{1}{T_1}\right)_{PABS}$$

order would then be expected because of the geometry of the molecule. The best model to fit the facts is one in which there are 2 *non-equivalent* binding sites, one for the PABS moiety and the other for phenyl. Molecules bound at either site exchange rapidly with free molecules, so that for each peak the observed relaxation is the weighted average of three different values. When phenylpropanol is added to a solution of sulphaphenazole and albumin, the PABS lines narrow and the phenyl line broadens. This result suggests that phenylpropanol selectively inhibits binding of the PABS fragment and supports the proposed model.

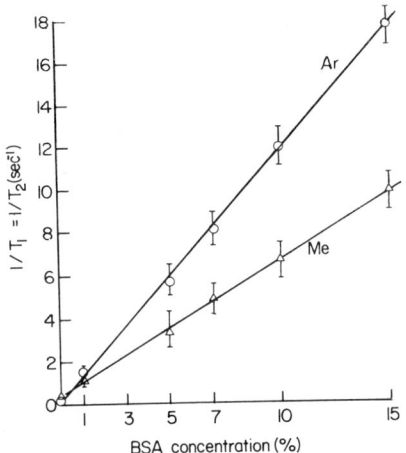

FIG. 8.26. Relaxation rates of 0·1 M sulphacetamide aromatic and CO<u>Me</u> PMR signals as a function of bovine serum albumin concentration in D_2O at pH 8·8 (Jardetzky and Wade-Jardetzky, 1965).

The study of specific molecular interactions by nuclear magnetic relaxation measurements has now been applied to a variety of systems, and brief notes on some of these are given below.

Adrenaline—Adenosine triphosphate

A weak complex between catecholamines and nucleotides is thought to be the storage form of sympathomimetic amines in the chromaffin granules of the adrenal medulla; relaxation studies were undertaken to find out which portions of the molecules are involved in the formation of the complex (Jardetzky, 1965). Selective broadening of the side chain methine and methylene protons of adrenaline **22** (0·3 M solution in D_2O, pH 1·5) is observed in the presence of 0·1 M adenosine triphosphate (**23**, ATP). The change is small when adrenaline is in excess and reaches a maximum when excess of the nucleotide is present. No similar selective change in the relaxation times is found in the PMR spectrum of ATP. It was concluded that the adrenaline–ATP complex involves specifically the attachment of the catecholamine side chain to one of the phosphate residues of the nucleotide. The phosphate side chain was chosen as the nucleotide site

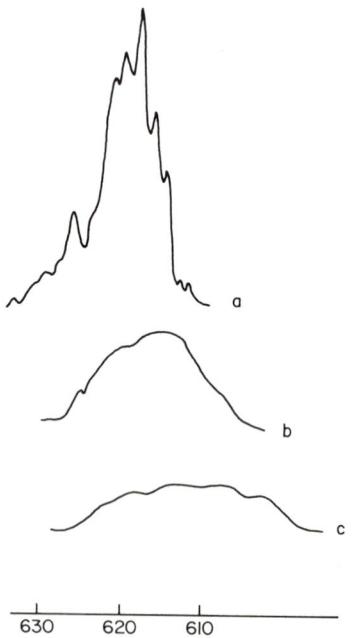

HO—⟨benzene ring⟩—CH(OH)CH₂N⁺H₂Me

22

23

because binding at this point leaves the rest of the molecule free. This point could possibly be corroborated by a study of the ^{31}P NMR spectrum of the complex. The PMR spectrum of adrenaline is also altered by the presence of mouse liver cells in a manner characteristic of a binding inter-action (Fischer and Jost, 1969). Large sample tubes are used so that signals from single protons present in 0·001 M concentration can be detected. The cells themselves produce a broad elevation of the baseline in the aromatic

630   620   610

Fig. 8.27. 90 MHz PMR signals of the aromatic protons of adrenaline in D₂O. Chemical shift scale is in Hz from DSS. (a) adrenaline salt; (b) liver cells with 0·01 M adrenaline; (c) liver cells with 0·005 M adrenaline (Fischer and Jost, 1969).

region but this does not interfere with measurements of the adrenaline phenyl peaks. Changes in signal width are greatest for the lowest adrenaline concentrations (fixed cell concentration) and widths approach those of free adrenaline as the concentration of the drug is increased (Fig. 8.27 shows broadening of the phenyl signals of adrenaline); these results are characteristic of binding of a large excess of small molecules to a fixed number of receptor sites and cannot be explained by a non-specific broadening mechanism.

Enzyme-substrate interactions

Kato's study of the hydrolysis of acetylcholine (Ach) by horse serum cholinesterase (HSchE) has already been described (Chapter 6, p. 228). This work has now been extended to an investigation of the Ach-cholinesterase interaction itself by relaxation time measurements (Kato, 1969). Relaxation times were obtained from line widths measured directly from the chart paper and corrected for instrumental line widths (no side band calibrations were employed). The technique was to mix $0 \cdot 1$ M Ach chloride with various concentrations of HSchE in a D_2O-phosphate buffer at pH $7 \cdot 0$, and to record the PMR spectrum of the mixture 8 sec later. Line width measurements were confined to the $^+NMe_3$ and $COMe$ singlets of the Ach spectrum. When Ach ($0 \cdot 1$ M) was added to the enzyme (150 mg/ml) there was a time-dependent broadening of the Ach spectral lines (Fig. 8.28). First, there was a sharp increase in the line widths of both $^+NMe_3$ and $COMe$ peaks; these lines then reached a maximum broadening after 1–10 min depending on the enzyme concentration. The line width then began to decrease at an approximately exponential rate. The broadening of the $^+NMe_3$ peak reached a maximum 2 min after mixing (conditions as in Fig. 8.28) while that of the $COMe$ peak occurred after 3 min. Similar transient line broadening was also observed for the Me peak of the acetate formed during hydrolysis (upfield of the $COMe$ signal of Ach). An increase in the enzyme concentration from 0 to 300 mg/ml raised the maximum degree of line broadening of the $^+NMe_3$ and $COMe$ signals by factors of 7 and 15 respectively, and it was inferred that the acetyl moiety was preferentially stabilized by interaction with the enzyme. This argument rests upon the accuracy of the $\Delta\nu_{1/2}$ values of the Ach peaks measured in the absence of enzyme. The maximum broadening of both the $^+NMe_3$ and $COMe$ peaks decreased as the Ach concentration increased, a result which shows the binding interaction to be of a type in which the enzyme active sites are saturable. The results of these experiments are interpreted as follows. The fact of the rate of initial broadening of the $^+NMe_3$ signal being greater than that of the $COMe$ resonance suggests that the quaternary head initially anchors the molecule to the anionic site of the enzyme and that this is followed by a more pronounced interaction between the acetate group of Ach and the esteratic site of the enzyme, leading to hydrolysis. Low concentrations of cholinesterase inhibitors such as physostigmine and neostigmine prevented the initial line broadening of the $^+NMe_3$ and COMe

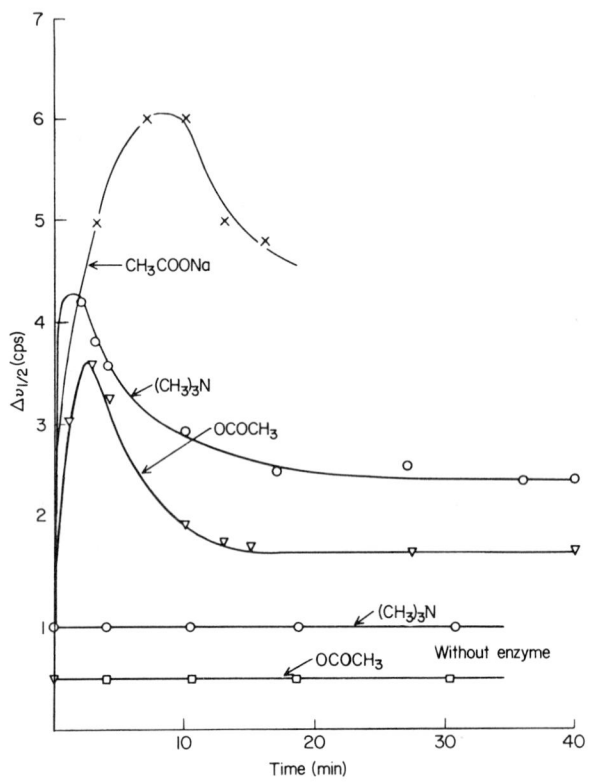

FIG. 8.28. Change in line width of $^+NMe_3$ and COMe PMR peaks of acetylcholine chloride (0·1 M) and the methyl peak of the newly formed acetate in the presence of cholinesterase (HSchE, 150 mg/ml) in D_2O-sodium phosphate buffer (0·1 M pH 7·0, 40°) (Kato, 1969).

signals (Fig. 8.28), whereas tetraethylpyrophosphate only prevented the broadening of the acetate peak. These results are evidence that the former drugs block both anionic and esteratic sites of the enzyme, whereas the phosphate blocks solely the esteratic site. All these conclusions are in agreement with current views upon the interactions of cholinesterases with their substrates and inhibitors (Barlow, 1964).

Enzyme-coenzyme interactions

In most studies of the binding of small molecules to enzymes, concentrations employed are much lower than those normally detectable by a single sweep NMR experiment, and the technique of multiple scanning and integration using a computer of average transients (CAT) has to be used. The spectrum of an 0·002 M solution of diphosphopyridine nucleotide (coenzyme I, DPN, structure given in Chapter 9, p. 361) in D_2O, derived from a total of 500 sweeps, shows selective broadening of its pyridine aromatic signals in the presence of 0·0005 M yeast alcohol dehydrogenase enzyme (YADH); this result was taken as evidence for the involvement of the nicotinamide ring in the enzyme-coenzyme interaction (Jardetzky *et al.*, 1963). Hollis (1967), however, using improved equipment, failed to duplicate this result. He found that all the PMR peaks corresponding to the protons of the nicotinamide and adenine rings were resolvable in a solution containing excess of DPN over enzyme binding sites, while adenine rather than pyridine peaks suffered the greater broadening (Fig. 8.29). It is known that

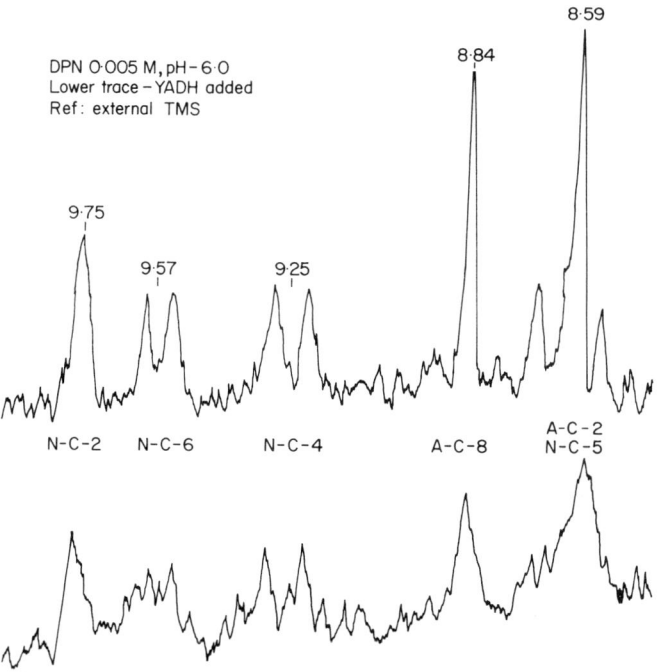

FIG. 8.29. 100 MHz PMR spectrum of 1·3 mM DPN in D_2O (pH 6·0). Upper trace recorded in the absence of enzyme; lower trace in the presence of 0·6 mM YADH. Sweep width 100 Hz in both cases (Hollis, 1967).

about 13% of DPN is bound at equilibrium (i.e. considerably more than is the case for penicillin G bound to albumin) and it follows from Equation (2) (p. 313) that $\Delta\nu_{1/2}$ values of the signals of bound species will be 30–40 Hz provided fast exchange between free and bound species is assumed. Values of this order are much less than those anticipated for bound molecules (line widths of 800–2200 Hz were estimated for the various protons of bound penicillin G, see p. 315), and Hollis argues that the data is more compatible with the case

$$\left(\frac{1}{T_2}\right)_{\text{free}} < \left(\frac{1}{T_2}\right)_{\text{exchange}} < \left(\frac{1}{T_2}\right)_{\text{bound}}$$

(see p. 313) in which relaxation rates of the bound species *do not* modify those of the free molecules. In such cases all free $\Delta\nu_{1/2}$ values depend on the exchange rate and should be similar, as is in fact observed experimentally taking into account the rather large experimental errors involved. The fact of the nicotinamide proton signals being little changed on addition of the enzyme is explained either by the fortuitous correspondence of $T_{2(\text{free})}$ (nicotinamide protons) and T_{exchange} or by $T_{2(\text{bound})}$ being equal to $T_{2(\text{free})}$ for the same protons. If the latter is true, the adenine nucleus must be the primary binding site. Selective broadening of the adenine peaks was also seen in the spectrum of the system YADH-reduced coenzyme I (DPNH). The pyridine C-4 methylene signal of DPNH (see p. 361) was also considerably broader after addition of the enzyme, and appeared to have lost intensity equivalent to about one proton. The enzyme had no effect on the methyl triplet of 1·4 mM of the substrate ethanol in D_2O. However when 4 mM of DPN was added the Me signal broadened and decreased in intensity by about 40%; the decrease was correspondingly less when only one equivalent of the coenzyme was present. These results suggest that ethanol is not taken up by YADH in the absence of its coenzyme.

Enzyme-inhibitor interactions

The uptake by α-chymotrypsin of tryptophan **24**, a reversible inhibitor of the proteolytic reactions of the enzyme, is an example of this class of

$$-CH_2CH(NH_2)CO_2H$$

24

relaxation study (Gerig, 1968). All lines of the 100 MHz spectra of the L-amino acid (44·5 mM in D_2O, 50 scans) broaden in the presence of native α-chymotrypsin, the effect being more pronounced for the side chain than for the aromatic proton signals. At a given ratio of enzyme to amino acid, the spectrum of the D-enantiomorph is slightly broader than that of the L-isomer; this result is consistent with the observation that the D-forms of several tryptophan-like molecules are more potent inhibitors of the chymotryptic reaction than are the L-isomers. The methionine-192 residue of α-chymotrypsin is believed to be resident in a key binding area for substrates based on aromatic amino acids. When this residue is alkylated with molecules such as N-bromoacetyl-3-trifluoromethylaniline (**25**), the binding of aromatic acid substrates is greatly diminished. It is also found that the line

$$\text{F}_3\text{C} \diagdown \hexagon \diagup -\text{NHCCH}_2\text{Br}$$
$$\underset{\text{O}}{\overset{\parallel}{}}$$

25

broadening effect of α-chymotrypsin upon the spectrum of tryptophan is abolished after the enzyme has been alkylated with **25**, and this result is evidence that tryptophan is bound to the enzyme at a location that is at or near the active site.

Hapten-antibody interactions

Joffe (1967) first showed that NMR spectroscopy could be used to detect the specific association of a hapten with its antibody. The 60 MHz spectrum (CAT-accumulation) of p-nitrophenyl-β-D-galactoside ($3 \cdot 2$ mM in a D_2O-phosphate buffer) shows distinct aromatic and sugar proton signals. When the anti-galactosyl antibody ($0 \cdot 125$ mM) is added the aromatic signals broaden so much that they are almost lost in the noise level while no significant changes are seen in the higher field signals. The same concentration of γ-globulin has no influence on the spectrum of the hapten.

From a pharmacological point of view, interest in relaxation studies lies in their potential application to gaining knowledge of the interaction of drug molecules with the receptors at which drug effects are mediated. So far no receptors have been isolated specifically but the use of interactions between antibodies and antigens as models for those of drugs and receptors is a promising approach (Metcalfe *et al.*, 1966), and a few relaxation studies of models of this type have already been reported.† An example is the investigation of the binding of choline derivatives to antibodies formed against the antigen obtained by reaction between diazotized 4-amino-phenoxycholine and bovine serum albumin (Burgen *et al.*, 1967; Metcalfe

† Recent work of Miledi *et al.* (1971) on the protein of *Torpedo* electric tissue gives promise to the early availability of the cholinergic receptor for PMR experiments.

et al., 1968). When the spectra of the phenylcholine ether analogues **26**a and b were examined in the presence of the antibody, selective broadening of the lines was observed; non-specific contributions were gauged by

$$\text{MeCN} \underset{\underset{\text{O}}{\|}}{\overset{\overset{\text{H}}{|}}{}} \text{—} \underset{}{\bigcirc} \text{—OCH}_2\text{CH}_2\overset{+}{\underset{\underset{\text{R}}{|}}{\overset{\overset{\text{Me}}{|}}{\text{N}}}}\text{—Me}$$

26

a: R = Me
b: R = CH$_2$Ph

recording the spectra in the presence of pooled γ-globulin from non-immunized animals. In both **26**a and b, the $1/T_1$ data showed that the quaternary head of the molecule was the most, and the acetyl group, the least stabilized, $(1/T_1)_{\text{bound}}$ values for **26**a, for example, of 230 sec^{-1} for NMe, 120 sec^{-1} for Ar and 80 sec^{-1} for C-Me were calculated from observed (antibody present) and free $1/T_1$ measurements and from the affinity constant of the complex (determined fluorometrically). In **26**b, the aromatic group attached to charged nitrogen was stabilized even more than the N-methyl groups. It is argued that these results are evidence that the binding site is pocket-shaped allowing the first bulky substituent (Me-CONHAr) to protrude from its orifice (and remaining relatively unrestricted in its movements) but requiring that the second bulky substituent ($^+$NMe$_2$R) be forced into its interior. This idea gains support from the fact that the N-benzyltrimethylammonium ion **27** shows a preferential stabiliza-

$$\text{PhCH}_2\text{—}\overset{+}{\text{N}}\overset{\nearrow\text{Me}}{\underset{\underset{\text{Me}}{\searrow}}{\searrow\text{Me}}}$$

27

tion òf the N-methyl groups over the aromatic substituent. It appears, therefore, that when the acetamidophenyl ether fragment is removed from **26**b, the best fit into the "pocket" is obtained by reversing the remainder of the molecule so that the N-benzyl residue protrudes towards the outside.

The same antibody also caused line broadening of the spectrum of

$$\text{MeCOCHMeCH}_2\overset{+}{\text{NMe}_3}$$
$$\underset{\text{O}}{\|}$$

28

(−)-β-methylacetylcholine (**28**) (and acetylcholine), but in this case line widths of all three types of methyl group increase by similar extents; this result suggests that the molecule is bound as a rigid unit in contrast to the binding of **26**a and b.

The finding that an acetyl methyl group is stabilized to almost the same degree as an N-methyl group if it is within the same distance from the latter as in acetylcholine, but not if it is further removed (as in **26**a or b) is of particular interest. It suggests that the antibody contains a specific site capable of binding the acetyl moiety, only if it is within a certain critical distance of the quaternary head. This site could not have arisen in response to the phenylcholine ether (used to form the antigen) because this compound lacks an acetyl group in the required position. It is possible, therefore, that the antibody binding site has a structure common to other binding sites for acetylcholine such as may occur in cholinesterase and the acetylcholine receptor.

Note: Recent reviews of relevance to this chapter have been published by McDonald and Phillips (1970) and Sheard and Bradbury (1970) on biopolymers, Burgen and Metcalfe (1970) on the applications of NMR to pharmacological problems and Roberts and Jardetzky (1970) on the NMR spectroscopy of amino acids, peptides and proteins. The last mentioned includes an account of the binding of small molecules to proteins.

References

Bak, B., Dambmann, C., Nicolaisen, F., Pedersen, E. J. and Bhacca, N. S. (1968). *J. Mol. Spectrosc.* **26**, 78.

Barlow, R. B. (1964). "Chemical Pharmacology", 2nd Ed. Methuen, London.

Becker, E. D., Bradley, R. B. and Watson, C. J. (1961). *J. Amer. Chem. Soc.* **83**, 3743.

Bovey, F. A. (1967). "NMR Data Tables for Organic Compounds", Vol. 1. Interscience, New York.

Bradbury, E. M. and Crane-Robinson, C. (1968). *Nature, Lond.* **220**, 1079.

Bradbury, E. M., Crane-Robinson, C., Goldman, H. and Rattle, H. W. E. (1968). *Nature, Lond.* **217**, 812.

Burgen, A. S. V. and Metcalfe, J. C. (1970). *J. Pharm. Pharmacol.* **22**, 153.

Burgen, A. S. V., Jardetzky, O., Metcalfe, J. C. and Wade-Jardetzky, N. G. (1967). *Proc. Nat. Acad. Sci. U.S.* **58**, 447.

Bystrov, V. F., Portnova, S. L., Tsetlin, V. I., Ivanov, V. T. and Ovchinnikov, Yu. A. (1969). *Tetrahedron* **25**, 493, and references cited therein.

Caughley, W. S., York, J. L. and Iber, P. K. (1967). *In* "Magnetic Resonance in Biological Systems" (A. Ehrenberg, B. G. Malström and T. Vänngård, eds). p. 25. Pergamon, Oxford.

Crespi, H. L., Rosenberg, R. M. and Katz, J. J. (1968). *Science, N.Y.* **161**, 795.

Crespi, H. L. and Katz, J. J. (1969). *Nature, Lond.* **224**, 560.

Crestfield, A. M., Stein, W. H. and Moore, S. (1963). *J. Biol. Chem.* **238**, 2413.

Cohen, J. S. (1969). *J. Clin. Pharmacol.* **9**, 72.
Davis, D. G., Mock, N. L., Laman, V. R. and Ho, C. (1969). *J. Mol. Biol.* **40**, 311.
Eaton, D. R. and Phillips, W. D. (1965). *Advan. Magn. Resonance* **1**, 103.
Feeney, J. (1969). "Annual Reports on the Progress of Chemistry for 1968", **65A**, p. 63. The Chemical Society, London.
Ferguson, R. C. and Phillips, W. D. (1967). *Science, N.Y.* **157**, 257.
Ferretti, J. A. (1967). *Chem. Commun.* 1030.
Fischer, J. J. and Jardetzky, O. (1965). *J. Amer. Chem. Soc.* **87**, 3237.
Fischer, J. J. and Jost, M. C. (1969). *Mol. Pharmacol.* **5**, 420.
Gerig, J. T. (1968). *J. Amer. Chem. Soc.* **90**, 2681.
Halpern, B., Nitecki, D. E. and Weinstein, B. (1967a). *Tetrahedron Lett.* 3075.
Halpern, B., Chew, L. F. and Weinstein, B. (1967b). *J. Amer. Chem. Soc.* **89**, 5051.
Heeschen, J. P. (1968). *Anal. Chem.* **40**, 560R.
Hollis, D. P. (1967). *Biochem.* **6**, 2080.
Hollis, D. P., McDonald, G. and Biltonen, R. L. (1967). *Proc. Nat. Acad. Sci. U.S.* **58**, 758.
Kato, G. (1969). *Mol. Pharmacol.* **5**, 148.
Kopple, K. D. and Marr, D. H. (1967). *J. Amer. Chem. Soc.* **89**, 6193.
Kowalsky, A. (1965). *Biochem.* **4**, 2382.
Kowalsky, A. and Cohn, M. (1964). *Ann. Rev. Biochem.* **33**, 481.
Jardetzky, O. (1964). *Advan. Chem. Phys.* Vol. 7, p. 499.
Jardetzky, O., Wade, N. G. and Fischer, J. J. (1963). *Nature, Lond.* **197**, 183.
Jardetzky, O. and Wade-Jardetzky, N. G. (1965). *Mol. Pharmacol.* **1**, 214.
Joffe, S. (1967). *Mol. Pharmacol.* **3**, 399.
Markley, J. L., Putter, I. and Jardetzky, O. (1968a). *Science, N.Y.* **161**, 1249.
Markley, J. L., Putter, I. and Jardetzky, O. (1968b). *Zeit. für anal. Chem.* **243**, 367.
Martin, J. S. (1970). Private communication.
McDonald, C. C. and Phillips, W. D. (1967). *J. Amer. Chem. Soc.* **89**, 6332.
McDonald, C. C. and Phillips, W. D. (1969). *J. Amer. Chem. Soc.* **91**, 1513.
McDonald, C. C. and Phillips, W. D. (1970). *In* "Biological Macromolecules", Vol. 3 (S. M. Timaskeff and G. D. Fasman, eds). Dekker, New York.
McTague, J. P., Ross, V. and Gibbs, J. H. (1964). *Biopolymers* **2**, 163.
Meadows, D. H. and Jardetzky, O. (1968). *Proc. Natl. Acad. Sci. U.S.* **61**, 406.
Meadows, D. H., Markley, J. L., Cohen, J. S. and Jardetzky, O. (1967). *Proc. Natl. Acad. Sci. U.S.* **58**, 1307.
Meadows, D. H., Jardetzky, O., Epand, R. M., Ruterjans, H. and Scheraga, H. A. (1968). *Proc. Natl. Acad. Sci. U.S.* **60**, 766.
Metcalfe, J. C., Marlow, H. F. and Burgen, A. S. V. (1966). *Nature, Lond.* **209**, 1142.
Metcalfe, J. C., Burgen, A. S. V. and Jardetzky, O. (1968). *In* "Molecular Associations in Biology", (B. Pullman, ed.). Academic Press, New York.
Miledi, R., Molinoff, P. and Potter, L. T. (1971). *Nature, Lond.* **229**, 554.
Nitecki, D. E., Halpern, B. and Westley, J. W. (1968). *J. Org. Chem.* **33**, 864.
Richards, F. M. and Vithayathil, P. J. (1959). *J. Biol. Chem.* **234**, 1459.
Roberts, G. C. K. and Jardetzky, O. (1970). *Advan. Protein Chem.* **24**, 447.
Roberts, G. C. K., Meadows, D. H. and Jardetzky, O. (1968). *Biochem.* **7**, 2053.
Rowe, J. J. M., Hinton, J. and Rowe, K. L. (1970). *Chem. Revs.* **70**, 1.
Sanders, J. K. M. and Williams, D. H. (1971). *J. Amer. Chem. Soc.* **93**, 641.
Saunders, M., Wishnia, A. and Kirkwood, J. G. (1957). *J. Amer. Chem. Soc.* **79**, 3289.

Sen, B. and Wu, W. C. (1969). *Anal. Chim. Acta* **46**, 37.

Sheard, B. and Bradbury, E. M. (1970). *In* "Progress in Biophysics and Molecular Biology", Vol. 20 (J. A. V. Butler and D. Noble, eds). Pergamon, Oxford.

Sheinblatt, M. (1966). *J. Amer. Chem. Soc.* **88**, 2845.

Sheinblatt, M. (1967). *In* "Magnetic Resonance in Biological Systems", (A. Ehrenberg, B. G. Malmström and T. Vänngård, eds). Pergamon, Oxford.

Sievers, R. E., Bayer, E. and Hunziker, P. (1969). *Nature, Lond.* **223**, 179.

Webb, R. G., Haskell, M. W. and Stammer, C. H. (1969). *J. Org. Chem.* **34**, 576.

Westley, J. W., Close, V. A., Nitecki, D. E. and Halpern, B. (1968). *Anal. Chem.* **40**, 1888.

Wüthrich, K., Shulman, R. G. and Yamane, T. (1968). *Proc. Nat. Acad. Sci. U.S.* **61**, 1199.

CHAPTER 9

Biochemical Aspects—2

CARBOHYDRATES

During the past decade PMR spectroscopy has contributed much to our understanding of sugar molecules, particularly so in the case of mono-saccharides. Two reviews (Hall, 1964; Inch, 1969) and a paper by Angyal (1969) on the composition and conformation of sugars in the solute state provide a good coverage of the field and illustrate the special value of the spectroscopic technique in solving problems of stereochemistry and equilibrium compositions of monosaccharides in solution. A few points of nomenclature must first be made clear before these aspects are illustrated in the present chapter.

In solution, monosaccharides based on both pyranose and furanose rings form mixtures of isomers which differ only in the configuration of the hemiacetal carbon atom (C-1); these isomers are termed *anomers*, and the C-1 proton the anomeric hydrogen atom. The anomeric centre is given the symbols α or β according to the following convention. For a D-sugar, when the hemiacetal ring is so orientated that the ring oxygen is in the rear and the anomeric carbon atom on the right hand side (as in **1** for pyranose sugars), the α form is the one which has the anomeric hydroxyl below the plane of

1

the ring and the β form that which has it above the plane. The preferred conformation of many pyranose sugars has an equatorially orientated 5-hydroxymethyl substituent, as in **2**, termed the normal (N) or CI form

2 CI

3 IC

(Eliel *et al.*, 1965). In rarer cases, when the axial 5-hydroxymethyl form **3** is favoured, the preferred conformation is termed the alternative (A) or IC form. The α/β anomer designation holds in either case and it is to be noted that R and β-OH are *cis*, while R and α-OH are *trans* in both CI and IC forms of hexoses.

Sugars must generally be examined as solutes in D_2O ($\equiv H_2O$) because of their highly polar nature, hence their spectra are complicated as a result of various isomerizations (e.g. mutarotation), described later; acetylated derivatives retain their integrity in solution and PMR studies of sugar acetates have mostly preceded those of the free sugars.

The anomeric protons of free and acetylated aldoses are clearly distinguishable in PMR spectra because of their low field position, a result of the deshielding influence of the oxygen atoms which flank the carbon atom (C-1) to which they are attached. Stereochemical evidence is derived from both the coupling constants and chemical shifts of these signals. Regarding the former, the spectra of monosaccharides provided some of the first examples which illustrated the dependence of vicinal coupling between two protons upon the dihedral angle between them (Chapter 3). It is now appreciated that a variety of variables affect the precise relationship between coupling constants and dihedral angles, hence the Karplus equation is best employed as a guide to conformation rather than a means of calculating accurate dihedral angles. If the C-1 and C-2 protons of the pyranose ring are *trans*, a J value in the range 7–10 Hz indicates a preference for a diaxial arrangement, while one in the range 2–4 Hz shows a diequatorial arrangement (or axial-equatorial for *cis* protons) to be favoured. Thus 40 MHz data for the penta-acetates of α-D-glucopyranose and β-D-allopyranose

4

5

doublet spacing of anomeric signal } 3 Hz   ($R = CH_2OAc$)   8 Hz support the preferred conformations **4** and **5** respectively for these two compounds (Lemieux *et al.*, 1958). This earlier study was later refined and extended by work at higher frequencies (Lemieux and Stevens, 1965). Line widths of the anomeric proton doublets derived from 100 MHz spectra give the most accurate $J_{1,2}$ values since those from 40 to 60 MHz spectra may give deceptive values as a result of virtual coupling effects (Chapter 6, p. 204). In several cases broad deformed doublets seen at 60 MHz were of increased width and symmetry in 100 MHz spectra. This occurred when the chemical shift difference (in Hz) between H_2 and H_3 approached the magnitude of $J_{2,3}$, as in β-D-galactopyranose pentaacetate (**6**); virtual

$\Delta\nu_{2,3}$ 0·2 ppm
      (12 Hz at 60 MHz)

$J_{2,3}$ 10 Hz

line separation $\Big\}$ 7·5 Hz (60 MHz)
of H-1 doublet   8·4 Hz (100 MHz)

coupling is reduced at the higher frequency because $\Delta\nu_{2,3}$ (Hz) then usually exceeds $J_{2,3}$ by a significant margin.

Stereochemical conclusions drawn from the PMR characteristics of anomeric protons may often be corroborated by data of other ring protons; to identify these signals, use is made of both spin decoupling and specific deuteration. For example, a multiplet near δ 4·12 in the spectrum of α-D-glucopyranose penta-acetate (**4**) is assigned to H-5 since it simplifies to a doublet (J 10 Hz) in the spectrum of the 6,6'-dideuterio analogue **7**,

7

with a coupling constant magnitude consistent with an H-4/H-5 diaxial arrangement.

An intermediate J_{vic} value (4–6 Hz) indicates that significant populations of both CI and IC conformations are present as in the case of β-D-ribo-pentopyranose tetraacetate **8** (J 4·7). At $-84°$C, the spectrum of this

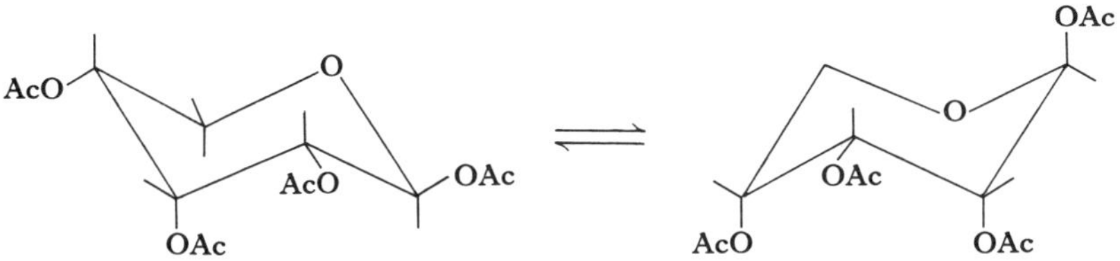

8

derivative in acetone-d$_6$ displays signals of both chair conformers in the approximate ratio of 2 ($J_{1,2}$ 1 Hz):1 ($J_{1,2}$ 8 Hz) (Bhacca and Horton, 1967); the J value observed at normal temperatures is therefore a time-averaged result due to rapid interconversion of the CI and IC conformers.

When the protons on C-1 and C-2 are *cis*, e.g. β-D-mannopyranose acetate (p. 337) the coupling constants of anomeric protons in CI and IC forms do not differ much, although it is generally found that $J_{1,2}$ is larger when the C-1 proton is equatorial ($J = 2.5$–3.5 Hz) than when it is axial ($J = 1.0$–1.5 Hz) (Lemieux and Stevens, 1966). In such cases, empirical rules that enable the chemical shift of any proton relative to β-D-glucopyranose penta-acetate (**9**, R = CH$_2$OAc) or (for pentoses) the corresponding β-xylose derivative (**9**, R = H) to be calculated in solvent CDCl$_3$, provide more definite answers to conformational problems.

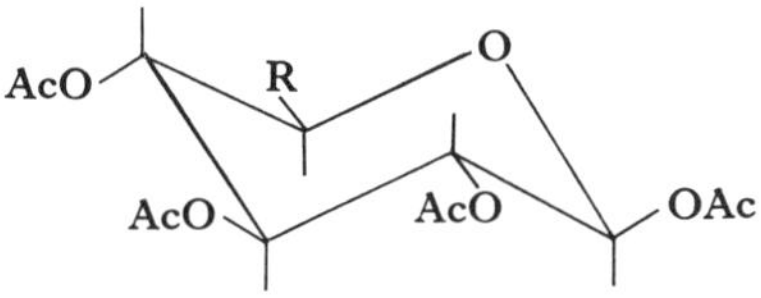

9

The rules are as follows:

1. If the proton under consideration has remained axial (all are axial in the model **9**, R = CH$_2$OAc);

(a) add 0·20 ppm for an axial acetoxy group at a neighbouring position;

(b) add 0·25 ppm for an opposition of the proton by an axial acetoxy group (H and OCOMe will have a 1,3 diaxial relationship).

2. If the proton under consideration has achieved an equatorial orientation;

(a) add 0·60 ppm because of the change from the environment of an axial hydrogen to that of an equatorial hydrogen;

(b) subtract 0·20 ppm for an axial acetoxy group at a neighbouring position;

(c) subtract 0·20 ppm for an axial acetoxy group at the next to neighbouring position. Use of these rules gives the calculated chemical shift on the δ scale, i.e. addition ≡ deshielding, and subtraction ≡ shielding.

The first rule depends on the fact that the changes a to b, and c to d both

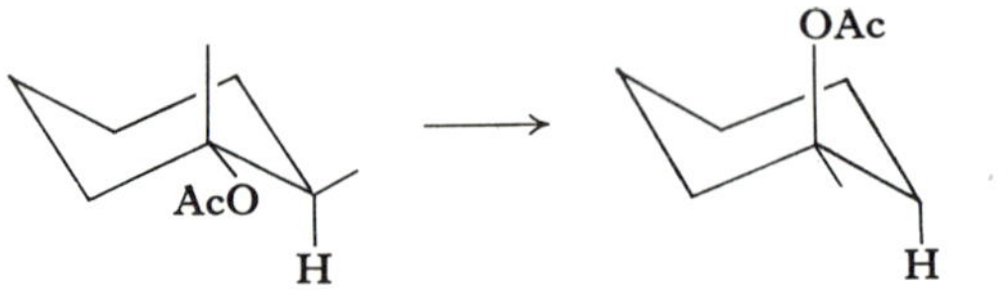

a

b

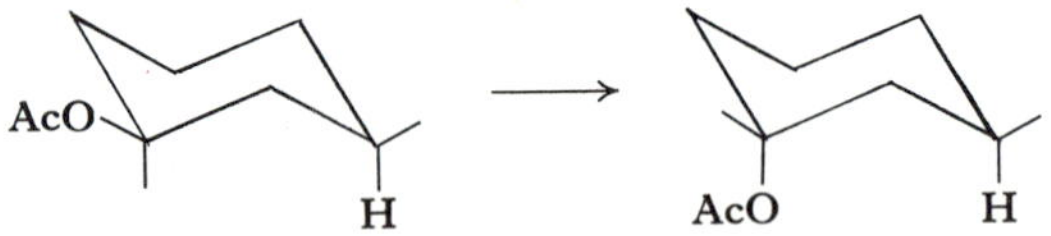

c

d

move the axial proton downfield, as illustrated by the following examples.

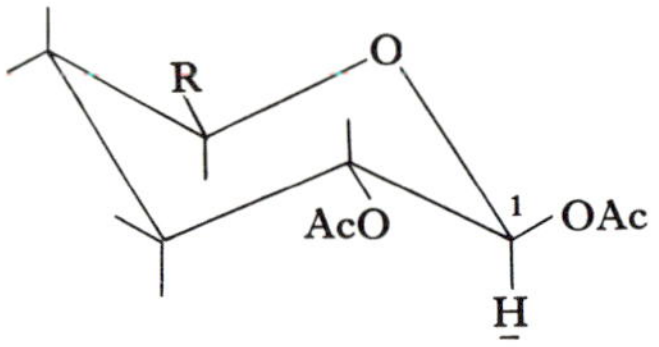

H-1 δ 5·75

β-D-glucopyranose penta-acetate

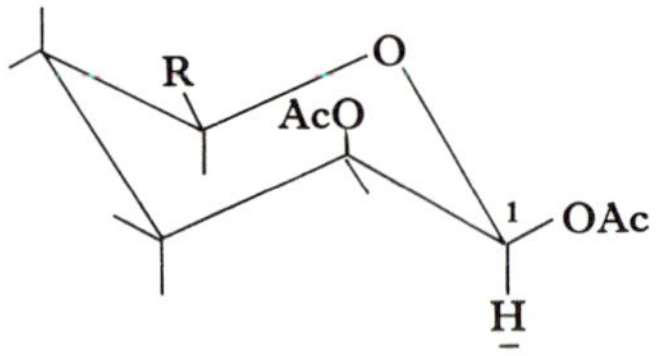

H-1 δ 5·93 Δν 0·18 ppm (to lower field)

β-D-mannopyranose penta-acetate

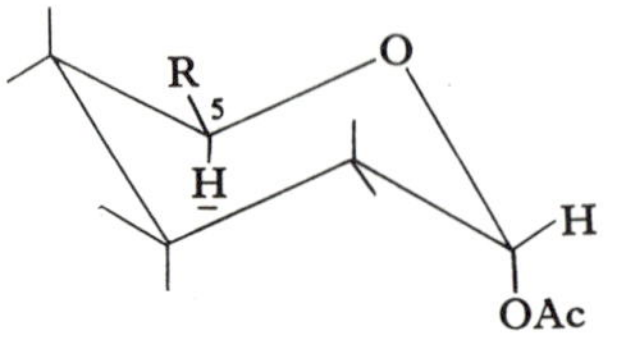

H-5 δ 4·12

α-D-glucopyranose penta-acetate

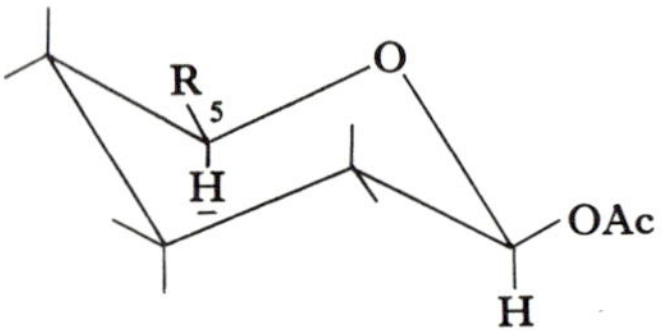

H-5 δ 3·88   Δν 0·24 ppm (to higher field)

β-D-glucopyranose penta-acetate

(partial structures: most common features not shown)

The last change is not unexpected in view of the known deshielding influence of an axial hydroxy group upon an opposed axial proton (Carr and Huitric, 1964, see p. 242).

The second rule follows from the greater deshielding of equatorial compared with axial protons in 6-membered chair conformations (see Chapter 3), and from the shielding effects of the changes e to f and g to h, as illustrated below.

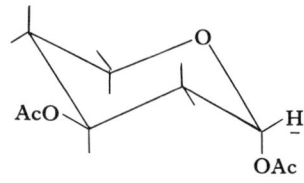

e	f
g	h
H-1 δ 6·33	H-1 δ 6·07 Δν 0·2 ppm (to higher field)
α-D-glucopyranose tetra-acetate	α-D-mannopyranose tetra-acetate
H-1 δ 6·27	H-1 δ 6·08 Δν 0·19 ppm (to higher field)
α-D-xylopyranose tetra-acetate	α-D-ribopyranose tetra-acetate

(partial structures: most common features not shown)

Thus an equatorial proton is more shielded when adjacent to axial than to equatorial OAc, while an axial proton is more shielded when adjacent to equatorial OAc. It is interesting that the same holds true in regard to the shielding effects of methyl substituents upon cyclohexane protons (Booth, 1966). Some examples which illustrate the value of these rules to stereochemical assignments are now given.

α-D-Altropyranose Penta-acetate

The CI and IC conformations of this derivative are shown below.† The

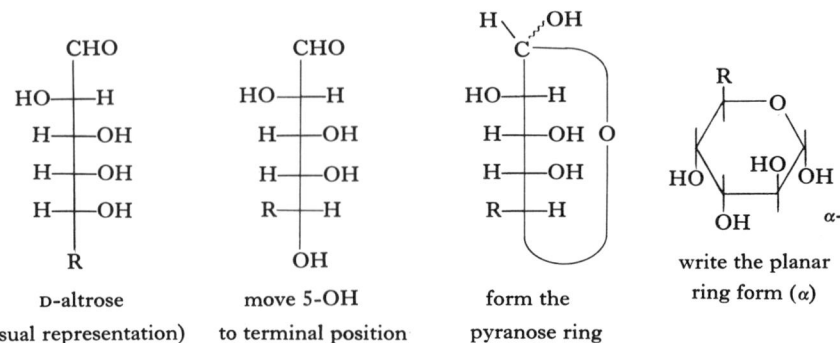

(R = CH₂OAc)

CI   IC

small value of $J_{1,2}$ (at 60 MHz the H-1 signal is a broad singlet) clearly favour CI, while observed and calculated H-1 and H-5 chemical shifts corroborate this assignment. Reference chemical shifts are those of H-1 ($\delta\, 5.76$) and H-5 ($\delta\, 3.9$) in β-D-glucopyranose penta-acetate (9, R=CH₂OAc) a compound in which all methine protons are axial, and all OAc groups are

† The following procedure, illustrated for D-altrose, is helpful for the conversion of a Fischer monosaccharide projection to a cyclic conformational representation.

equatorial. Taking CI as the preferred conformation, H-1 has achieved an equatorial orientation, therefore:

Add   0·6   (rule 2a)
Subtract   0·2   (axial OAc at C-2, rule 2b)
Subtract   0·2   (axial OAc at C-3, rule 2c)
Net addition   0·2
Reference value   5·76
Calculated value   5·96
Observed value   6·02   (All δ values)
H-5 has remained axial
Add 2 × 0·25 = 0·5 (opposition from two axial OAc, rule 1b)
Reference value   3·9
Calculated value   4·4
Observed value   4·5

α- and β-Mannopyranose Acetates

In this example, $J_{1,2}$ values fail to differentiate between the two anomers since the line width of the anomer proton signal is narrow in both cases. Chemical shift calculations clearly show that the β-derivative has the preferred conformation **10**, and the α-anomer, **11**.

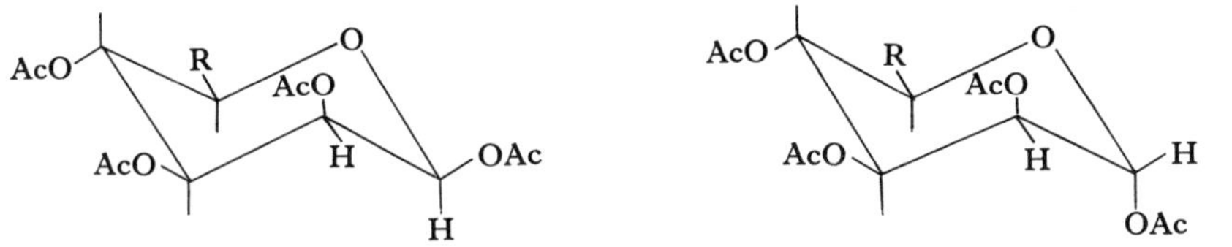

β-(CI)   α-(CI)
10   **11**

(R = CH₂OAc)

H-1 has remained axial   H-1 has become equatorial
Add   0·2   (rule 1a)   Add   0·6   (rule 2a)
Reference value 5·76   Subtract   0·2   (rule 2b)
Calculated value 5·96   Net addition   0·4
Observed value   5·93   Reference value 5·76
                                          Calculated value 6·16
                                          Observed value   6·09
H-5 axial   H-5 axial
Calculated value 3·9   Add   0·25   (rule 1c)
  (as standard)   Reference value 3·9
Observed value   3·9   Calculated value 4·15
                                          Observed value   4·1

PMR studies of hexoses bear out the conclusion that the conformational preferences of these sugars are determined largely by the bulkyl hydroxy-methyl substituent. Another, although smaller factor, is the preference of the C-1 hydroxyl group for the axial position. An explanation of this so-called anomeric effect is that repulsive interactions between dipoles involving C—O bonds of ring oxygen and C-1-hydroxyl are more pronounced when the last group has an equatorial orientation (cf. **12** and **13**). The anomeric

<table>
<tr><td align="center">**12**</td><td align="center"></td></tr>
<tr><td></td><td align="center">**13**</td></tr>
</table>

effect increases when the hydroxy group is methylated and becomes the governing factor in pentoses (CH$_2$OH absent) when OH is replaced by a halogen. The spectrum of tri-O-acetyl-β-D-xylopyranose chloride (Fig. 9.1), for example, fails to display the large vicinal coupling constants anticipated for the CI conformation **15**; instead, all the couplings are small (J <4 Hz) and establish the IC form **14** as the favoured conformation (Holland *et al.*, 1967). Remarkably, all 4 substituents in this conformation adopt preferred axial orientations. The possibility that the chloro derivative

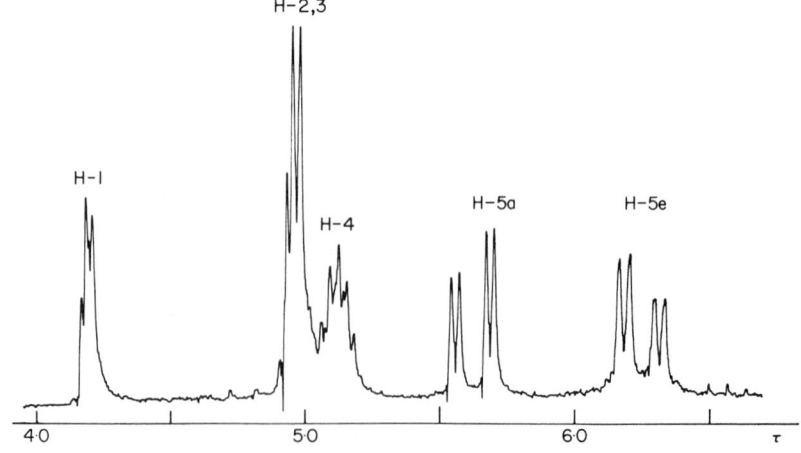

Fig. 9.1. Part of the 100 MHz PMR spectrum of tri-O-acetyl-β-D-xylopyranosyl chloride in CDCl$_3$ (Holland *et al.*, 1967).

adopts a conformation of the flexible type (e.g. skew-boat) was not considered probable because all possible structures of this type would be expected to give either a large $J_{4,5}$ coupling or a large $J_{1,2}$ coupling. The C-3 proton is long-range coupled to both C-1 and C-5e protons, probably as a result

of the coupled protons being linked by a near-planar W pathway (see Chapter 3, p. 91). Reversal of the usual H-5e and H-5a chemical shift relationship is explained by the axial proton being *syn* diaxial to deshielding chlorine and acetoxy substituents. Similar conformational preferences are found for fully esterified pentopyranosyl fluorides (Hall and Manville, 1969) which may be studied by both 'H and ^{19}F NMR. The favoured conformation **16** for α-D-xylopyranosyl fluoride, and **17** for the β-anomer

are supported both by H_1—H_2 and F—H_2 coupling constant magnitudes. The angular dependence of vicinal ^{19}F—'H coupling upon the FCCH dihedral angle appears to be similar to that described by Karplus for vicinal 'H—'H coupling. Hence a higher J_{F,H_2} value for **16** ($\phi = 180°$) is expected than for **17** ($\phi = 60°$), as is in fact observed.

PMR evidence may also assist studies of acyclic derivatives of sugars; the case of osazone structure, for long a controversial issue, is given as an example (Khadem *et al.*, 1965). Of 3 structures proposed for D-lyxohexose-phenylosazone, the acyclic form **18** contains 2 imino protons while each

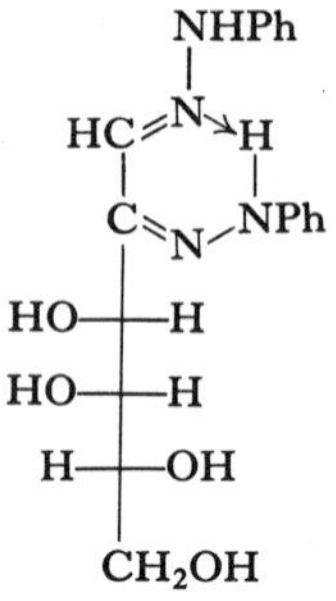

NHPh
|
HC=N·H
| |
C=N·NPh
|
HO——H
|
HO——H
|
H——OH
|
CH$_2$OH

18

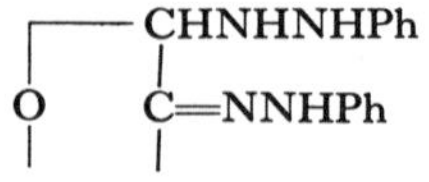

    CHNHNHPh
    |
O   C=NNHPh
    |

19

HC=NNHPh
|
CNHNHPh
|
O

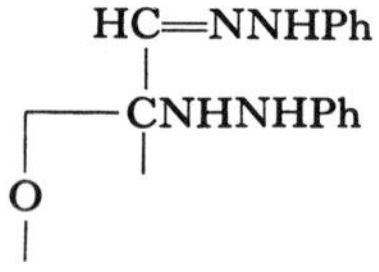

20

of the 2 cyclic variants **19** and **20** contain 3. The cyclic structures are untenable, however, because the spectrum of the osazone in DMSO shows only 2 one-proton singlets that can be assigned to NH protons; these are very low field (τ −0·55 and +2·4) and disappear when D$_2$O is present. Proton signals due to the 4 OH substituents are assigned to higher field signals on the basis of comparison with the spectrum of the analogue **21**

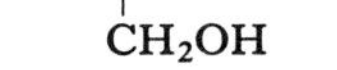

HC=N
| NPh
C=N
|
HO——H
|
HO——H
|
H——OH
|
CH$_2$OH

21

MeNPh
|
HC=N·H
| |
C=N·NPh
|
AcO——H
|
AcO——H
|
H——OAc
|
CH$_2$OAc

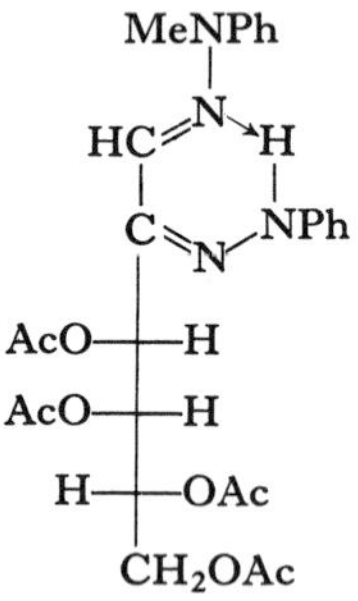

22

which contains no imine protons. Two low field singlets (τ −2·48 and +1·58), also attributed to NH protons, are seen in the spectrum of the fully acetyl-ated derivative of the osazone. A single NH resonance appears in the spectrum of the mixed osazone **22** and its chemical shift is close to the lower field NH signal of the symmetrical derivative. The latter signal is therefore assigned to the NH proton of the C-2 phenylhydrazone moiety of **18**. Structure **18** requires that this proton be coordinated to nitrogen of the C-1 residue and the fact of its very low field position supports this formulation.

Equilibrium Composition of Sugars in Solution

A pentose or hexose sugar is a single molecular species in the crystalline state. When it is dissolved in water (or D_2O), however, isomerization occurs and an equilibrium mixture results which may be composed of at least 5 compounds. These are the α-pyranose, the β-pyranose, the α-furanose, the β-furanose and the acyclic aldehyde form; the population of the last component is usually very small and may be neglected. These changes are commonly detected by optical rotational variations but they may also be

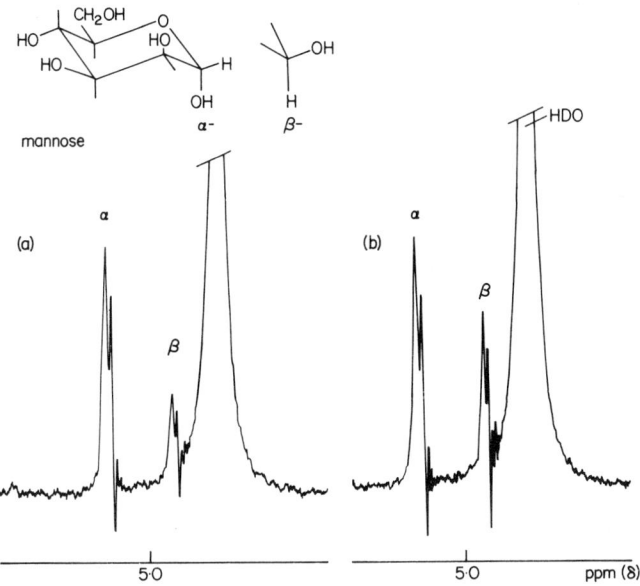

Fig. 9.2. 60 MHz anomeric PMR signals of: (a) a freshly prepared solution of an α-mannose sample in D_2O; (b) the same solution stored for 24 hours. *Note*: the significant proportion of β-anomer present in the fresh solution is indicative of either rapid mutarotation or a lack of anomeric purity of the sample. The latter is probably the case because the relative intensities of the anomeric signals of a fresh solution in DMSO-d_6 (a solvent in which mutarotation is slow) were the same as in (a).

monitored by observing changes in the anomeric signals in spectra of sugars dissolved in D_2O. Thus a solution of mannose recorded shortly after preparation showed a major low field (α) and a minor high field (β) anomeric signal (Fig. 9.2). In the spectrum recorded a day later, the β- had grown at the expense of the α-signal and their relative intensities were 67 (α):33 (β), a ratio which agrees closely with that derived from optical rotational measurements. At 60 MHz, only the anomeric protons are readily observed

(Lenz and Heeschen, 1961) but at 100 MHz other protons may be resolved and their characteristics used to corroborate isomeric assignments and to differentiate between pyranose and furanose forms (Lemieux and Stevens, 1966). Thus, the 100 MHz spectra of D-xylose, D-lyxose and D-arabinose in D_2O at equilibrium display only 2 anomeric proton signals; this shows that structures adopted in solution are restricted to 2 species, either a pyranose or a furanose pair (no case of single pyranose-single furanose pair has been reported). The PMR parameters of the anomeric signals provide the initial clue to structure, particularly when one member displays a typical a/a coupling constant as in the case of D-xylose (β, $J = 7\cdot4$ Hz) and D-arabinose (α, $J = 7\cdot2$ Hz) diagnostic of pyranose rings. Confirmation may follow

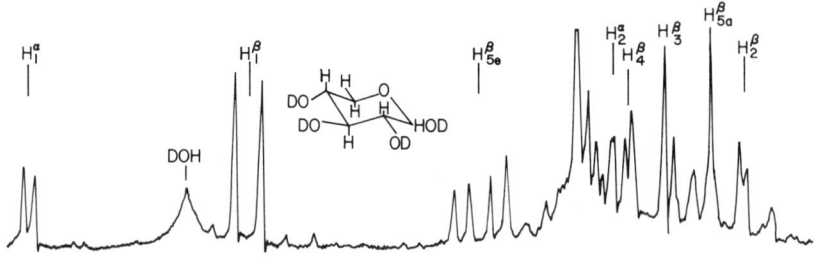

FIG. 9.3. 100 MHz PMR spectrum of D-xylose at equilibrium in D_2O at 35° (Lemieux and Stevens, 1966).

from data upon other signals as illustrated in detail for the case of D-xylose. In the equilibrium spectrum (Fig. 9.3), the band δ 3·5–3·9 is due to protons of the α-anomer, as seen from the spectrum of a freshly prepared solution; H_2^α falls in this region at δ 3·63 became spin decoupling at this frequency causes the H_1^α, signal to collapse. The quartet at δ 4 is assigned to a β-proton (its intensity corresponds with that of the H_1^β, signal) and its spacings of 4·3 and 10·4 Hz indicate it to arise from the equatorial atom attached to C-5

β-

23

since this should display a large geminal and a small vicinal (e/a) couplings in terms of the CI chair conformation **23**; further proof comes from the signal at δ 4 being a doublet in the spectrum of the 5-deuterio analogue. The position of the axial proton at C-4 is fixed by identifying the resonance position at which irradiation causes the δ 4 quartet (H_{5e}^β) to collapse to a doublet *which retains the larger (geminal) coupling*; this occurs at δ 3·58. The intensities of the signals to higher field of δ 3·5 correspond to three protons based on the intensity of H_1^β (α-signals are absent from this region as already mentioned), thus the resonance area of signals due to H_2^β, H_3^β and H_{5a}^β is established. The signal near δ 3·36 must be due to H_{5a}^β because it collapses when H_{5e}^β (δ 4 quartet) is irradiated. The quartet at δ 3·26 (J 7·4 and 8·5 Hz) changes to a doublet when H_1^β is irradiated, and hence is identified as due to H_2^β. It follows that the remaining triplet (δ 3·49) is due to H_3^β. The approximate coupling constants, listed below, clearly establish that the β-anomer of D-xylose exists in the near CI-chair conformation **23** in solution.

(J values in Hz for the β-anomer of D-xylose in D_2O)

$$
\begin{array}{ll}
H_1-H_2 & 7\cdot4 \\
H_2-H_3 & {\sim}8\cdot5 \\
H_3-H_4 & {\sim}8\cdot5 \\
H_4-H_{5a} & {\sim}10 \\
H_4-H_{5e} & 4\cdot3 \\
H_{5a}-H_{5e} & 10\cdot4
\end{array}
$$

The spectrum of D-ribose in D_2O at equilibrium displays 4 anomeric signals (Fig. 9.4) of total intensity 1, present in the ratio (moving upfield) of 6:18:56:20; both pyranose and furanose species must therefore be

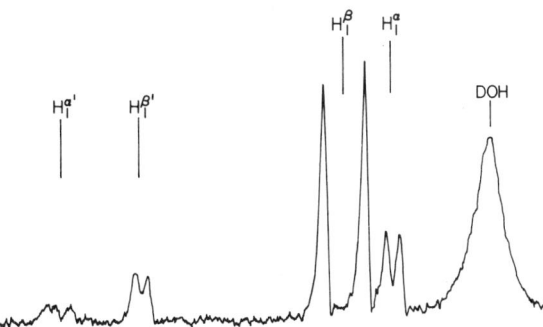

Fɪɢ. 9.4. 100 MHz anomeric PMR signals of D-ribose at equilibrium in D_2O at 35° (Lemieux and Stevens, 1966).

present. To obtain a model for the chemical shifts of the anomeric protons of ribofuranoses, the spectrum of 5-O-methyl-D-ribose was examined, a compound which may only form a 5-membered ring (see Scheme I). Its

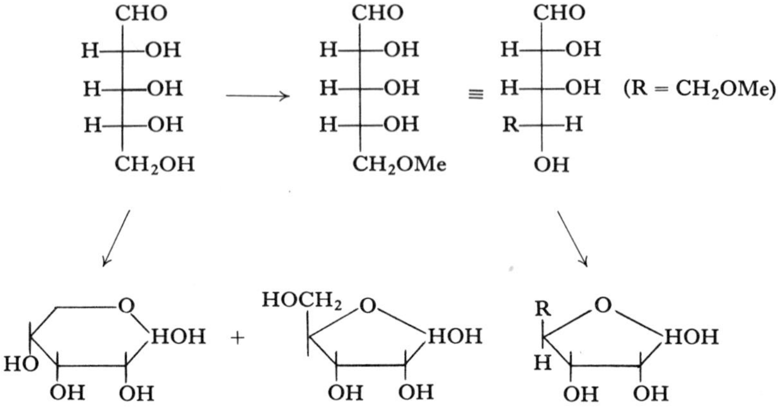

Scheme I

spectrum showed anomeric signals at δ 5·51 ($J \sim 3$ Hz) and 5·40 ($J \sim 1·5$ Hz) in the ratio 1:2·6. The higher field signal was assigned to the β-anomer on the grounds of its greater intensity (the *trans* 1,2 diOH should be more stable than the *cis* arrangement) and lower $J_{1,2}$ value. In general, derivatives of β-ribofuranose have $J_{1,2}$ values of about 1 Hz as compared to 4–5 Hz for the α-anomers. From these results, the lower pair of anomeric signals (δ 5·42 and 5·30) in the spectrum of D-ribose itself were assigned to the furanose forms. The major anomeric signal (δ 4·99, J 6·4 Hz) was assigned to β- rather than α-D-ribopyranose on account of its $J_{1,2}$ value being typical of a/a coupling in 6-membered ring sugars. Its unusually low field position is accounted for by the deshielding influence of the axial hydroxyl group at

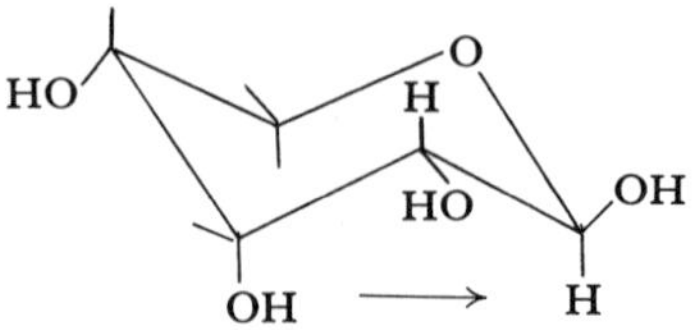

24   deshields

C-3 (**24**), as discussed shortly. Chemical shifts and J_{vic} values of the other ring protons of the major component were established by decoupling experiments akin to those outlined in the case of D-xylose and these data were consistent with β-D-ribopyranose having the CI conformation **24**. The highest field anomeric signal (δ 4·91, *J* 2·1 Hz) was assigned to α-D-ribopyranose and it is shown below that its chemical shift is close to the predicted value for a IC conformation.

Furanose species may also be detected in certain hexose sugars by PMR, the spectrum of the rare sugar idose being a particularly good example (Fig. 9.5).

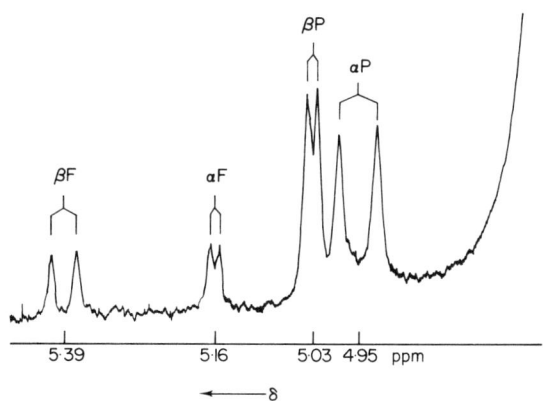

FIG. 9.5. 100 MHz anomeric PMR signals of D-idose in D₂O at 60°; αF, βF signify α- and β-furanose, αP, βP denote α- and β-pyranose signals (Angyal, 1969).

Rules for calculating the chemical shifts of protons relative to corresponding ones in β-D-xylopyranose and β-D-glucopyranose, analogous to those already described for acetylated derivatives, have been advanced on the basis of spectral analyses of aldoses in D₂O (Lemieux and Stevens, 1966). These are as follows:

1. If the proton under consideration is in an equatorial orientation, add 0·6 ppm.

2. If the proton under consideration is in an axial orientation,

(a) add 0·3 ppm for each neighbouring axial hydroxyl group and
(b) add 0·35 ppm for each axial hydroxyl group which is in opposition to the axial proton.

These rules are simpler than in the previous case because whereas an axial acetyl group has a shielding influence on an equatorial proton 1 or 2 bonds removed due to the anisotropy of the carbonyl function, similarly placed hydroxyl groups have no appreciable effect.

Application of these rules to α-D-ribopyranose supports the IC form as its preferred conformation, as follows:

	CI		IC
	H-1 equatorial		H-1 axial
Add	0·6 (rule 1)		Add 0·3 (rule 1a)
Reference value	4·65		4·65
Calculated value	5·25		4·95
Observed value	4·91		4·91

When dimethylsulphoxide (DMSO-H_6 or d_6) is used as solvent for PMR studies of sugars, signals due to the hydroxyl groups are observed and much valuable information is thereby made available (Casu *et al.*, 1964, 1965). The signals of secondary OH groups appear as doublets and primary as triplets, as a result of HOCH coupling; this is only detected in hydrogen bonding solvents such as DMSO which make the rate of proton exchange at oxygen NMR-slow (see Chapter 1, p. 28). The spectrum of α-D-glucose in DMSO-d_6 is a typical example (Fig. 9.6); mutarotation is slow in this solvent so signals due to a single anomer are seen provided the crystalline sugar is isomerically pure. The low field doublet (1 proton) near δ 6·2, and 4 of the 5 protons within the band δ 4·2–5·2 are due to OH as is clear from the reductions in signal intensity that follow the addition of D_2O to the solution. Complete D/H exchange is difficult to achieve because of the high proportion of exchangeable protons in a sugar molecule. Figure 9.6 shows a spectrum in which exchange is almost complete; the lowest field doublet is much reduced in intensity while the chief remaining feature of the 5 proton band is the narrow doublet near δ 5. The last signal is due to the anomeric C-1 proton and it forms a triplet in anhydrous DMSO as a result of coupling with the C-1 hydroxyl hydrogen atom. The OH resonances of the band δ 4·2–5·2 form 3 overlapping doublets (due to O_2H, O_3H and O_4H) and a triplet near δ 4·3 due to the primary CH_2OH. These signals are well resolved in the 100 MHz spectrum [Fig. 9.6(c)]. The lowest field doublet (∼ δ6·2) is known to be due to O_1H because this signal is absent in the spectrum of α-methylglucoside. Data upon the anomeric OH protons of a series of sugars have been tabulated (Casu *et al.*, 1964). In most cases,

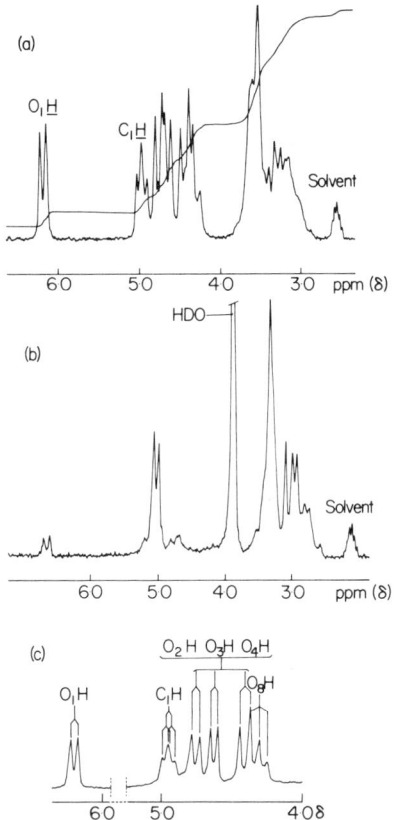

FIG. 9.6. PMR spectra of α-D-glucose in DMSO-d$_6$: (a) 60 MHz spectrum; (b) 60 MHz spectrum in DMSO-d$_6$-D$_2$O; (c) 100 MHz spectrum in DMSO-d$_6$ (last is due to Casu *et al.*, 1965).

α-anomers absorb in the range δ 6·30–6·04 and display coupling constants of the order 4·0–4·5 Hz, the corresponding data for β-anomers being δ 6·6–6·42 and J 6·0–7·0 Hz. These differences are intelligible in terms of the anomeric OH group being axial in α, and equatorial in β-anomers. The lower field positions of the β-OH signals are probably due to: (i) the greater deshielding of an equatorial as compared with an axial environment in alicyclic 6-membered rings, and (ii) a greater deshielding contribution due to hydrogen bonding of OH to the solvent; the more exposed equatorial β-groups will bond more strongly with DMSO than the axial groups. With regard to the coupling constant differences, work on cyclic mono-hydric alcohols has shown that equatorial OH protons couple more strongly with adjacent hydrogen atoms than do axial OH protons because the conformation **25** (HCOH dihedral angle 180°) is more favoured in the former case (Chapter 1, p. 29). The higher range of coupling constants observed for β-anomers is therefore in agreement with these findings. Sugars with configurations at C-2 different from that of glucose, such as mannose (**26**), show values that do not fit the mentioned range for anomeric O$\underline{H}$ chemical shifts although ^3J values still conform.

Perlin (1966) has given further demonstrations of the utility of DMSO in the PMR of sugars. Partially acylated sugars, for example, may be characterized neatly by a combination of OH and CH data. Thus, the spectrum of a tri-O-acetyl-β-D-mannose isomer (Fig. 9.7) shows two O$\underline{H}$ doublets (both absent in DMSO-D$_2$O), hence both free hydroxyls must be secondary (free CH$_2$OH would produce a triplet O$\underline{H}$ signal). The lower field doublet has a chemical shift typical of the anomeric O$\underline{H}$, so C$_1$—OH is free. When D$_2$O is added the 2 O$\underline{H}$ doublets vanish while the doublet of doublets Y becomes a narrow doublet; Y must therefore be due to a $\underline{H}$—C—OH proton and is identified as the anomeric C-1 proton because of its position and the fact that its larger spacing (7·4 Hz) corresponds with that of the

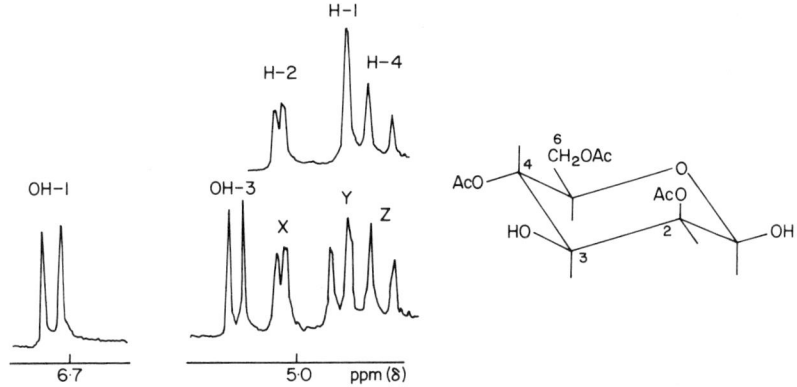

FIG. 9.7. Part of the 100 MHz spectra of 2,4,6-tri-O-acetyl-β-D-mannose in DMSO-d_6 (bottom) and DMSO-d_6-D$_2$O (top) (Perlin, 1966).

C_1—OH signal. Signals X and Z, unchanged in the presence of D$_2$O, must be due to H—COCOMe functions and their low field positions are a result of the acylation shift (see p. 30). Signal X is weakly coupled to Y (C_1—H) and is thus assigned to C_2—H. Z does not have a spacing in common with X and cannot therefore be due to the C-3 proton; by elimination it must be assigned to C_4—H. The C_3—H signal is not subject to the acylation shift and appears upfield of the group shown in Fig. 9.7. This analysis confirms the derivative as the 2,4,6-triacetate.

Furanose forms are more favoured in DMSO than in D$_2$O. The spectrum of a freshly prepared solution of arabinose anomers (mutarotated in water), for example, displays 2 major and 1 minor C_1—OH signal, the last being attributed to a small amount of furanose. The same signal is much more pronounced in the spectrum recorded 4 days later. Mackie and Perlin (1966) suggest that water may stabilize pyranoses preferentially by hydrogen bonding; bulk increases due to OH solvent interactions will destabilize furanoses because of the eclipsing of substituents that obtains in 5-membered rings. In support, 2,3-di-O-methyl-D-arabinose, which has less hydroxyl groups available for hydrogen bonding, produces at least 3 H-1 signals in D$_2$O, whereas only 2 significant signals are found for arabinose itself. The intensity of the weakest signal shows that about 17% of furanose is present. A similar value (20%) is given by the C_1—OH signals observed immediately after the mixture is transferred from water to DMSO (Fig. 9.8). Assignment of the furanose signal is corroborated by a corresponding upfield triplet caused by the primary carbinol group (CH$_2$OH) present only in 5-ring pentoses. In the spectrum recorded 2 weeks later, the furanose content has risen to 65% [Fig. 9.8(b)].

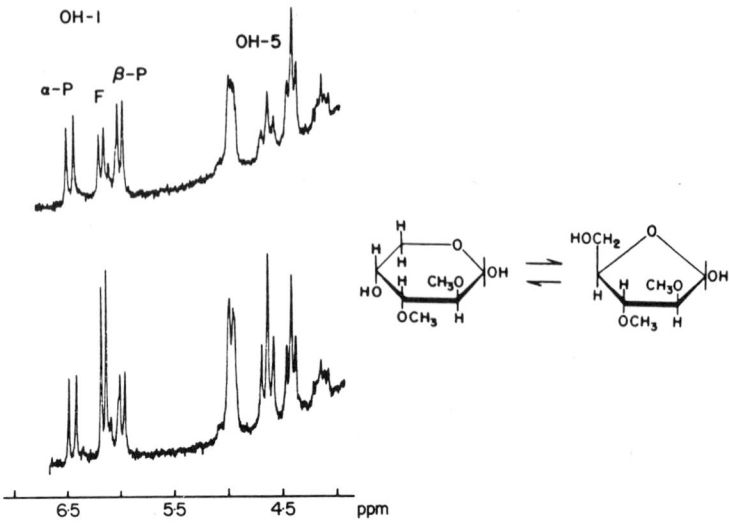

FIG. 9.8. Part of the 100 MHz spectra of 2,3-di-O-methyl-D-arabinose in DMSO-d_6. Top: spectrum of syrupy product directly after transfer from an aqueous solution; bottom: spectrum recorded 2 weeks later (Mackie and Perlin, 1966).

The PMR features of a series of disaccharides and higher polyglucoses in DMSO-d_6 have been reported (Casu *et al.*, 1966, 1968). The 2 forms of maltose may be differentiated, as usual, by their anomeric OH signals (α, τ 3·7, J 4·5 Hz; β, τ 3·4, J 6·4 Hz). The spectra of both anomers show a 2-proton signal which is low field of the usual non-anomeric hydroxyl proton range of τ 5–6. These low field signals are assigned to the protons of the "internal" O_2H and $O_3{'}H$ groups of the α 1,4-linked glucose units.

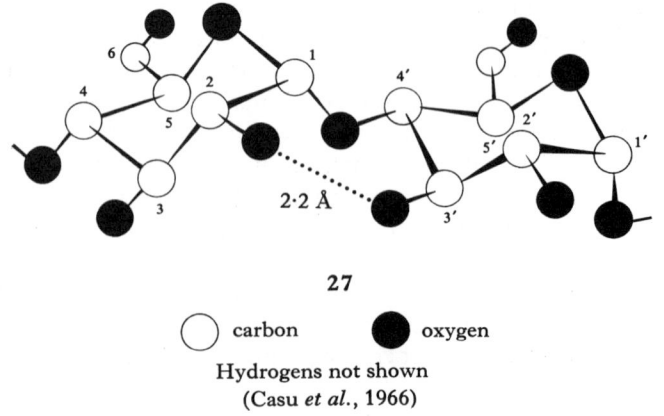

27

◯ carbon   ● oxygen

Hydrogens not shown
(Casu *et al.*, 1966)

In the model **27** built with CI rings, these two OH groups face each other and are close enough to permit formation of strong intramolecular hydrogen bonds with consequent deshielding of the protons in question. The other hydroxyl protons are also subject to deshielding by hydrogen bond formation but in these cases the bonding (intermolecular to solvent DMSO-d₆) is considered to be weaker. This hypothesis is supported by the fact that the intensities of the low field absorption of the non-anomeric hydroxyls of maltotriose (3 CI) units corresponds to 4 hydroxyls per molecule. Close approach of 2 hydroxyls as in **27** is not possible in 1,6-linked saccharides, and sugars of this type like gentiobiose and melibiose do not show low field

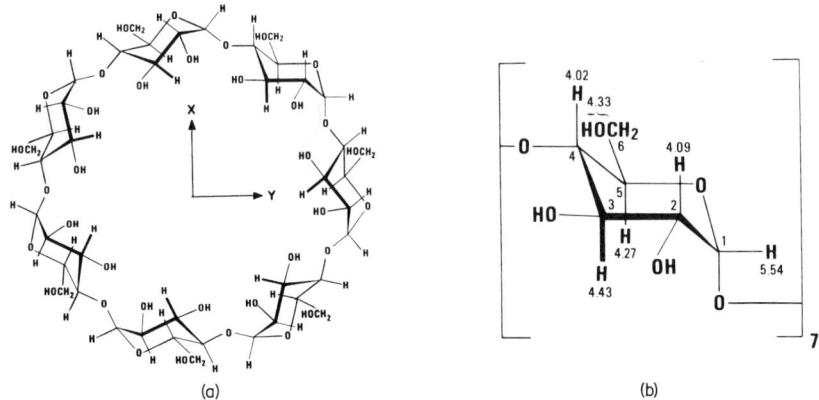

(a)   (b)

FIG. 9.9. (a) Cyclohepta-amylose molecule. Z-Axis is parallel to the cyclohepta-amylose cavity, i.e. normal to the plane of the paper; (b) recorded chemical shifts (δ scale) for various cyclohepta-amylose protons (Demarko and Thakkar, 1970).

non-anomeric OH absorptions. The spectra of amylose and cyclodextrins also display O$\underline{\text{H}}$ signals in the range τ 4–5, consistent with these molecules being constituted from repeating CI units as in **27**. The lowest field signal in this region is seen in the spectrum of β-cyclodextrin (indicative of intramolecular hydrogen bonding being strongest in this molecule) and this is explained in terms of a limited rotation of the glucose units of the macrocycle (see Fig. 9.9). The spectrum of β-cellobiose [4-O(β-D-glucopyranosyl) β-D-glucose] in DMSO-d₆ has a 1-proton absorption signal in the range typical of intramolecularly hydrogen bound O$\underline{\text{H}}$, and a conformation which involves a single internal H-bond between 3-O$\overline{\text{H}}$ of one ring and the ring oxygen atom of the other has been proposed based on this finding (**28**).

28

Cyclodextrins are of interest as models for enzymes and proteins and it has been suggested that they form inclusion complexes with organic substrates in aqueous solution (Demarko and Thakkar, 1970 and references cited therein). A PMR study has now verified the inclusion of a variety of aromatic substrates within the cavity of a cyclodextrin. X-ray and PMR

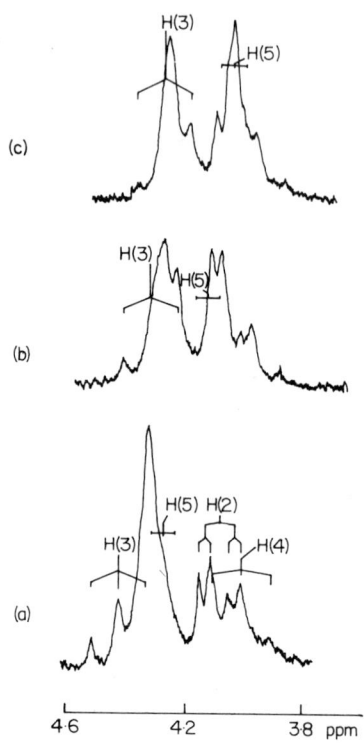

FIG. 9.10. Part of the 100 MHz PMR spectra of cyclohepta-amylose in D_2O (1.79×10^{-2} M) at $30 \pm 1°$ containing various amounts of *p*-hydroxy benzoic acid: molar ratio, substrate: cyclohepta-amylose: (a) no substrate, (b) 0.77, (c) 3.09 (Demarko and Thakkar, 1970). *Note*: Signal assignments are based on (i) analysis of individual splitting patterns and *J* values at 220 MHz; (ii) decoupling experiments and (iii) expected chemical shift behaviour, i.e. H-3 and H-5, being 1,3-diaxial to the C-1 axial oxygen should resonate at lower field than H-2 and H-4.

studies (some of which have just been described) have established the CI chair conformation of the constituent glucose units in cyclodextrins. Cycloamylose thus has primary and secondary hydroxy groups crowning opposite ends of its torus, H-3 and H-5 directed toward its interior and H-1, H-2 and H-4 located on its exterior (Fig. 9.9). Models show that a substituted aromatic molecule should just fit the cyclohepta-amylose cavity provided the long axis of the "guest" be aligned parallel to the cavity axis. It was thus expected, in the light of the screening environment associated with aromatic molecules, that if inclusion occurs, protons located within or near the cavity (e.g. H-3, H-5 or H-6) should be strongly shielded. On the other hand, if association takes place at the exterior of the torus, H-1, H-2 and H-4 should be the more strongly affected. Changes in the 100 MHz spectrum of cyclohepta-amylose in D_2O ($1\cdot79 \times 10^{-2}$ M) following the addition of increasing amounts of p-hydroxybenzoic acid bear out the former prediction. Thus, the H-3 and H-5 signals move upfield as the concentration of aromatic substrate increases while the remaining resonance lines (H-1, H-2, H-4 and H-6 experience only marginal upfield shifts) (Fig. 9.10). Inclusion complex formation was demonstrated for a variety of aromatic substrates, characterized by the chemical shift changes $\varDelta\delta$ [$\varDelta\delta = \delta_{free}$ (no substrate) $- \delta_{sat}$ (solution saturated with substrate)] for the various proton signals, e.g., acetylsalicyclic acid $\varDelta\delta$ ppm: H-1, $+0\cdot03$; H-2, $+0\cdot03$; H-3, $+0\cdot14$; H-4, $+0\cdot03$; H-5, $+0\cdot20$; H-6, $+0\cdot06$. No evidence for complex formation was obtained for substrates that had bulky substituents such as 2,6-di t-butyl phenol or were very soluble in water, e.g. catechol and resorcinol.

Furanoses

In pyranose sugars, only two principle conformations (CI and IC) need to be considered in the majority of cases. In furanose derivatives, however, the 5-membered ring has numerous conformations of similar free energies separated by low energy barriers (Eliel *et al.*, 1965), hence their stereochemistry is more difficult to establish. This is unfortunate from a biochemical point of view because sugar moieties of nucleosides and nucleotides exist in the furanoid form. In spite of these problems, many PMR studies of furanose derivatives have been reported and conformational and configurational assignments made from the data obtained (Hall, 1964; Inch, 1969).

The earliest papers provided evidence that the 5-membered ring of natural ribofuranoses is not planar (Jardetzky, 1960, 1962). Thus analysis of the 60 MHz spectrum of adenosine-2′-monophosphate (2′-AMP) gives the vicinal coupling constants shown alongside formula **29**. If the ribose ring of

$J_{1,2}$ 6·0 Hz
$J_{2,3}$ 5·2 Hz
$J_{3,4}$ 3·4 Hz

Ad = adenine

29

planar
arrangement

29 were planar, $J_{1,2}$ and $J_{3,4}$ should both have values close to 2 Hz on the basis of equal dihedral angles of 120° and direct application of the Karplus $\cos^2\phi/J$ relationship as shown above; also $J_{2,3}$ ($\phi = 0°$ in the eclipsed conformation) should be about 8 Hz. It is now recognized that substituent effects modify the extents of vicinal coupling and that only approximate J_{vic} values may be derived from the original Karplus relationship; nevertheless, it is clear that the experimental values shown in **29** do not conform to a planar model. Amongst spectra of a series of β-ribose nucleosides and nucleotides in D_2O, spacings of the doublet due to the anomeric proton H_1 (identified by its low field position relative to the other ribose proton signals) are larger for purine (4–6 Hz) than for pyrimidine derivatives (<3 Hz), and this result suggests that the two classes differ in conformation. Jardetzky

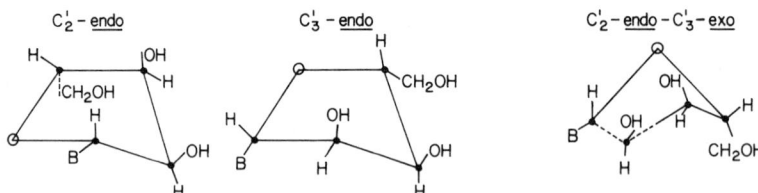

Fig. 9.11. In these conformations the carbon atom(s) specified is moved out of the plane of the remaining ring atoms; in *endo* forms the displaced atom is directed towards the same side of the ribose ring as the β C_1-heterocyclic substituent and the C_4—C_5 bonds, while in the *exo* form its direction is away from these bonds. B denotes the heterocyclic base (Jardetzky, 1960).

(1960, 1962) and Smith and Jardetzky (1968) measured the dihedral angles of wire models of maximally puckered furanose rings, e.g. C_2-*endo.*, C_3-*endo* and C_2-*endo*-C_3-*exo*, (Fig. 9.11) and calculated the theoretical coupling constants corresponding to each conformation. Preferred forms were proposed after comparisons of calculated and observed data but in most cases it was not possible to restrict the choice to a single conformer. The example of cyclic 3′,5′-AMP is a special one because its conformation is governed by the energetically favoured chair conformation of the 6-membered ring formed from the atoms C_3, C_4, C_5, O_5, P and O_3. In the model **30**, the H_1—H_2 dihedral angle has a value (105°) which corresponds

30

with a very small coupling constant, and evidence that the conformation is maintained in solution is provided by the observation of a $J_{1,2}$ value of less than 1 Hz.

In considering the possibility of assigning configurations to anomeric nucleosides and nucleotides by PMR, Lemieux and Lineback (1963) pointed out the uncertainties of interpreting the J values of anomeric proton signals as a result of the H_1—H_2 dihedral angle variations in flexible 5-ring systems (*cis* 0° to 45° corresponding to J values of 3·5 to 8 Hz, and *trans* 75° to 165° corresponding to values of 0 to 8 Hz). On the basis of these overlapping ranges of J values, calculated from the original Karplus expression, assignments may only be made safely when J is less than 3·5 Hz. In 4 pairs of α-β methyl furanosides, e.g. **31** $J_{1,2}$ values of 2 Hz or less were found in each of the 4 1,2 *trans* derivatives while the $J_{1,2}$ *cis* results ranged from 4 to 4·5 Hz (Capon and Thacker, 1964).

α-methylglucofuranoside (R = H, R′ = OMe)
$J_{1,2}$ 4·2 Hz

β-anomer (R = OMe, R′ = H)
$J_{1,2}$ 0
(solvent, pyridine)

31

A full analysis of the PMR spectra of a pair of isomeric nucleosides may allow configurational assignments to be made as is demonstrated by the case of the deoxyribose derivatives, thymidine and its anomer (Lemieux, 1961).

Th = thymine

thymidine

anomer

Spacings of the H_1', H_2', H_2'' and H_3' signals of the anomer of thymidine (60 MHz spectrum in D_2O) establish the couplings $J_{H_1', H_2'}$ ~4 Hz and $J_{H_1', H_2''}$ ~ 7 Hz, while H_3' is coupled by about 4 Hz to one of the H_2 protons and 7 Hz to the other. Considering the H_2—H_3 values: if H_3 is projected

32   **33**

between H_2' and H_2'' as in **32**, J values of 3·6 and 0·6 Hz follow from the Karplus relationship. A very low J value is not obtained experimentally, hence this arrangement is eliminated. The form **33**, however, satisfies the experimental values. If the anomer has the β-configuration, the arrangement requires that H_1 is projected between H_2' and H_2'' (**34**) whence J values of about 7 and 0·5 Hz are anticipated.† Only one of the experimental $J_{1,2}$ values (7 and 4 Hz) corresponds in this case, thus the anomer cannot possess this configuration. The α-arrangement (**35**), on the other hand, leads to predicted J values which are close to both observed values.

34   35   36

It follows that thymidine itself is the β-anomer. The first order analysis of the CH_2CH system of the α-form just described is valid because there is a significant chemical shift difference between all protons concerned. In the spectrum of thymidine itself, however, the H_2' and H_2'' chemical shifts are almost the same. Under these circumstances the two vicinal coupling constants tend to be averaged rather than have values characteristic of the two HCCH dihedral angle relationships; failure to recognize complications resulting from averaging of coupling constants in cases like this can lead to errors in the interpretation of PMR spectra and the use of coupling constant values for conformational assignments (Trager *et al.*, 1967). Thus the anomeric PMR signal of thymidine is a triplet rather than a quartet but its spacings (7 Hz) may not be used as direct evidence of a particular conformation. In this particular case, Lemieux considered that the magnitude of the average value requires that H_1' be strongly coupled (J 5 Hz) with H_2' since the coupling of H_1' with H_2'' should not be greater than 9, and proposed the arrangement **36** in which H_1' defines dihedral angles of about 150° and 30° with H_2' and H_2''. Both conformations suggested for thymidine and its anomer are forms in which all substituents are staggered rather than eclipsed, and hence are favourable on energetic grounds.

The PMR spectra of a variety of anomeric 2'-deoxy-D-ribofuranosyl nucleotides have been reported and in every case the β-anomeric proton

† These relationships are most readily appreciated by the use of flexible models, e.g. Framework Molecular Models. Dreiding models are less satisfactory because they do not permit the "freezing" of any particular flexible conformation of a 5-membered ring.

forms a "pseudo triplet" (*J* 6·5–7 Hz) and the α, a doublet of doublets with spacings of about 7 and 3–4 Hz (Jardetzky, 1961; Robins and Robins, 1965). Thus, the conformation of the deoxyribose ring of these nucleosides is not affected to any appreciable amount by changing the heterocyclic base, and configurational assignments to purine *deoxyribose* nucleosides may be made with reasonable confidence on the basis of the PMR spacings of the anomeric proton.

In contrast, the spacings of C-1 proton signals of *ribo* furanosyl nucleosides do not aid the stereochemical assignment of a pair of anomers because both signals are doublets showing similar separations (α-anomers: 4–7·5 Hz, β-anomers: 5–6 Hz) (Nishimura and Shimizu, 1965). However, it is found that the C-1 signals of *cis* isomers **37** come to resonance at lower field than

37 *cis* H₁, H₂

38 *trans* H₁, H₂

1-D-ribofuranosyluracil
α, −143†, *J* 4·0 Hz
β, −129, *J* 4·0 Hz

α (*cis*)

β (*trans*)

1-D-lyxofuranosylthymidine
α, −131, *J* 7·0 Hz
β, −144, *J* 6·0 Hz

α (*trans*)

β (*cis*)

† Chemical shift in Hz from dioxane at 60 MHz (Nishimura and Shimizu, 1965).

those of *trans* isomers **38**, the α/β chemical shift difference being about 15 Hz at 60 MHz. In deoxyribofuranosyl nucleosides the anomeric proton signals have similar resonance positions, hence the chemical shift difference in question must be due to the 2-hydroxyl substituent. Axial hydroxyl deshields an adjacent *trans* proton in 6-membered-ring sugars (see p. 343) and probably has the same influence in the *cis* isomers **37** where similar H/OH geometry obtains.

Data upon both α- and β-anomers are strictly required if firm stereochemical conclusions are to be drawn using any of the above methods. Cushley *et al.* (1967) have described a procedure involving acetylated derivatives in which only one isomer is needed. Acetylated pyrimidine nucleosides exist in a preferred conformation in which the 5,6 double bond of the pyrimidine nucleus is *endo* to the 5-membered sugar ring (Fric *et al.*, 1966) as shown in **39** for α-ribo and α-arabino derivatives. The *cis* acetoxy group in **39a** is found in the cone of positive shielding of the 5,6 double bond. This is substantiated by the fact that the C_2-acetoxy resonance appears

39

a: R = OCOMe, R′ = H (arabino)

b: R = H, R′ = OCOMe (ribo)

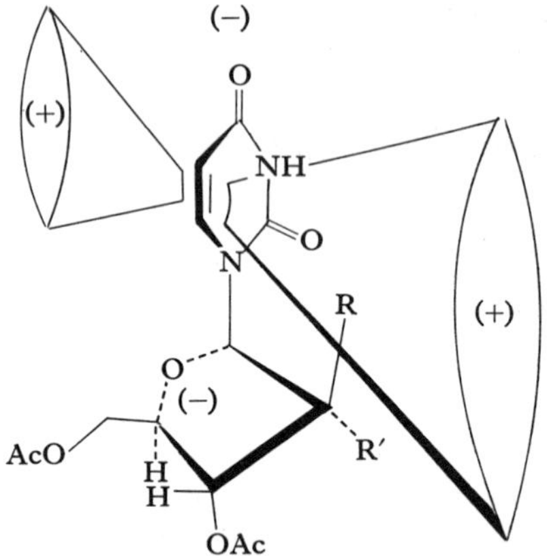

well upfield of the C_3 and C_5 signals in the two *cis* C_2—OCOMe/C_1-heterocycle nucleosides examined. In addition, the C_2—OCOMe resonance in the same derivatives moves downfield by 0·05–0·1 ppm when the compounds are hydrogenated, e.g. **40**; reduction of the 5,6 double bond is readily achieved by shaking the nucleoside with hydrogen at atmospheric pressure in the presence of 5% rhodium on alumina. In nucleoside-dihydro-

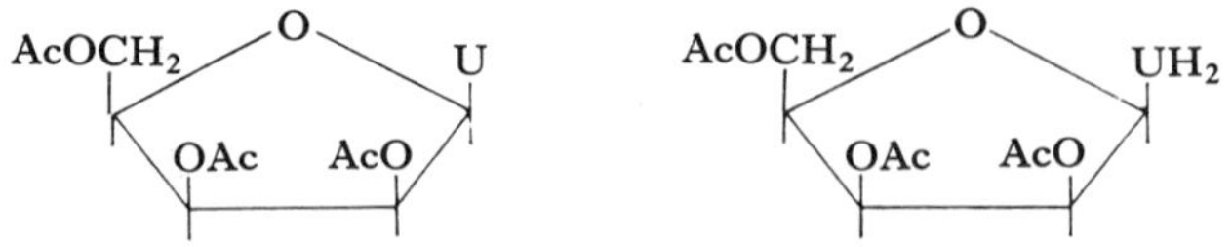

C_2-OCO<u>Me</u> τ 8·07
chemical shift
Lyxofuranosides

τ 7·97 (DMSO-d_6 solvent)

U = uracil, UH_2 = dihydrouracil

40

nucleoside pairs in which a *trans* relationship exists between the C_1 and C_2 substituents, either an upfield shift is seen in the C_2 acetoxy resonance or no appreciable effect, e.g. **41**. The authors point out that observation of changes in the 2-acetoxy signal of an anomer of unestablished geometry

	TH₂
AcOCH₂ O T	

C₂-OCOMe
chemical shift   τ 7·92   τ 7·93

Ribofuranosides   T = Thymine, TH₂ = dihydrothymine

41

should therefore allow its configurational assignment, but give no specific examples; they emphasize that pure conformers are not required for observation of a shift but simply a substantial population in a preferred conformation.

Hart and Davis (1969) have sought evidence about the conformation of pyrimidine nucleotides by application of the nuclear Overhauser effect

42
anti
in reference to pyrimidyl

43
syn
2C=O and ribose ring

(NOE, see p. 102). In **42**, the authors consider that H-6 can be close enough to all sugar protons with the exception of H-4′ and protons attached to oxygen to allow an NOE. A sole interaction with H-5′ and/or H-5″ would indicate a predominant *anti* conformation **42**, while an exclusive interaction with H-1′ would indicate that the *syn* conformation **43** is preferred. In the event, enhancement of the H-6 signal followed the irradiation of more than 1 signal (see results for uridine on p. 360) and it was concluded that the *syn* conformation as well as intermediate-range conformations have significant populations in addition to the usually accepted *anti* form.

Positional, as opposed to stereochemical, isomers of nucleosides may also be characterized by PMR spectroscopy, as in the case of the two isomers isolated from reaction between 2,3,5-tri-O-benzyl-D-arabinofuranosyl chloride and N⁶-benzoyladenine (Darnall and Townsend, 1966). The major product was 9-β-D-arabinofuranosyladenine **44** and the minor was

NOE results for uridine

Solvent	Proton(s) irradiated	Per cent enhancement[a] of H-6 signal
DMSO-d_6	H-1′	10
	H-2′, 3′	13
	H-5′	10
D_2O (75) plus Pyridine[b] (25)	H-1′	6
	H-2′, 3′, 4′	18
	H-5′	12

[a] A positive enhancement >4% is taken as evidence for a NOE.
[b] Added to change relative chemical shifts of H-1′ and H-5 so that H-1′ could be irradiated independently of H-5.

presumed to be the corresponding α-anomer. The 60 MHz PMR spectrum of **44** in DMSO-d_6 displayed closely placed singlets for the 2- and 8-protons of the adenine ring ($\Delta\nu$ 6 Hz). These signals should have shown a similar

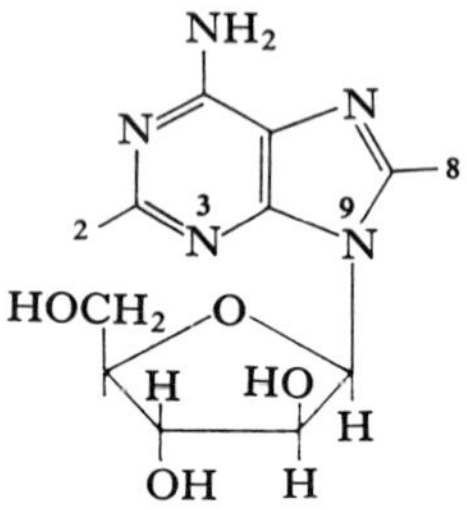

44

degree of separation in the spectrum of the minor isomer if it were the α-anomer. In fact, the $\Delta\nu$ value was 40 Hz and this result identified the minor

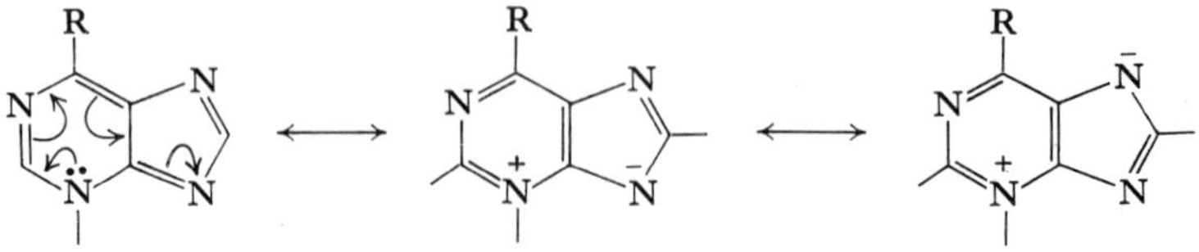

45   **46**   **47**

component as a 3-adenine derivative **45** (Townsend *et al.*, 1964). In derivatives of this type, contributions from the resonance forms **46** and **47** deshield the 2-H and shield the 8-H proton of the adenine ring and account for the large difference in chemical shift. Finally the minor isomer was assigned a β-anomeric configuration from the chemical shift of its anomeric proton signal. The fact that its C-1 proton resonance (δ 6·67) was lower field than that of the known 3-β-D-ribofuranosyladenine **48** (δ 5·95), a *trans* 1,2-

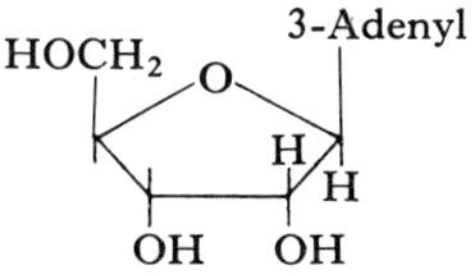

48 (*trans*) 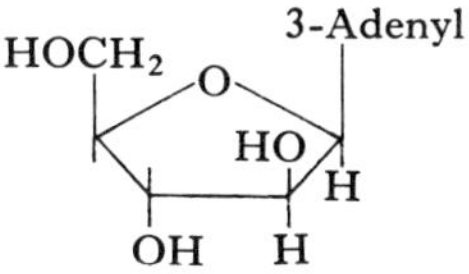   49 (*cis*)

substituted furanoside, showed that it must be the *cis* derivative **49** in accord with the results of Nishimura and Shimizu (1965) as previously discussed.

Pyridine Dinucleotides

Extensive PMR studies of coenzymes I and II, known as diphospho-pyridine nucleotide (DPN, **50**, R = H) and triphosphopyridine nucleotide

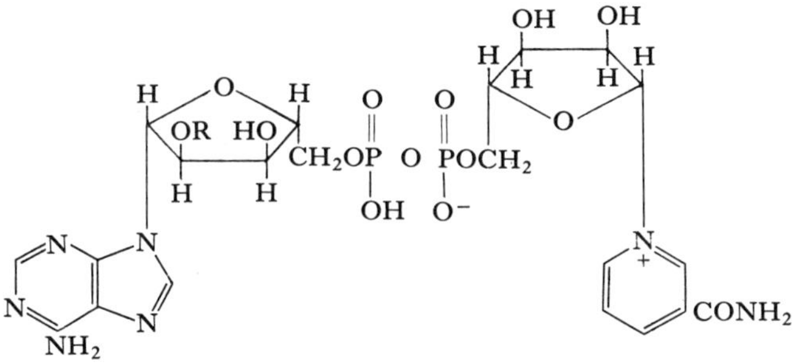

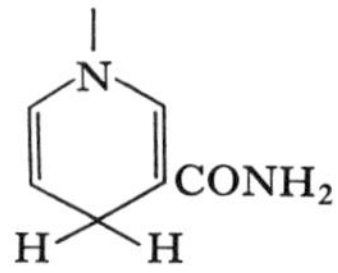

50

DPN   R = H
TPN   R = PO₃H₂

(TPN, **50**, R = PO₃H₂) respectively, have been made in order to obtain evidence about the conformations of these molecules which play such an important role in the oxidation-reduction processes of living organisms. In the case of the reduced form of coenzyme I (DPNH), an added interest in this work is the problem of the specific recognition by enzymes of one of the C-4 protons of the dihydronicotinamide ring. All stable conformations of the dinucleotides may be regarded as special cases of 1 of 3 conformational classes, namely, open chain **51**a, hydrogen-bonded **51**b in which the 2 heterocyclic rings lie in the same plane, and folded **51**c in which one heterocyclic is stacked in a plane parallel to that of the other. In the last arrange-

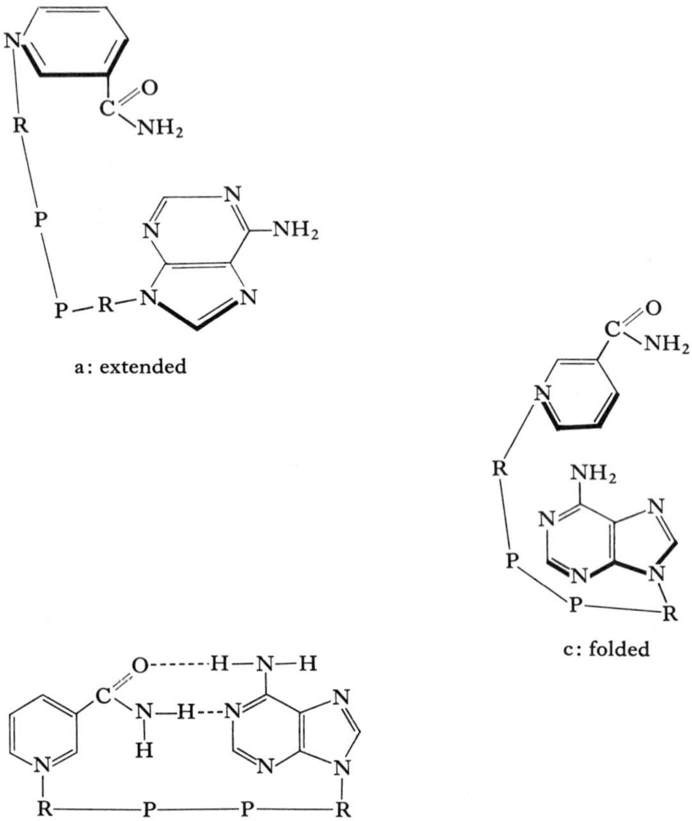

a: extended

c: folded

b: hydrogen-bonded

51

ment, the protons of the pyridyl ring lie in the diamagnetic shielding zone of the adenine aromatic nucleus, while the C-2 and C-8 protons of adenine fall in the same zone of the nicotinamide substituent. Hence, a demonstration that the PMR signals of the heterocyclic protons of DPN and DPNH have unusually high field positions, provides evidence supporting conformations of the folded variety. The first clue suggestive of stacked conformations being favoured, was provided by a comparison of the spectra of DPNH in D_2O and N-benzyldihydronicotinamide **52** in chloroform (Meyer *et al.*, 1962). Upfield shifts of the pyridyl signals of DPNH relative to those

52

of **52** were observed, but the possibility of shifts being caused by solvent and substituent effects was not investigated. Firmer evidence was provided by Jardetzky and Wade-Jardetzky (1966) who compared the spectrum of DPN (the natural β-anomer) with that of an equivalent mixture of nico-

53

54

tinamide mononucleotide (**53**, NMN) and adenosine 5'-phosphate (**54**), i.e. the constituent mononucleotide halves of the coenzyme; the solvent was D_2O in both cases. The aromatic peaks and the H'-1 peak of the ribose units shifted to higher fields as compared to those of the mixture of mononucleotides. The shifts were not large (5–15 Hz at 60 MHz) and varied from signal to signal in the spectrum. These results indicate a specific structural difference between DPN and the mononucleotide mixture, affecting different protons to different extents, which could arise from the formation of either intra- or inter-molecular complexes, or a combination of both. To eliminate the possibility of intermolecular contributions, the dependence of chemical shifts upon pH and concentration were examined. In the mononucleotides the shifts of the adenine peaks reflected ionization of the N-1 basic centre (pK_a 3·8), plots of the C-2 and C-8 chemical shifts against pH having the typical S shape of titration curves with inflections near pH 4†

† The apparent pH value in D_2O is measured by means of a pH meter, and henceforth the term pD replaces pH in this account. In some, but not all, of the data the correction pD = meter reading plus 0·4 has been applied (Glascoe and Long, 1960).

(see Chapter 2, p. 68); pyridyl proton signals of NMN were independent of pH over the accessible range (the nucleotide precipitated below pD 5) and were assumed to be so at lower pH values since no ionizable centre comparable with that of adenine is present. In the dinucleotides, however, *both* the adenine and pyridine shifts showed a transition near the pK_a of adenine. Thus a change in the ionization state of the adenine ring is reflected in a change in the chemical shifts of the pyridyl protons, and this result is good evidence that the two rings are close together. Shifts of the pyridyl signal to low field below pD 4 could be due to a dissociation of the pyridine-adenine complex brought about by a repulsive interaction between the 2, now positively charged, rings (see later). The intramolecular nature of the complex was further corroborated by the lack of a concentration dependence of chemical shifts of the dinucleotide over the range 0·005–1·0 molar.

Similar evidence and conclusions about dinucleotide conformations were drawn from a study of the spectra of α- and β-DPN and adenosine diphosphoribose, ADPR (the coenzyme minus its pyridyl moiety) (Sarma *et al.*, 1968b). At pD 7·0 the adenine C-8, C-2 and C′-1 (ribose) proton signals in both anomeric forms of DPN were shifted to higher fields compared with those in ADPR, the shifts being larger for the β-isomer (Table 9.1).

TABLE 9.1

Compound		Adenine peaks[a]		
		C-8	C-2	C′-1 (ribose)
ADPR	pH 7	512·0	495·0	370·8
	pH 2	520·2	510·0	372·3
αDPN	pH 7	508·7	488·7	363·5
	pH 2	518·8	508·5	368·8
βDPN	pH 7	504·0	486·0	363·0
	pH 2	516·7	508·0	369·3

[a] Evidence for these and related assignments is discussed later.
All measurements made on 0·15 M solutions in D_2O at 38°; chemical shifts in Hz from DSS at 60 MHz.

In all cases a change in pD from 7·0 to 2·0 resulted in downfield shifts, but shifts were more pronounced in the case of the dinucleotide signals (due to adenine protonation plus unfolding effects as mentioned above) than those of ADPR in which deshielding by N-1 protonation is the sole effect operating. The protonation shifts are known from data upon ADPR,

hence the unfolding contribution to the dinucleotide shifts may be calculated; these prove to be greater for the C-2 proton of adenine (α 4·8;
β 7·0 Hz) than for the C-8 proton (α 1·9; β 4·5 Hz). This difference implies
that in both α- and β-DPN, the pyridine ring is in close juxtaposition with
the pyrimidine portion of the adenine ring rather than the imidazole nucleus
of this base (see p. 370).

The chemical shifts of the adenine C-8, C-2 and C'-1 protons were
unaffected when the 3-carboxamido substituent of β-DPN was replaced by
COMe, CHO, NH_2, CONHMe and CONHEt. This would not be expected
if the primary force of interaction between the two basic moieties were of a
hydrogen-bonded nature, and is evidence against the hydrogen-bonded
conformation. When the 3-substituent was $CONMe_2$, the adenine signals
were slightly lower field (2–3 Hz) than in DPN, and this was taken as
evidence that the substituent groups on the amide nitrogen could hinder
folding of the molecule if they were sufficiently bulky.

PMR data also supports the idea that the conformations of the reduced
coenzyme β-DPNH is a folded one in aqueous solution. If such is the case,
however, the 2 hydrogens at C-4 of the dihydronicotinamide moiety should
be magnetically non-equivalent because one is close to, and the other
removed from, the adenine nucleus of the coenzyme. The resonance signal
of these protons, both at 60 and 100 MHz (Hollis, 1967), appears as a broad
singlet, however (small coupling to the C-5 proton is not resolved), indica-

Folded   Folded
55   **56**

tive of the 2 protons having an identical environment as might arise as a
result of rapid equilibrium between folded conformations of the right-
handed and left-handed type (**55** and **56**). Further evidence upon this aspect
of the problem is provided by studies at 220 MHz, now outlined.

Comparison of the 220 MHz spectra of β-NMN **53** and β-DPN (Fig. 9.12)
provides a striking illustration of the upfield chemical shifts of the pyridyl
signals of β-DPN in comparison with their resonance positions in the
mononucleotide (Sarma and Kaplan, 1969a). Several dinucleotides were
examined and, with the exceptions of the acetylpyridine analogues (AcPy)-
DPN and (AcPy)TPN, the C-2 proton of the pyridine ring was found to
suffer a much larger upfield shift than that at the C-6 position, e.g., β-DPN,
C-2 56·5 Hz, C-6 35·5 Hz at 220 MHz. Thus, adenine shields the C-2

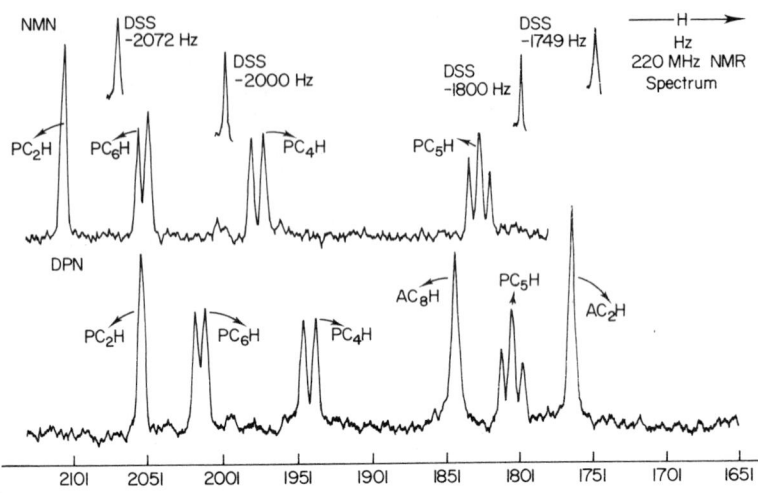

Fig. 9.12. 220 MHz PMR spectra of β-NMN **53** (top) and β-DPN (bottom) taken at sweepwidth 500 Hz with DSS as internal standard (Sarma and Kaplan, 1969a). *Note*: assignment of pyridyl proton signals is based upon relative chemical shifts and first order splitting patterns: C_2—H is lowest field (flanked by $^+$N and C=O) and a singlet, C_6—H and C_4—H both doublets (*ortho* coupled to C_5—H) with the former (adjacent to $^+$N) at lower field, and C_5—H a triplet (2 *ortho* couplings) at highest field (it is removed from the electronegative features).

proton more than the C-6 proton in the folded conformation of the dinucleotide and must therefore interact specifically with that part of the pyridine nucleus which contains the amide side chain. Examination of molecular models shows that such interactions could prevail only if the pyridine ring assumes a *syn* conformation with respect to the D-ribofuranose ring (**57**). In the cases of the acetyl pyridine analogues of β-DPN and TPN

$$CONH_2$$

syn NMN (partial structure)

57

(but not their reduced forms), the C-6 protons underwent the greater deshielding, e.g. (AcPy)DPN, C-2 27 Hz, C-6 69 Hz. It was concluded, therefore, that adenine associates preferentially with that half of the pyridine nucleus which lacks the 3-substituent in these molecules; in support, very little shielding of the CO$\underline{Me}$ group was seen on passing from the mono to the dinucleotide. Models show that an interaction of this type requires the *anti* pyridine-ribose conformation in which C$_2$—H is directed *away* from the ribose ring. Evidence about the *syn* and *anti* conformations of the mononucleotides may be obtained by studying the effect of ionization of the secondary phosphate (i.e. the second ionization of the 5′-phosphate group). The original report concerned a variety of purine and pyrimidine derivatives (Schweizer *et al.*, 1968; Danyluk and Hruska, 1968) and the case of adenosine 5′-phosphate (5′-AMP) will shortly be described because of its relevance to the structure of DPN.

First, however, a digression is made to discuss the question of assigning the PMR signals of purine bases since this information is vital to any conclusions about nucleotide conformations that are based on chemical shift arguments. The spectrum of purine **58** itself in D$_2$O displays a trio of singlets placed (moving upfield) at 555, 545 and 527 Hz respectively from external hexamethyldisiloxane at 60 MHz (Bullock and Jardetzky, 1964). The highest field signal is due to the C-8 proton since it is absent in the spectrum of 8-deuteriopurine obtained from 4,5-diaminopyrimidine and deuterioformic acid (see **59**), while the lowest field signal is due to the C-6

58

proton because it does not appear in the spectrum of the 6-deuterio analogue **60**, prepared as shown from 6-chloropurine. The central singlet (545 Hz)

59

60

of the purine spectrum must therefore be due to the C-2 proton. The spectrum of the product derived from 6-chloropurine also shows a fall in the intensity of the highest field signal, indicative of D/H exchange occurring at the C-8 site. This is confirmed by the fact that the C-8 proton signal is completely absent from the spectrum of a solution of purine in D_2O kept at 105° for 4 hours; the C-2 and C-6 proton signals are unaffected by this

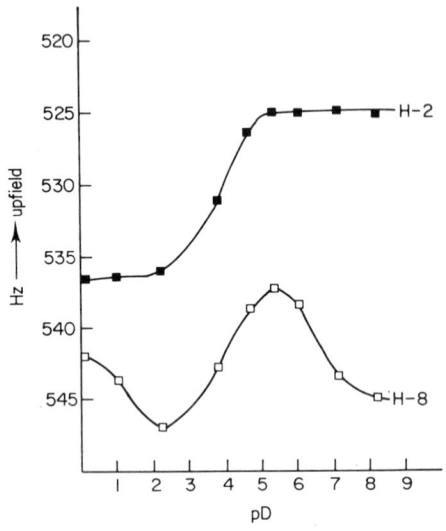

FIG. 9.13. 60 MHz chemical shifts (in Hz relative to external TMS) of the adenine C-2 and C-8 protons of adenosine 5′-monophosphate versus pD of the solution (adapted from Danyluk and Hruska, 1968).

treatment (Schweizer *et al.*, 1964). The exchangeable nature of the C-8, but not the C-2 proton, therefore provides a ready means of differentiating the two signals in adenine derivatives. Chemical shift comparisons are unreliable in this regard; thus the C-8 proton signal of purine is upfield, while that of the adenine moiety of 5′-AMP (absent after D_2O exchange) is downfield of the corresponding C-2 signal (Schweizer *et al.*, 1968). Spectral assignments to other nucleotide bases are described later in this chapter.

Discussion of the influence of ionization changes upon the adenine PMR signals of 5′-AMP may now be resumed. In a spectrum of 5′-AMP the chemical shifts of the adenine protons both move to higher field as the pD changes from 4·5 to 6·0 (Fig. 9.13) as a result of the shielding which follows removal of a proton from the basic centre of adenine. Over the range pD 6·0 to 8·7 the secondary phosphate ionizes and this is reflected in the increased shielding of the methylene protons of C-5′ of the ribose moiety—

this shielding is not seen in the corresponding monomethylester in which the phosphate group

$$\left[(5')CH_2OP \diagdown \begin{matrix} O \\ \bar{O} \\ OMe \end{matrix} \right]$$

cannot release a second proton. The same pD change brings about a substantial deshielding of the C-8 proton of adenine but has only a small effect on the C-2 proton of the same ring (Fig. 9.13). Again, corresponding signals of the monomethyl ester do not suffer these effects. The results show that the phosphate function must be closer to the C-8 proton than the C-2 adenine proton and thus support the *anti* arrangement **61** as preferred conformation of 5'-AMP. Schweizer *et al.* (1968) suggest that the de-

61

shielding action of the diionized phosphate group upon the C-8 proton is brought about by a direct electrostatic field effect which polarizes the H—C$_8$ bond (this proton already has acidic character as shown by the fact that it can be replaced by deuterium). A mechanism of this type is in contrast with that of electron release by the phosphate dianion towards C-5' and consequent shielding of protons attached to this atom.

Sarma and Kaplan (1969a) made a similar study of pD effects upon the aromatic PMR signals of β-NMN (the nicotinamide component of DPN), α-NMN and (AcPy)MN. Thus, in β-NMN at pD 8·0, the chemical shifts of the C-2 and C-6 protons were both shifted downfield compared with their values at pD 4·5 with the C-2 proton being the more affected, while the C-4 and C-5 proton signals in this and other examples suffered little or no change (Table 9.2).

These results are evidence that β-NMN has the preferred *syn* conformation **57**. In (AcPy)MN, however, ionization of the secondary phosphate deshielded the C-6 proton but not the C-2 proton, results in support of the *anti* conformation for this nucleotide. The entire pyridyl substituent

TABLE 9.2

**Downfield shift in Hz (220 MHz data) of pyridyl
protons on raising pH from 4·5 to 8·0**

Nucleotide	C-2	C-6	C-4	C-5
β-NMN	27·5	10·6	0	0
α-NMN	3·5	3·0	1·5	3·0
(AcPy)MN	−2·0	37·0	−6·0	4·5

is directed away from the 5′-phosphate group in α-NMN and, as expected,
none of the pyridyl protons were influenced significantly by a change in
the phosphate ionization state.

With knowledge of the likely conformations of the constituent mononucleotides of DPN to hand, Sarma and Kaplan (1969a) proposed overlap
models for the base pairs in the dinucleotide which are consistent with
chemical shift data. The overlapping structure Fig. 9.14(a) was drawn from

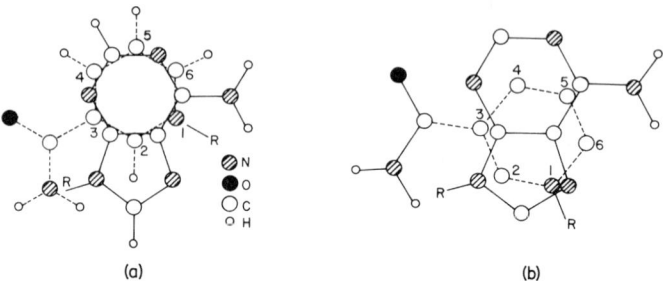

(a)   (b)

Fig. 9.14. (a) The *anti-syn* overlap between base pairs in α-DPN, β-DPN and
TPN (see text). The solid lines represent adenine and the broken lines the pyridine
ring which lies below the adenine. (b) The *anti-syn* structure of the stacked conformation in β-DPNH (Sarma and Kaplan, 1969a).

skeletal models in which the adenine moiety was held *anti* and the nicotinamide in *syn* conformations to their respective ribose substituents. This
arrangement would cause the adenine to shield the C-2 proton of the
nicotinamide ring more than that at C-6, as required by the experimental
data, and also allows for the fact that base pair stacking involves juxtaposition
of the pyridine nucleus with the pyrimidine rather than imidazole half of
the adenine ring. The overlap model derived from the *anti-anti* structure
best describes the stacked conformation of the analogues (AcPy)DPN
and (AcPy)TPN since it requires that the more shielded C-6 proton takes
the place occupied by the C-2 proton in Fig. 9.14(a).

TABLE 9.3

Pyridyl protons. Chemical shifts in Hz from DSS at 220 MHz

	C-2	C-6	C-5	C-4	
β-NMNH	1555·0	1072·0	1084·0	642·7	642·7
β-DPDH	1510·5	1051·5	1041·0	560·5	581·9
Upfield shift	45·5	20·5	43·0	82·2	60·8

Turning now to reduced dinucleotides, chemical shift comparisons show that the C-2, C-5, and particularly the C-4 pyridyl protons are strongly shielded in β-DPNH (Table 9.3) and the overlap shown in Fig. 9.14(b) best accounts for these results. Comparing Fig. 9.14(a) (DPN) and (b) (DPNH), it is to be noted that in the latter the C-4 proton is moved deeper into the shielding zone of the adenine nucleus, while the C-6 proton (less shielded than in DPN) is moved further away from the same base.

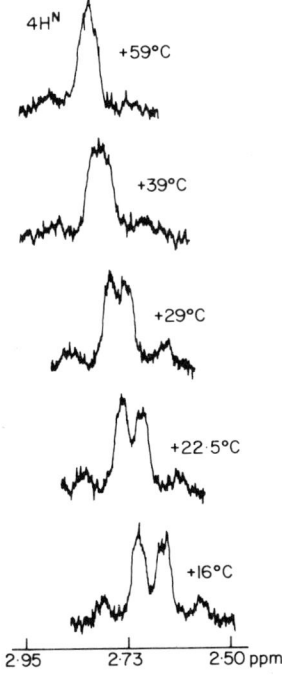

FIG. 9.15. Temperature dependence of the 220 MHz C-4 methylene PMR signal of DPNH in D$_2$O, pD = 9·0 (Patel, 1969).

The anomaly of the 2 C-4 pyridyl protons in the folded DPNH conformation having the same magnetic environment as judged by their giving
rise to a single resonance line in 60 and 100 MHz spectra (already mentioned) has now been solved by 220 MHz data (Sarma and Kaplan, 1969b;
Patel, 1969). In the spectrum of DPNH recorded at the highest frequency,
the single C-4 methylene resonance of earlier spectra is clearly split in two
and the typical quartet pattern of an AB system may be discerned; the
chemical shift difference between the signals ($\Delta = 15$ Hz) is small even at
220 MHz—hence, failure to detect it at lower frequencies is understandable.
In a concurrent study at the same frequency, Patel (1969) also reported the
C-4 methylene signal of DPNH as an AB quartet, and he found the resolution
of the signal was much improved on lowering the temperature (Fig. 9.15).
To account for this, he proposed that a rapid equilibrium of open and one
folded conformation of DPNH favours the open form (chemical non-
equivalence but magnetic equivalence of the C-4 methylene protons) at
higher temperatures and the folded form (chemical and magnetic non-
equivalence of the same protons) at lower temperatures. Another possibility,
also consistent with the data, might be a *slow* equilibrium between the 2
folded conformations, and evidence that such is indeed the case has been
obtained by studies on dihydropyridine monodeuterated DPNH with the
absolute configuration R (**62**) at C-4 (Sarma and Kaplan, 1970a). If exchange

62

between the RH and LH helices (**63**) of this derivative is NMR-fast, the
PMR signal of the remaining C-4 proton will be a singlet of chemical shift

63

representing a weighted average of the C-4 proton environment in the two forms. If the exchange is NMR-slow, 2 distinct signals (one due to the C-4 proton of the RH helix and the other to that of the LH form) will appear in the spectrum. The latter result is in fact obtained [Fig. 9.16(a)] in support of a slow helix exchange rate. In the spectrum recorded at 74° when the dinucleotide has unfolded, a single C-4 methylene resonance line is found *but at lower field* since these protons are no longer shielded by the adenine nucleus. Another feature of the C-4 proton signal of the deuterated analogue

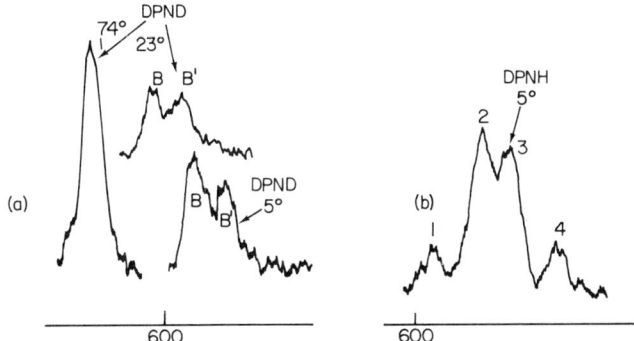

FIG. 9.16. 220 MHz PMR signals of (a), the C-4 proton of (R)-DPND at 5°, 23° and 74°; (b), the C-4 methylene protons of DPNH at 5° (Sarma and Kaplan, 1970a).

(**62**) is its lack of symmetry at 5° and 23°, the area of B being about 15% larger than that of B' [Fig. 9.16(a)]. This result shows that there is a conformational preference for one of the helices, and a careful examination of space filling models reveals subtle differences which may give rise to a difference in the energy content of the 2 forms. The C-4 methylene resonance signal of DPNH (and TPNH) itself is clearly asymmetric at 5° [Fig. 9.16(b)] and it is concluded that the AB quartet observed results from the overlap of the 2 expected AB quartets from the 2 unequally populated, non-equivalent helical forms of DPNH.

Further evidence for the existence of significant populations of both types of folded conformer is provided by data upon N-ethyl-N-methylnicotinamide dinucleotide (**64**) (Sarma *et al.*, 1970). The spectrum of N-methyl-N-ethylnicotinamide itself displays duplicate N-methyl and N-ethyl PMR

CONMe(Et)

ADPR

64

signals of equal intensity at 35°, due to the forms **65** and **66**, which coalesce

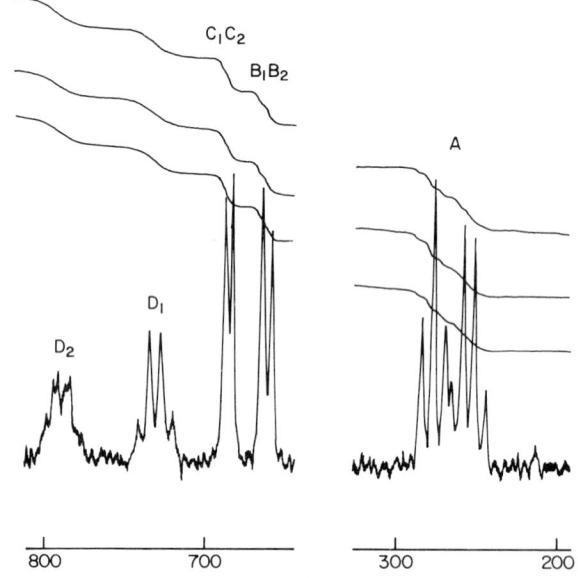

65   **66**

at 90° when rotation about the C—N bond (of partial double bond character) becomes NMR-fast. In the corresponding dinucleotide (**64**), however,

Fig. 9.17. 220 MHz spectrum of N-methyl-N-ethylnicotinamide-adenine dinucleotide (**64**) in D_2O showing the N—$CH_2\underline{Me}$ (A), N—$\underline{CH_2}Me$ (D) and N—$\underline{Me}$ (C, B) signals. The chemical shift scale is in Hz from DSS (Sarma *et al.*, 1970).

signals indicative of 4 non-equivalent N-methyl and 4 non-equivalent N-ethyl groups are seen (Fig. 9.17) which may only be accounted for if the 2 helix forms of **64** are present in comparable amounts.

The clear demonstration by PMR data of the fact that the C-4 methylene protons of reduced dinucleotides differ in their environments in the 2 helical forms almost certainly provides an explanation for the differing behaviour of these protons towards chemical and enzymatic agents (Sarma and Kaplan, 1970a).

Finally, attention has been given to the question of changes in the conformation of the ribose backbone of dinucleotides that occur when DPN is reduced (Sarma and Kaplan, 1970b). In this case evidence is derived from PMR signals due to the adenine and pyridine protons linked to C'-1 of the ribose fragments which are well separated in the spectrum of both DPN and DPNH (Fig. 9.18). In the oxidized form, the lower field

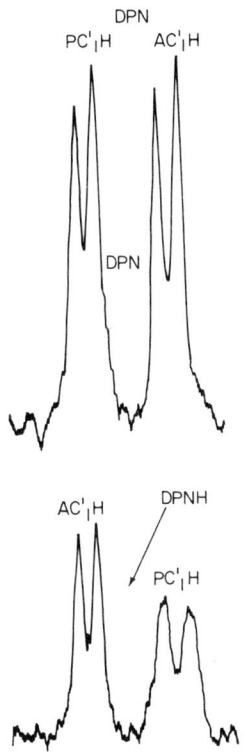

FIG. 9.18. 220 MHz PMR signals of the C-1' (ribose) protons of the pyridine (P) and adenine (A) moieties of DPN and DPNH (Sarma and Kaplan, 1970b).

signal is assigned to the pyridyl moiety because the full positive charge on pyridyl N-1 should have a greater deshielding influence on an adjacent C'-1 proton than the uncharged N-9 atom of adenine. When the coenzyme is in the reduced form, however, the relative positions of the C_1'-H signals are considered to be reversed for the following reasons: (i) In DPNH both nitrogens linked to the ribose fragments are uncharged and the deshielding influence of the adenine atom should now be the greater since it is part of a more highly conjugated system; (ii) the adenosine C_1'-H signal should be lower field in DPNH than in DPN because the shielding contribution it receives from the non-conjugated dihydropyridine nucleus of the reduced form will be less than that provided by the fully aromatic pyridyl unit of the oxidized coenzyme.

Selective broadening of the higher field C_1'-H signal in DPNH provides evidence in support of these assignments (Fig. 9.18). This result is attributed to a change in the relaxation rate of nitrogen which is more likely to be suffered by the pyridyl nitrogen than the adenine N-9 atom when DPN is reduced.

Taking these assignments as valid, it is to be noted that, whereas $J_{1',2'}$ of the adenine ribose C'-1 proton of DPN changes little after reduction, that of the pyridyl ribose moiety increases from about 5 Hz to 8·5 Hz, similar findings being obtained in the case of TPN. This result is considered more likely to be due to a change in the dihedral angle between the Py—C_1'—H and C_2'—H ribose protons than to the effect of the change $=\overset{+}{N}= \rightarrow \ \equiv N$ on $J_{1',2'}$, and is thus good evidence for a distinct conformational difference between oxidized and reduced pyridine dinucleotides.

Chemical shift data upon flavine adenine dinucleotide (FAD, **67**) and corresponding mononucleotide fragments similarly provide evidence for base-pair stacking in preferred conformations of this compound [Sarma *et al.*, 1968a (60 MHz); Kotowycz *et al.*, 1969 (220 MHz)]. The conformational arguments (not detailed here) rest as usual upon the correct assignment of proton signals arising from the heterocyclic bases, and identification of the C-5 and C-8 methine protons of the isoalloxazine ring of FAD presents certain problems which deserve mention in this section.

Flavin adenine dinucleotide (FAD)

67

Electron density calculations (Pullman and Pullman, 1959) showed that C-8 has a higher π-electron density than C-5 in this ring system, as may

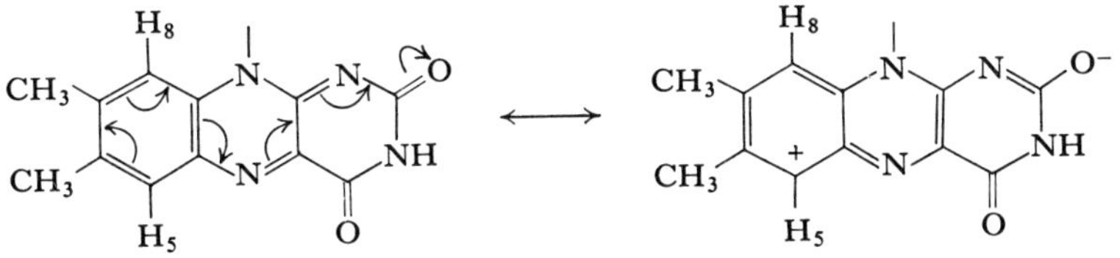

68

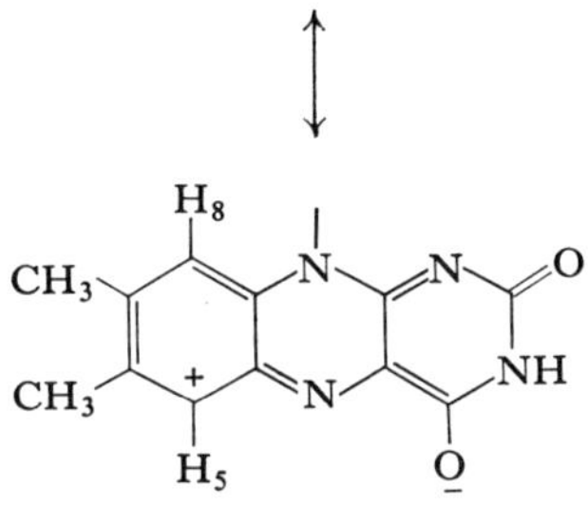

69

be depicted in the resonance forms **68** and **69**. On these grounds the lower field isoalloxazine PMR signal may be assigned to the C-5 proton and the higher field resonance to the C-8 proton. These assignments are supported by the fact that only the low field signal persists in the spectrum of 8-deuteriolumiflavine **70**, prepared as shown. Uncertainties in this approach are revealed, however, by data obtained in the 220 MHz study of FAD

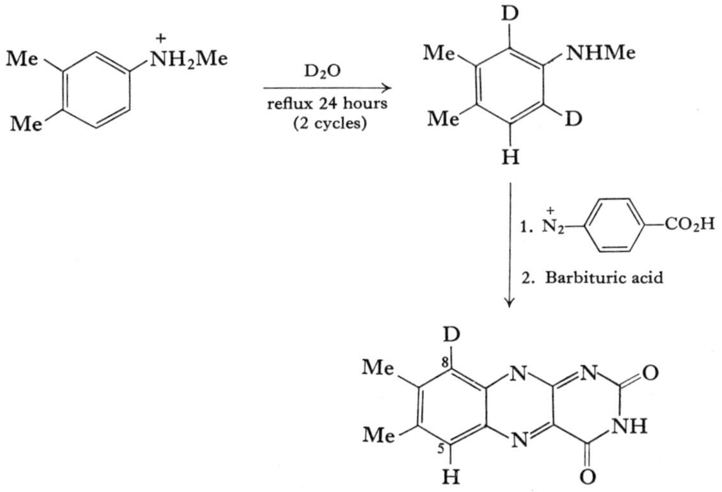

70

and riboflavine-5′-phosphate (FMN) since it is found that the relative positions of the 2 isoalloxazine methine signals are both concentration and temperature dependent. Thus the FMN signals become equivalent in D_2O-dioxane at infinitive dilution while those of FAD converge from a point of maximum separation at $[M]$ = zero to chemical shift equivalence at a molarity of about 0·035 (Fig. 9.19), results which are unexpected from electron density considerations alone. Identification of the isoalloxazine methine signals of FAD was finally made on the basis of chemical shifts

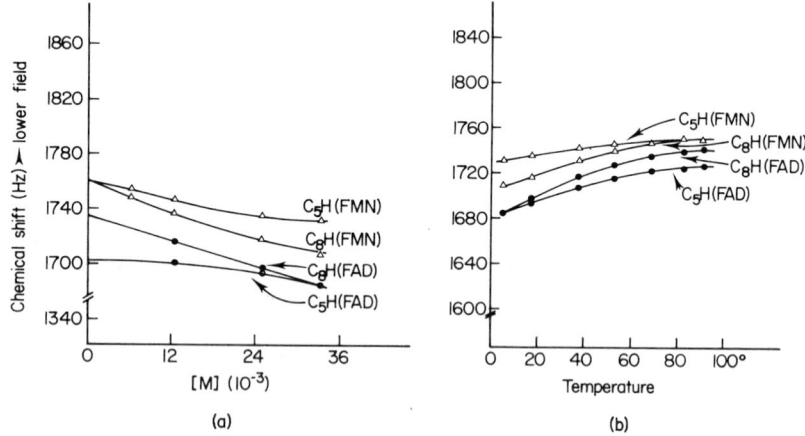

Fig. 9.19. Concentration (a) and temperature dependence (b) of the C-5 and C-8 proton chemical shifts of flavin mononucleotide (FMN) and FAD (67) in D_2O-dioxane-d-8 (70 : 30 v/v) at pH 7·0 (Kotowycz et al., 1969).

versus temperature plots. In FMN, the C-5 and C-8 signals (unambiguously assigned on chemical evidence already described) differ in their temperature dependence as shown in Fig. 9.19, that of the C-8 proton moving to low field at the greater rate as the temperature is raised. The isoalloxazine signals of FAD also differ in their temperature dependence and, on the assumption that the rates of change of corresponding signals of FMN and FAD are similar, the FAD signal most affected by a temperature increase is assigned to the C-8 proton, and that least affected to the C-5 proton.

Polynucleotides

PMR studies of polymeric forms of nucleotides such as the deoxy (DNA) and ribonucleic acids (RNA) are comparable to those of proteins in the following respects: (i) spectra of the native forms of the biopolymers exhibit, at best, overlapping multiplets that are highly broadened as a result of the long correlation times which obtain in the rigid helical forms of these molecules; (ii) spectral features are far better resolved, or may only be seen in fact, in single strand molecules obtained by heating the native forms above their "melting" temperatures; and (iii) assignment of the resonances in the spectra of "melted" forms may be based on the PMR characteristics of the constituent mononucleotides which are linked in the polymer via 3′,5′-phosphodiester bridges.

The case of DNA, which was first examined by PMR spectroscopy at 60 MHz and later at 220 MHz (McDonald *et al.*, 1964, 1967), is discussed here as an example. The DNA samples, derived from various sources, were first dissolved in neutral D_2O to pre-exchange labile protons and then freeze-dried. The deuterated material was dissolved in fresh D_2O (15–30 mg/ml) and the spectrum of the solution recorded at room temperature to determine whether any interfering impurities were present other than the residual HDO. DNA at this temperature, if it is in the double-stranded helical form, does not exhibit high-resolution PMR as the resonances are so broadened by anisotropic dipole-dipole interactions that they are not detected. When the solution is heated to above 75° sharp resonance lines characteristic of single-stranded DNA emerge from the broad background absorption. These resonances do not gradually sharpen but appear abruptly as the complex is "melted" and this result shows that the portions that are and are not "melted" do not form a rapidly equilibrating system.

Single stranded calf thymus DNA in neutral D_2O at 90° exhibits PMR absorption at 220 MHz in 5 discrete spectra regions:

1. δ 7·3 to 8·3, includes protons at C-2 and C-8 of adenine, C-8 of guanine, and C-6 of cytosine and thymine;
2. δ 5·8 to 6·4, includes protons at C-5 of cytosine (shielded by adjacent NH_2 group) and C-1′ of the deoxyribose fragments;
3. δ 4·6 to 5·1, due to the C-3′ protons (deoxyribose);
4. δ 2·1 to 2·8, due to the C-2′ protons (deoxyribose);
5. δ 1·6 to 1·9, due to the two partially resolved methyl resonances of thymine (T-Me) (71).

71

Signals due to protons at C-4′ and C-5′ are obscured at 90° by the HDO resonance.

Observation of two resonance bands for the T-Me signal of separation 28 Hz at 220 MHz (Fig. 9.20) shows that there are two magnetically non-equivalent environments for the thymine residues of DNA; further, the intensity difference between the bands shows that one environment is populated more than the other. Duplicate T—Me signals are also found in samples derived from other sources, e.g. bacteriophage and salmon sperm, with the same chemical shifts (δ 1·83 and 1·71) but different intensity ratios ($I_{1·71}/I_{1·83}$). The magnetic environment of T—Me will be influenced by

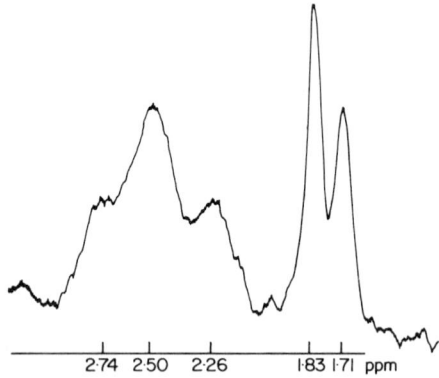

FIG. 9.20. High-field regions of the 220 MHz PMR spectrum of calf thymus DNA in D_2O (20 mg/ml) pD 7·0, 93° (an average of 20 spectra) (McDonald *et al.*, 1967).

neighbouring bases especially if base stacking occurs. Clues to the nature of this neighbour-base dependent environment for thymine methyl groups were found upon examination of the PMR spectra of 3′-5′ linked deoxy-ribonucleoside monophosphate dimers containing thymine. The spectra of most dimers showed T—Me signals close to the low field thymine methyl band of DNA; however, dimers of the type (72) in which thymine has a purine neighbour attached to the sugar moiety having a free hydroxyl at C-5′, showed higher field resonances which moved to lower field when the temperature was raised, e.g., T—Me signal of ApT, δ 1·61 (20°), 1·71 (50°), 1·78 (90°). These differences are attributed to the differential shielding resulting from purine-pyrimidine as opposed to pyrimidine-pyrimidine stacking interactions; there is evidence that these are stronger in the former case (Ts'o *et al.*, 1963). Stacking probably occurs also in dimers of type **72** in which thymine and the purine base are interchanged, but the favoured conformation is such that the methyl group of thymine is too far from the purine ring to be influenced by the purine ring-currents. It is

72

concluded that the T-Me resonance of DNA at δ 1·71 arises from thymine bases that are in a stacked relationship with neighbour purine bases such that the thymine methyl groups are in ring-current fields of the purine rings. In addition, all the thymine-purine conformational relationships that are effective in shifting the T-Me resonance are maintained at 90° since there is no change in position or intensity of this resonance as the temperature is decreased. The stacking arrangement of thymine with neighbour purines in single-stranded DNA is believed to be the same as that in dinucleotides of type **72** and therefore the DNA resonance at δ 1·71 arises only from methyl protons of thymine that have a purine at the C-5' neighbour position. The intensity ratio $I_{1·71}/I_{1·83}$ then provides a measure of the ratio of purine to pyrimidine bases arranged in this manner and the fact that values of unity are not obtained shows that the frequency of occurrence of the various possible 5'-neighbours of thymine in double-stranded DNA is not random.

Smith *et al.* (1969) have pointed out the restricted scope of PMR spectroscopic studies of polynucleotides compared with that of proteins, due principally to the fact that DNA and RNA contain only 4 major bases, i.e., adenine, cytosine, guanine and thymine (DNA) or uracil (RNA) whose proton chemical shifts are quite similar. In a study of the relatively low molecular weight polymer, transfer RNA (tRNA) they noted a disappointing lack of improvement in resolution on going from 60 to 100 MHz for unfractionated tRNA or going to 220 MHz and using purified alanine tRNA when spectra were recorded at the normal operating temperatures. In the 220 MHz spectrum, only a peak attributable to protons at the C-8 position of adenine could be assigned with certainty (the C-8 proton resonance is the lowest field of all PMR signals of the common bases). The methyl signals of thymidine and of certain rare bases present in tRNA such as 1-methylguanosine could, however, be located between δ 3·1 and

1·9 when extensive time-averaging of spectra was employed. The methyl bands narrowed and split into several components when the temperature was raised, and these effects were most pronounced in the "melting" range of 50–65° previously observed optically. The results indicate that the rare nucleotides of tRNA are located in highly organized regions of the molecule.

References

Angyal, S. J. (1969). *Angew. Chem. Int. Ed.* **8**, 157.
Bhacca, N. S. and Horton, D. (1967). *J. Amer. Chem. Soc.* **89**, 5993.
Booth, H. (1966). *Tetrahedron* **22**, 615.
Bullock, F. J. and Jardetzky, O. (1964). *J. Org. Chem.* **29**, 1988.
Capon, B. and Thacker, D. (1964). *Proc. Chem. Soc.* 369.
Carr, J. B. and Huitric, A. C. (1964). *J. Org. Chem.* **29**, 2506.
Casu, B., Reggiani, M., Gallo, G. G. and Vigevani, A. (1964). *Tetrahedron Lett.* 2839.
Casu, B., Reggiani, M., Gallo, G. G. and Vigevani, A. (1965). *Tetrahedron Lett.* 2253.
Casu, B., Reggiani, M., Gallo, G. G. and Vigevani, A. (1966). *Tetrahedron* **22**, 3061.
Casu, B., Reggiani, M., Gallo, G. G. and Vigevani, A. (1968). *Tetrahedron* **24**, 803.
Cushley, R. J., Watanabe, K. A. and Fox, J. J. (1967). *J. Amer. Soc.* **89**, 394.
Danyluk, S. S. and Hruska, F. E. (1968). *Biochem.* **7**, 1038.
Darnall, K. R. and Townsend, L. B. (1966). *J. Hetero. Chem.* **3**, 371.
Demarko, P. V. and Thakkar, A. L. (1970). *Chem. Commun.* 2.
Eliel, E. L., Allinger, N. L., Angyal, S. J. and Morrison, G. A. (1965). "Conformational Analysis". Wiley, New York.
Fric, I., Smejkal, J. and Farkas, J. (1966). *Tetrahedron Lett.* 75.
Glascoe, P. K. and Long, F. A. (1960). *J. Phys. Chem.* **64**, 188.
Hall, L. D. (1964). *Adv. Carbohyd. Chem.* **19**, 51.
Hall, L. D. and Manville, J. F. (1969). *Can. J. Chem.* **47**, 19.
Hart, P. A. and Davis, J. P. (1969). *Biochem. Biophys. Res. Commun.* **34**, 733.
Holland, C. V., Horton, D., and Jewell, J. S. (1967). *J. Org. Chem.* **32**, 1818.
Hollis, D. P. (1967). *Biochem.* **6**, 2080.
Inch, T. D. (1969). *In* "Annual Review of NMR Spectroscopy" (E. F. Mooney, ed.), Vol. 2, p. 35. Academic Press, London.
Jardetzky, C. D. (1960). *J. Amer. Chem. Soc.* **82**, 229.
Jardetzky, C. D. (1961). *J. Amer. Chem. Soc.* **83**, 2919.
Jardetzky, C. D. (1962). *J. Amer. Chem. Soc.* **84**, 62.
Jardetzky, O. and Wade-Jardetzky, N. G. (1966). *J. Biol. Chem.* **241**, 85.
Khadem, H. El., Wolfrom, M. L. and Horton, D. (1965). *J. Org. Chem.* **30**, 838.
Kotowycz, G., Teng, N., Klein, M. P. and Calvin, M. (1969). *J. Biol. Chem.* **244**, 5656.
Lemieux, R. U. (1961). *Can. J. Chem.* **39**, 116.
Lemieux, R. U. and Lineback, D. R. (1963). *Ann. Rev. Biochem.* **32**, 156.
Lemieux, R. U. and Stevens, J. D. (1965). *Can. J. Chem.* **43**, 2059.

Lemieux, R. U. and Stevens, J. D. (1966). *Can. J. Chem.* **44**, 249.

Lemieux, R. U., Kullnig, R. K., Bernstein, H. J. and Schneider, W. G. (1958). *J. Amer. Chem. Soc.* **80**, 6098.

Lenz, R. W. and Heeschen, J. P. (1961). *J. Polymer Sci.* **51**, 247.

Mackie, W. and Perlin, A. S. (1966). *Can. J. Chem.* **44**, 2039.

McDonald, C. C., Phillips, W. D. and Penman, S. (1964). *Science, N.Y.* **144**, 1234.

McDonald, C. C., Phillips, W. D. and Lazar, J. (1967). *J. Amer. Chem. Soc.* **89**, 4166.

Meyer, W. L., Mahler, H. R. and Baker, R. H. Jr. (1962). *Biochem. Biophys. Acta* **64**, 353.

Nishimura, T. and Shimizu, B. (1965). *Chem. Pharm. Bull.* **13**, 803.

Patel, D. J. (1969). *Nature, Lond.* **221**, 1239.

Perlin, A. S. (1966). *Can. J. Chem.* **44**, 539.

Pullman, B. and Pullman, A. (1959). *Proc. Natl. Acad. Sci. U.S.* **45**, 136.

Robins, M. J. and Robins, R. K. (1965). *J. Amer. Chem. Soc.* **87**, 4934.

Sarma, R. H. and Kaplan, N. O. (1969a). *Biochem. Biophys. Res. Commun.* **36**, 780.

Sarma, R. H. and Kaplan, N. O. (1969b). *J. Biol. Chem.* **244**, 771.

Sarma, R. H. and Kaplan, N. O. (1970a). *Biochem.* **9**, 539.

Sarma, R. H. and Kaplan, N. O. (1970b). *Biochem.* **9**, 557.

Sarma, R. H., Dannies, P. and Kaplan, N. O. (1968a). *Biochem.* **7**, 4359.

Sarma, R. H., Moore, M. and Kaplan, N. O. (1970). *Biochem.* **9**, 549.

Sarma, R. H., Ross, V. and Kaplan, N. O. (1968b). *Biochem.* **7**, 3052.

Schweizer, M. P., Chan, S. I., Helmkamp, G. K. and Ts'o, P. O. P. (1964). *J. Amer. Chem. Soc.* **86**, 696.

Schweizer, M. P., Brown, A. D., Ts'o, P. O. P. and Hollis, D. P. (1968). *J. Amer. Chem. Soc.* **90**, 1042.

Smith, I. C. P., Yamane, T., and Shulman, R. G. (1969). *Can. J. Biochem.* **47**, 480.

Smith, M. and Jardetzky, C. D. (1968). *J. Molec. Spect.* **28**, 70.

Trager, W. F., Nist, B. J. and Huitric, A. C. (1967). *J. Pharm. Sci.* **56**, 698 and references cited therein.

Ts'o, P. O. P., Melvin, I. S. and Olsen, A. C. (1963). *J. Amer. Chem. Soc.* **85**, 1289.

Appendix

NOTES ON SOME SOLVENT AND HYDROGEN BONDING EFFECTS†

Dimethylsulphoxide

Deuterated dimethylsulphoxide (DMSO-d_6) is a valuable solvent for use in NMR spectroscopy because of its power of dissolving a wide range of compounds. It is a particularly good solvent for polar substances, e.g. hydrohalide and methohalide salts of bases, which are often only sparingly soluble in CDCl$_3$. The protonated solvent, i.e. $(CH_3)_2SO$, serves perfectly well provided only low field signals such as those due to aromatic and vinylic signals need to be recorded but not otherwise because the intense solvent signal together with its satellite and spinning side bands blanket out the region δ 1–4. In most cases, therefore, the deuterated solvent, $(CD_3)_2SO$, needs to be employed. The spectrum of the usual commercial product, of about 99·5% isotopic purity, displays a solvent signal due to residual DMSO-d_5 which forms a characteristic multiplet formed as a result of

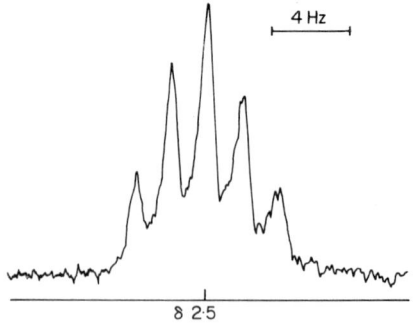

4 Hz

δ 2·5

Fig. A.1. Residual DMSO-d_5 60 MHz PMR signal of deuterated dimethyl-sulphoxide, recorded at a sweep width of 100 Hz.

† Unreferenced work in this section refers to unpublished results of the author and his colleagues.

coupling between one proton and two geminal deuterium atoms (Fig. A.1). This signal may be troublesome, especially in work with dilute solutions involving time-averaging techniques, and solvent with an isotopic purity of 100% has recently become available (Diaprep, Atlanta). The use of DMSO-d$_6$ for the study of hydroxyl proton resonances has been described in Chapters 1 (general) and 9 (sugars). The ability of the solvent to fix an exchangeable proton is well illustrated by the case of the spectrum of the 2-methylene pyrrolidine **1**; in CDCl$_3$ the vinylic signals are both extensively

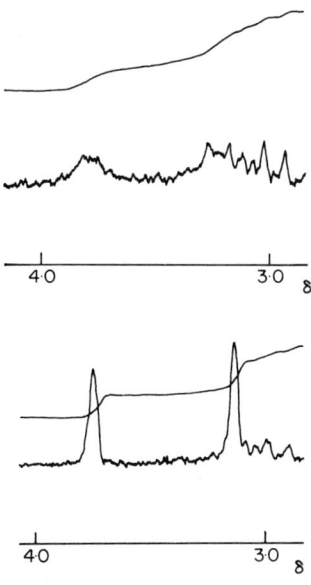

1

broadened indicative of exchange, but in DMSO-d$_6$ the signals are markedly narrower (Fig. A.2) since exchange between the two quite distinct vinylic environments (one *cis* and the other *trans* to the CPh$_2$ group) is impeded presumably by hydrogen bonding of the form $=$C—H$\cdots\bar{\text{O}}$—$\overset{+}{\text{S}}$Me$_2$.

Fig. A.2. 60 MHz vinylic PMR signals of 3,3-diphenyl-5-methyl-2-methylene-pyrrolidine (**1**) in CDCl$_3$ (top) and DMSO-d$_6$ (bottom).

Use of DMSO-d$_6$ for determining isomer purity and/or stereochemistry is illustrated by the next 2 examples.

Reaction between 2-methylcyclohexanone and potassium cyanide-hydrochloric acid gives a mixture of cyanohydrins **2**. The solid which

2   **3** (major)   **4** (minor)

separates has a PMR spectrum in DMSO-d$_6$ which displays a single sharp OH resonance at δ 6·32. The spectrum of the mother liquors, similarly recorded, shows two O$\underline{H}$ resonances, one at the previous chemical shift and the other at higher field (δ 6·1). These results (i) establish the purity of the solid (major) component, and (ii) provide evidence that major product has an equatorial OH orientation (and hence is more extensively hydrogen bonded to DMSO-d$_6$ leading to a lower field position) and the minor, an axial orientation as in **3** and **4** respectively. The configurations of the products also follow since reversal of these assignments requires the major isomer to have an axial methyl orientation which is unlikely on the grounds of the greater conformational free energy value of Me as compared with that of CN (Eliel *et al.*, 1965). The OH resonances arc not clearly resolved in spectra of the cyanohydrins recorded in CCl$_4$.

5   **6**

In the isomeric 4-piperidinol methiodides **5** and **6**, the lower field position of the two OH resonances of the salts in DMSO-d$_6$ similarly serves to identify the isomer with an equatorial hydroxy group (**5**) (Fig. A.3). The difference is quite small but the assignment is confirmed by the larger separation of the lower field signal, both resonances being doublets (*J* 5·5 and 4·2 Hz respectively) as a result of HOCH coupling (see p. 29). Note that the OH signals are only clearly resolved in spectra of the piperidinols when the basic centre (which promotes exchange of the OH proton even in DMSO-d$_6$) is neutralized by quaternization (see p. 28).

The spectra of a particular compound recorded in 2 different solvents usually differ in greater or lesser degree. When specific rather than general differences are observed, i.e. when only certain signals are significantly affected by the solvent change, solvent shift data may provide valuable clues to the structure, and particularly the stereochemistry, of the compound in question. In this regard, the effect of replacing solvents such as CCl$_4$ or

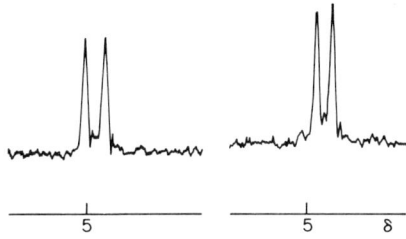

FIG. A.3. 60 MHz hydroxyl PMR signals of the methiodide of α-1,3-dimethyl-piperidin-4-ol (**5**) (left-hand figure) and the corresponding β-isomer (**6**) (right-hand figure) in DMSO-d$_6$.

CDCl$_3$ by benzene have been most extensively studied and several examples of ASIS have been given in this book (reviews: Laszlo, 1967; Ronayne and Williams, 1969). Solvent effects are explained in terms of the geometry of probable collision complexes and consequent ring current effects of the aromatic solvent molecules upon the various proton features of the solute.

Less attention has been paid to the solvent effects of DMSO-d$_6$ as such. The following examples show them to exist and demonstrate their possible use in structural investigations (all data refer to spectra recorded at 60 MHz).

The 3-methyl doublet signal of the PMR spectrum of α-prodinol (**7a**)

OH

Me~N Ar
Me

α-
7

OH

N Ar
Me

β-
8

a: Ar = Ph
b: Ar = o-tolyl

moves upfield by 9·5 Hz when solvent CDCl$_3$ is replaced by DMSO-d$_6$; the shift of the corresponding signal of the β-isomer (3-Me axial, **8a**) is smaller (6 Hz). The significance of this result is supported by similar results for other diastereoisomeric pairs, e.g. **7b** 3Me shift; α, 6·3 Hz; β, 4 Hz, while the same order of shift is found for the 2-methyl group of the

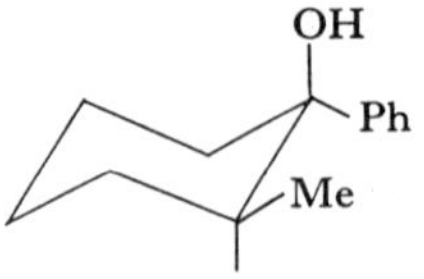

3-Me shift (CDCl$_3$ → DMSO-d$_6$)

5·5 Hz

9

non-basic analogue of α-prodinol (**9**). These results may be accounted for

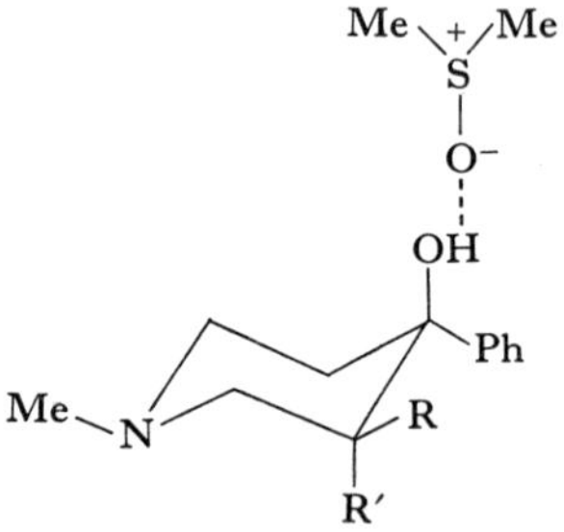

10

in terms of a collision complex of type **10** stabilized by hydrogen bonding. Accepting the view of the anisotropy of the sulphoxide bond being analogous to that of a carbon-carbon triple bond, as discussed on p. 266, it follows that both an equatorial (R) and an axial (R′) methyl substituent in **10** would fall within the shielding zone of the sulphoxide bond in this model, with the closer equatorial group being the more affected.

The chemical shifts of methyl adjacent to the C-4 centre of the γ- and β-isomers of 1,2,5-trimethyl-4-phenylpiperidinol **11** in CDCl$_3$ are close

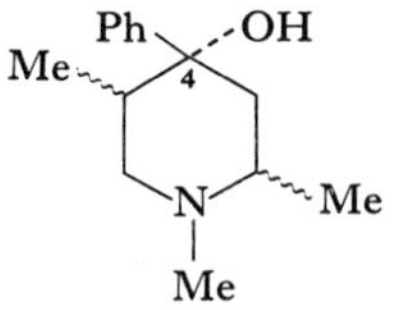

5-Me chemical shift (Hz from TMS at 60 MHz)

γ 37·5 (CDCl$_3$); 29·5 (DMSO-d$_6$)

β 45 (CDCl$_3$); 41 (DMSO-d$_6$)

11

to those of 3Me in α- and β-prodinol respectively and the signals shift upfield in similar manners when CDCl$_3$ is replaced by DMSO-d$_6$. These results support the equatorial orientation of methyl adjacent to C-4 in the γ-isomer, and the axial orientation of the same substituent in the β isomer as in **12** and **13** respectively.

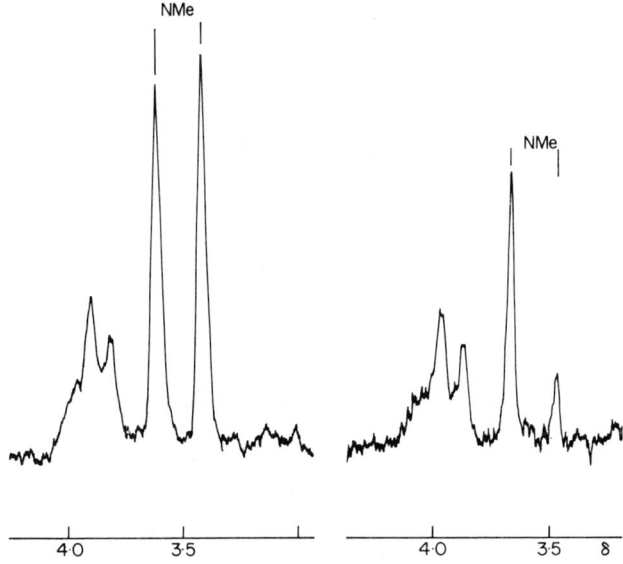

Another solvent effect of DMSO-d_6 is revealed by a study of the methiodides of 1-methylpiperidine derivatives. The spectrum of the methiodide of 1-methyl-4-phenylpiperidine (**14**) in CDCl$_3$ exhibits 2 well

spaced N—Me signals of separation 15 Hz at 60 MHz. The higher field signal is assigned to the axial group since its intensity is reduced in the spectrum of the N-trideuteromethiodide (total product), for reasons

Fig. A.4. 60 MHz N-methyl signals of the methiodide (left-hand figure) and trideuteromethiodide (right-hand figure) of 1-methyl-4-phenylpiperidine in CDCl$_3$.

explained in Chapter 4 (p. 159) (Fig. A.4). When $CDCl_3$ is replaced by DMSO-d_6, however, both signals move upfield and converge to a single line. DMSO-d_6 must therefore shield both N—Me groups, its influence upon the equatorial group (shifted by 27·5 Hz) being almost twice that upon the axial group (shifted by 14·5 Hz). Similar results are obtained for

	Chemical shift (Hz at 60 MHz)		
	$CDCl_3$	DMSO-d_6	$\varDelta$
e-NMe	219	190	29 Hz
a-NMe	200*	183	17 Hz

15

* (Reduced intensity in spectrum of corresponding CD_3I salt, Stien *et al.*, 1968).

the 4-*t*-butyl analogue **15**. It is clear, therefore, that DMSO-d_6 associates with the quaternary derivatives in such a manner that the deshielding influence of the charged nitrogen centre upon the N-methyl protons is reduced. A possible model is shown in **16**. The shielding action of this

16

solvent may be a result of its dispersing the positive charge around the nitrogen atom, shown distributed on the nitrogen substituents by analogy with molecular orbital calculations of charge distribution in the ammonium ion (Kier, 1968). Additionally, the N-methyl groups may fall within the diamagnetic shielding zone of the sulphoxide bond if collision complexes of the type **16** are favoured. The fact that the equatorial group suffers the greater shielding may be explained by its more ready access to solvent molecules. In the case of the methiodide of 1,2 dimethylpiperidine (**17**) the equatorial group is still shielded more than a-NMe by the solvent change $CDCl_3 \rightarrow$ DMSO-d_6 but the difference is less, possibly due to a hindering influence of the 2-methyl substituent upon solvation.

	Chemical shift (Hz at 60 MHz)		
	CDCl$_3$	DMSO-d$_6$	Δ
e-NMe	208	189	19
a-NMe	191	176·5	14·5

17

D$_2$O also has a shielding effect upon the N-methyl signals of quaternary salts (the NMe signals of **14** and **15** are close to the chemical shifts found with DMSO-d$_6$ as solvent). It is evident from these results that due consideration of solvent should be made in the interpretation of the spectra of quaternary salts of cyclic derivatives especially if comparisons between spectra recorded in polar and non-polar solvents are being made. For example, the reversal of the relative resonance positions of the a and e-NMe signals in *cis* and *trans* 4-t-butyl-N-benzyl-N-methylpiperidinium chlorides

18

(**18**) when CDCl$_3$ is replaced by D$_2$O (axial NMe higher field in CDCl$_3$ but lower field in D$_2$O) (Bottini and O'Rell, 1967) may be explained by the more extensive solvation of the equatorial group.

Another factor which brings about significant changes in a PMR spectrum following a solvent replacement is a solvent-induced conformational change. A good example of this phenomenon is the case of the substituted cyclohexylamine derivative **19**a shown in Fig. A.5 (Whelton, 1969). The positions of deuterium substitution makes possible observation of the H$_1$ and H$_5$ signals as clear doublets, that at lower field being assigned to H$_1$ because this proton signal is subject to an acylation shift (p. 30). The spectrum of the compound in DMSO-d$_6$ is consistent with the chair conformation **19**a with the *o*-tolyl substituent equatorial and the $\overset{+}{H}NMe_2$ group axial since the H$_1$ signal displays a typical a/a coupling (9 Hz) and H$_5$, an e/e value (3·9 Hz). In formic acid-D$_2$O (95:5), however, a striking reversal of these couplings occurs (Fig. A.5, bottom), and this result shows that the inverted chair form of **19**a with the aromatic substituent axial and the ammonium group equatorial must be preferred in this solvent; in this conformation H$_1$ and H$_2$ have an e/e and H$_4$ and H$_5$ an a/a relationship, in agreement with the coupling constants of the H_1 and H_5 signals. Presumably the steric requirements of the solvated protonated dimethylamino group (axial in **19a**) are greater in $HCO_2H\text{-}D_2O$ than in $DMSO\text{-}d_6$ and govern the conformational preference in the former solvent. The $J_{1,2}$ and $J_{4,5}$ values

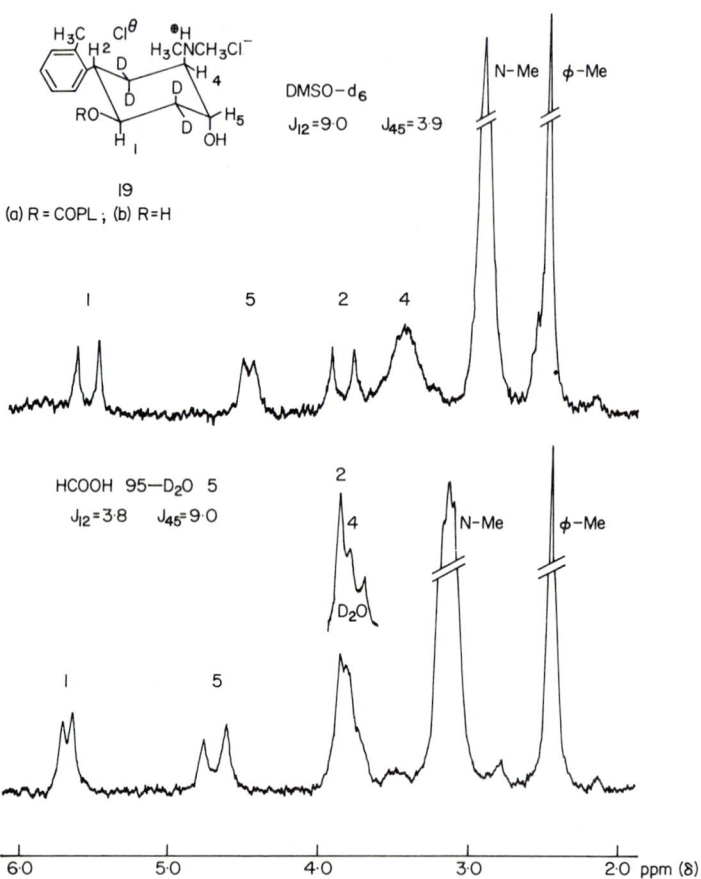

FIG. A.5. Part of the 60 MHz PMR spectrum of the cyclohexylamine derivative **19a** in $DMSO\text{-}d_6$ (top) and $HCO_2H\text{-}D_2O$ (95:5) (bottom) (Whelton, 1969).

of H_1 and H_5 in the spectrum of the corresponding diol (**19b**) do not differ greatly in $DMSO\text{-}d_6$ and have values close to a $J_{a/a}\text{-}J_{e/e}$ average ($J_{1,2}$ 5·6 Hz; $J_{4,5}$ 6·5 Hz); this result is evidence that both chair conformers and/or flexible forms (skew boats) of **19b** have comparable populations in this solute condition. The J values for the diol in $HCO_2H\text{-}D_2O$ ($J_{1,2}$ 5·0; $J_{4,5}$ 8·2) indicate a change in the conformational equilibrium.

In the above examples, evidence for conformational change rests securely upon coupling constant data. Conclusions about conformational changes drawn from chemical shift evidence may be equivocal however, as the following example demonstrates. In CDCl$_3$, the 3-methyl signal of β-prodine hydrochloride **21** is 18 Hz lower field than the corresponding

20	**21**
ν_{Me} 44 Hz (CDCl$_3$)	62 Hz (CDCl$_3$)
43·5 Hz (D$_2$O)	43·5 Hz (D$_2$O)

(Hz from TMS at 60 MHz)

signal of the α-isomer **20** at 60 MHz, and this difference has been interpreted in terms of the β-3-Me group being closer to the deshielding $^+$N feature of the molecule (see p. 203). In D$_2$O the β-3-Me resonance moves upfield and has a chemical shift coincident with that of the α-signal (Casy, 1968) (see **20**, **21**). This result could be due to (i) the deshielding influence of protonated nitrogen being reduced when it is solvated by D$_2$O; (ii) a shielding change due to break up of ion-pairs ($\overset{+}{N}H$ -- $\overset{-}{C}l$), more likely to exist in CDCl$_3$ than in D$_2$O (see p. 60); (iii) a conformational change, e.g. to a skew boat, which moves the β-3-Me group away from the charged nitrogen centre, or a combination of two or more of these factors. The fact that the N-methyl resonances of these esters are little altered after the solvent change is evidence for the conformational factor; unfortunately coupling constant evidence is not available because the ring protons signals form a complex band that cannot be resolved.

Pyridine

Pyridine serves as an alternative to benzene as an aromatic solvent capable of inducing changes in chemical shifts relative to values in CDCl$_3$ and CCl$_4$. More recently it has found application in the study of polar molecules containing an hydroxyl function to which pyridine will hydrogen bond (Demarco *et al.*, 1968). In many alcohols unexpectedly large solvent shifts ($\Delta = \nu_{CDCl_3} - \nu_{C_5D_5N}$) are observed for protons and proton groups orientated in certain ways towards the hydroxyl group, and their magnitudes provide useful stereochemical information. The following are examples of the various relationships.

1,3-*Diaxial deshielding*

Single protons or methyl groups occupying positions 1,3-diaxial to OH experience deshielding effects of the order of 0·2–0·4 ppm in pyridine relative to $CDCl_3$. Two examples are 5α-androstane-2β,17β-diol **22**, and

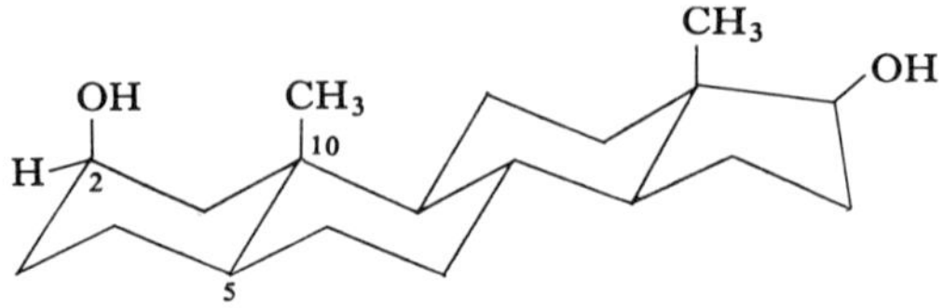

22

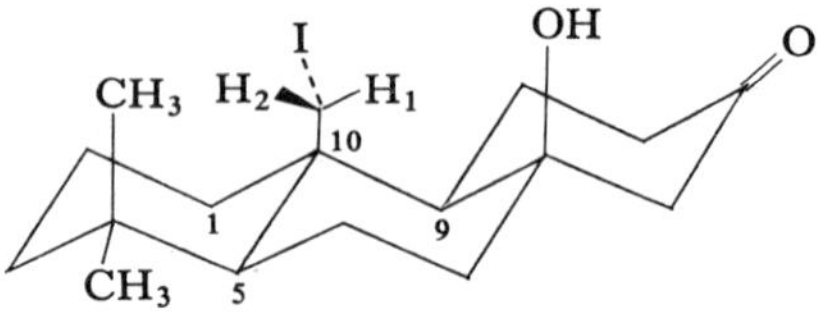

23

the 10-iodomethylsteroid **23**. The PMR spectrum of **23** in pyridine displays two low-field 1-proton doublets ($J = 11$ Hz) at δ 5·17 and 3·73 which are assigned to protons H_1 and H_2 respectively (Wenkert and Mylari, 1967). The large chemical shift difference between H_1 and H_2 is rationalized if the iodomethyl group has the preferred conformation shown in **23** in which non-bonded interactions between iodine and the β-axial substituents in rings A and B are minimal. In this conformation H_1 is much closer to the C_8 axial OH than is H_2, and it will thus experience a greater deshielding effect from the pyridine molecules coordinated to the OH function.

Vicinal deshielding

Protons and methyl groups vicinally situated to an OH function are deshielded, and in extent dependent on the dihedral angle (φ) between the proton (or methyl group) and the OH function. Only small *Δ* values are observed when the angle is large, e.g. **24**, but large shifts occur when φ is 60°, e.g. **25**.

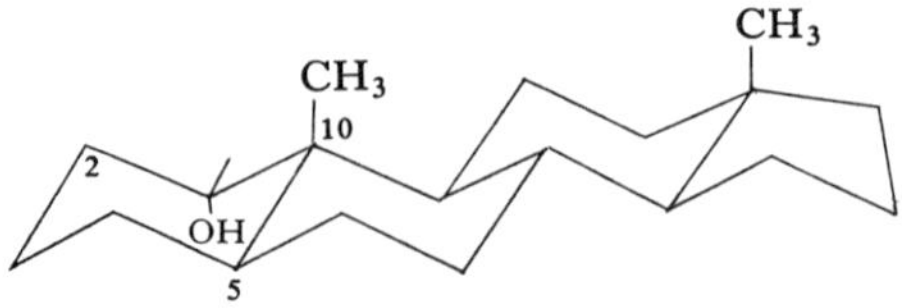

24

H-19   *Δ* − 0·03 (φ = 180°)

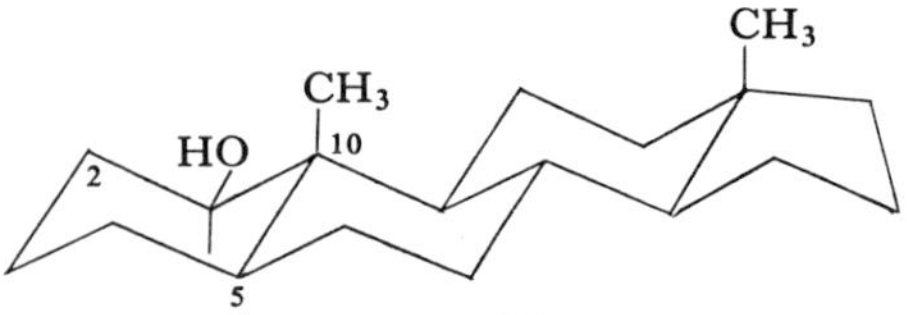

25

H-19   *Δ* − 0·22 (φ = 60°)

Geminal deshielding

Protons or methyl groups situated on a carbon atom bearing an OH function are deshielded by 0·15–0·25 ppm in pyridine relative to CDCl₃. The last 2 effects are verified in the isomeric piperidinols **26** and **27**.

α	β
26	**27**
3-Me Δ −0·13	Δ −0·14
4-H (not resolved	Δ −0·12
in CDCl₃)	

Both α- and β-3-Me signals move downfield in similar extent when CDCl₃ is replaced by pyridine (ϕ is 60° in both cases). The same solvent change moves the 4-H methine signals, which are diagnostic of the conformation (see p. 242), away from the main ring proton resonance band and hence resolvable. Unexpectedly, Δ values for the 3-Me signals of α- and β-prodinol (**28** and **29**, R = H) are similar even though the relevant β-dihedral

α	β	
28	**29**	**30**
3-Me Δ −0·2 (R = H)	Δ −0·19 (R = H)	
Δ 0 (R = COMe)	Δ −0·01 (R = COMe)	

angle is 180° in the chair form **29**. This result may mean that β-prodinol has a preferred skew boat conformation **30** in pyridine; in this arrangement 3-Me is moved closer to the OH group. An alternative explanation is that association between the hydroxyl group and pyridine in the β-alcohol modifies the preferred orientation of the 4-phenyl group with respect to the piperidine ring, a change which makes the shielding influence of 4-phenyl upon the axial 3-methyl group different than its effect in CDCl₃ and CCl₄ (Chapter 6, p. 202). The fact that the 3-Me Δ values are negligible

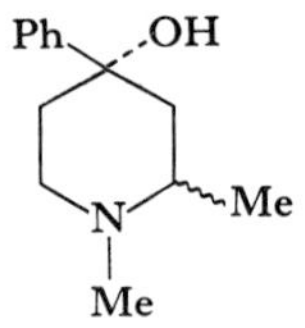

31 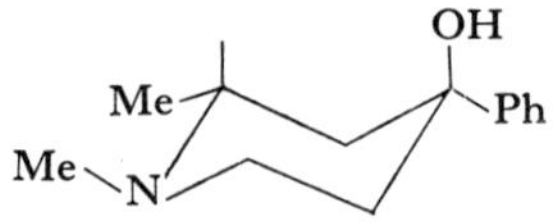   32 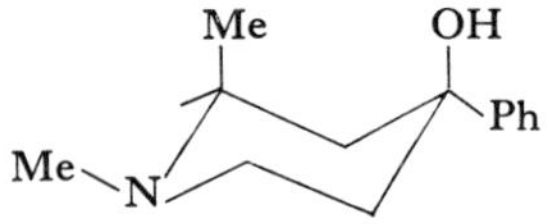   33

for the corresponding acetoxy esters of the prodinols emphasizes the importance of a hydrogen bonding association in bringing about these shifts. Neither 2-methyl signal of the isomeric piperidinols **31** shows a significant pyridine-induced shift; this result is anticipated for the 3-Me isomer **32**

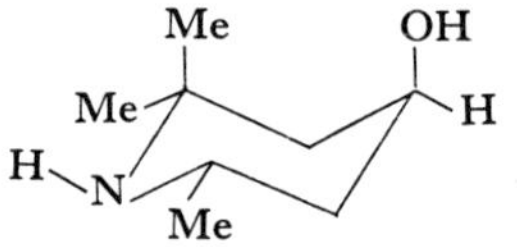

34 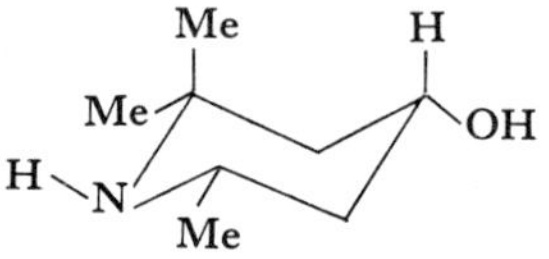   35

but not for isomer **33** in which the 2-Me and 4-OH groups have a 1,3-diaxial relationship in the chair form. The expected shift is seen in the axial 2-Me signal of the isomeric 2,2,6-trimethyl-4-piperidinol **34** with an axial OH group; this signal is lower field of the other two methyl signals in CDCl₃ because of its proximity to the OH group (see p. 242) and is moved even lower field in pyridine (Fig. A.6 upper); in the e-OH isomer **35** all 3 Me signals have similar resonance positions and are little affected by the same solvent change (Fig. A.6, bottom). Results with one of the 2-methyl-4-phenylpiperidinols **31** are thus anomalous and, again, provide evidence that the axial isomer **33** undergoes a conformational change in pyridine which moves the OH group away from the 2-methyl substituent. In the trimethyl analogue **34**, a conformational change is unlikely since it would place the 2,6-equatorial methyl substituents in unfavourable orientations.

Pyridine solvent shifts have provided evidence about the relative stereochemistry of the 3 erythromycin aglycones **36a** and **36b** (C-9 epimers) and

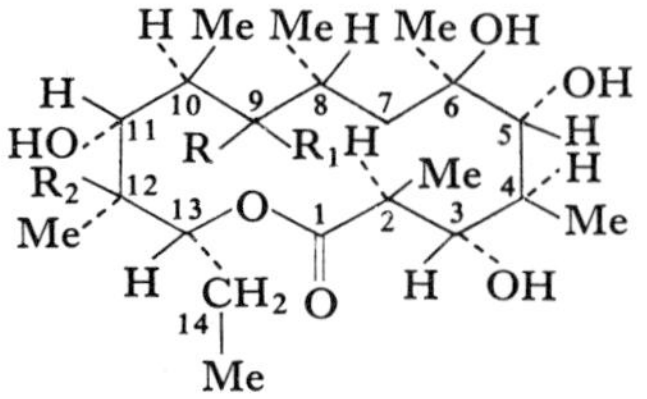

a: R = OH, R₁ = R₂ = H

b: R = H, R₁ = OH, R₂ = H

c: R = H, R₁ = R₂ = OH

36

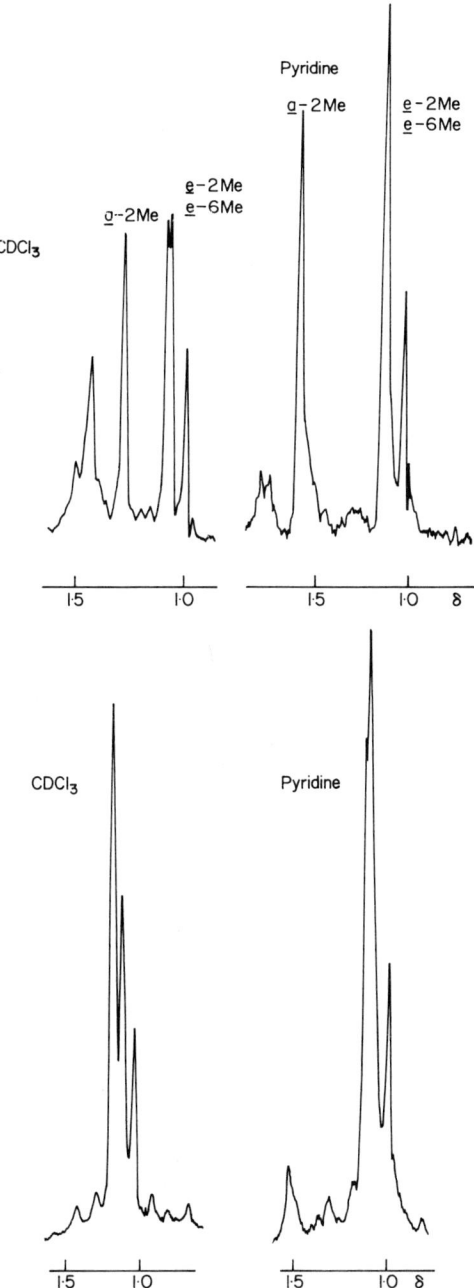

F<small>IG</small>. A.6. Methyl resonance region of the 60 MHz PMR spectra of *trans* (4-OH/6-Me) 2,2,6-trimethyl-4-piperidinol (**34**) (upper) and the corresponding *cis* isomer **35** (lower) in CDCl₃ and pyridine.

36c (Demarko, 1969a). Signal assignments to the individual protons and methyl groups were made by a detailed study not described here (Demarko, 1969b). Solvent shift data established that all the aglycones adopt the same approximate conformation in solution. Constant Δ values were noted for protons and methyl groups situated at positions 2, 3, 4, 5 and 6, hence their spatial position relative to nearby OH substituents must be the same in **36a**, b and c. Identical Δ values for H-9, H-11 and H-13 in **36b** and **36c** show that these 2 compounds share the same C-9 OH configuration which differs from that of **36a**. The large Δ values for H-11 and H-13 in **36b** and

Δ Values

Compound	H-9	H-11	H-13
36a	-0.66	-0.46	-0.56
36b	-0.38	-0.66	-0.76
36c	-0.40	-0.64	-0.80

36c are evidence that these protons bear a 1,3-diaxial and pseudo 1,3-diaxial relationship to the C-9 OH function (see **37**). Similar reasoning accounts for the greater deshielding observed for H-9 in **36a** than in the other 2 compounds. Models show that when H-9 takes the place occupied by 9-OH in **37**, it is close to the C-6 hydroxyl group.

(partial structure)

37

The use of PMR spectroscopy as a tool for the study of hydrogen bonding phenomena is now well established and the topic has been extensively reviewed (Pimental and McClellan, 1960; Laszlo, 1967; Murthy and Rao, 1968). Although the PMR technique does not allow the separate detection of individual species, e.g. free, inter- and intramolecular bonded OH as is possible by use of infra-red spectroscopy, it does permit the study of hydrogen bonding under a wide variety of solvent, concentration and temperature conditions. When a proton is hydrogen bound its chemical shift moves downfield. The chief cause of this deshielding is believed to be a distortion of its electronic structure (Pople *et al.*, 1959). Thus, in **38**, the electrostatic field of the hydrogen bond tends to draw H towards Y and to repel the X—H bonding electrons towards X; as a result the electron density

$$\overset{+}{\text{X—H}} \cdots \overset{-}{\text{Y}}$$

38

around H is reduced and the proton is deshielded (cf. explanation of the deshielding influence of the diionized phosphate group upon the C-8 proton of adenine in 5′-AMP, p. 368). Another factor influencing the chemical shift of a hydrogen bound proton is the magnetic anisotropy of the donor molecule Y, as mentioned earlier in the case of DMSO-d_6. This contribution may be positive or negative but is always outweighed by the X—H bond polarization factor. The sensitivity of the extent of hydrogen bonding to temperature and concentration is the reason for the variations seen in the resonance signal of acidic protons such as those of hydroxyl and amino groups, as already discussed in Chapter 1. These resonances are less sensitive to the same variables when spectra are recorded in hydrogen bonding solvents such as DMSO-d_6 and pyridine. The chemical shift of a free OH proton (aliphatic) determined by plotting the chemical shift of an alcohol against concentration and extrapolating to zero, is about the same as a methyl proton ($\sim \delta$ 1), while the OH chemical shift under hydrogen bonding conditions falls in range δ 1–4. Hydrogen bonding solvents generally move the O<u>H</u> resonance to lower field as compared with values in solvents such as CCl_4 and CDCl (see p. 27). Even lower field

Hydroxyl proton resonances (δ) in various solvents

	CCl_4	DMSO-d_6	Benzene	Pyridine
MeOH (distilled from lime) O<u>H</u>	3·93 broad quartet ($J \sim$5 Hz)	4·03 sharp quartet (J 5·5 Hz)	Within Me signal, broad singlet δ 3·25	5·43 broad singlet
![cyclohexanol structure] cis-trans mixture	2·5 v. broad band	4·33 (J 4·8 Hz) 4·05 (J 3·5 Hz) sharp doublets[a]	Within ring protons band above δ 2	4·8 broad band

[a] Lower field signal due to *trans* and higher to *cis* isomer.

OH signals are seen when intramolecular bonding operates, and in such cases change to a bonding solvent often has little effect; the case of 5-hydroxy-1,4-naphthoquinone **39** is an example (Casu *et al.*, 1966).

ν_{OH} 11·86 (CDCl$_3$)
11·75 (DMSO-d$_6$)

39

Much of the PMR studies of hydrogen bonding that have been reported concerns association between donors and acceptors and the gaining of quantitative data upon self-associated systems (monomer-dimer-polymer equilibria), and the reader is directed to the reviews mentioned above for further information upon these aspects. The relative strengths of solute-solvent hydrogen bonds in alcohols of related structure, as gauged from OH chemical shifts, provides information about the steric environment of hydroxyl functions and may aid stereochemical assignments. Thus the chemical shifts of the OH resonance of cyclohexanol and *trans* *t*-butylcyclo-hexanol in CCl$_4$ (linearly related to concentration over the range 0·015 to 0·002 molar) are lower field than those of *cis* t-butylcyclohexanol (Fig. A.7); this result reflects the conformational preference for equatorial OH (more accessible for hydrogen bonding) in the first 2 compounds. Differences in

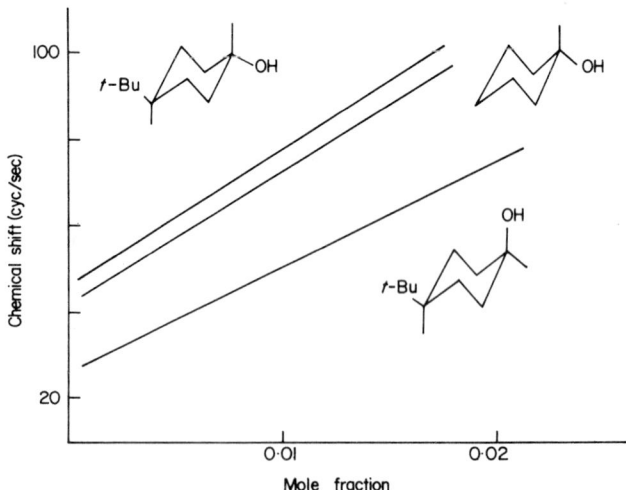

FIG. A.7. Chemical shift dependence of hydroxyl proton with concentration at 40° in CCl$_4$. Chemical shifts in Hz from TMS at 60 MHz (adapted from Ouellette, 1964).

the slopes of Fig. A.7 are also of stereochemical significance; the smaller slope of the *cis* derivative shows that the OH chemical shift of the axial alcohol is less concentration dependent than that of its epimer, a result of its being the less hydrogen bonded of the two isomers. Evidence about intramolecular hydrogen bonding provides valuable information about the preferred conformation of a compound since it establishes the relative orientation of donor and acceptor features within the molecule. The following examples add to those already given in this book.

The anti-blood clotting agent dicoumarol is considered to have the

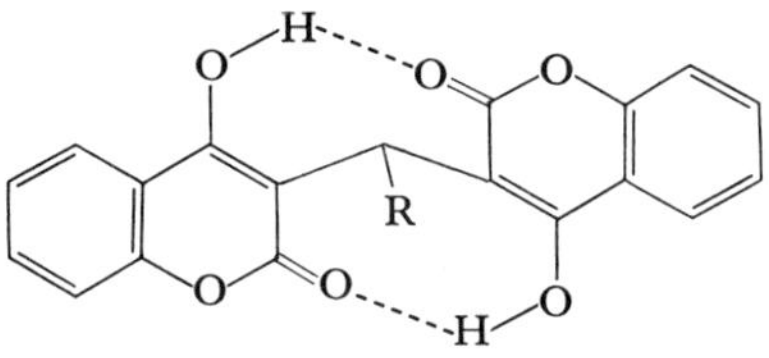

40

preferred conformation **40** (R = H) in CDCl$_3$ on the basis of its PMR spectrum displaying a low field singlet (δ 11·31), of intensity 2 protons, which is independent of concentration and is removed on the addition of a trace of D$_2$O (Hutchinson and Tomlinson, 1969). When R in **40** is an alkyl or aryl group, each OH proton gives rise to a separate signal since R is not symmetrically disposed toward the two hydroxyl protons.

The alkaloids sparteine and α-isosparteine possess a skeleton built of two quinolizidine rings; in α-isosparteine both nitrogen ring junctions are *trans* while in sparteine one is *cis* and the other *trans* (Krueger and Skolik, 1967). There is clear evidence for the all-chair conformation of α-isosparteine but arguments in favour of both conformers **41** (all-chair) and **42**

41

42

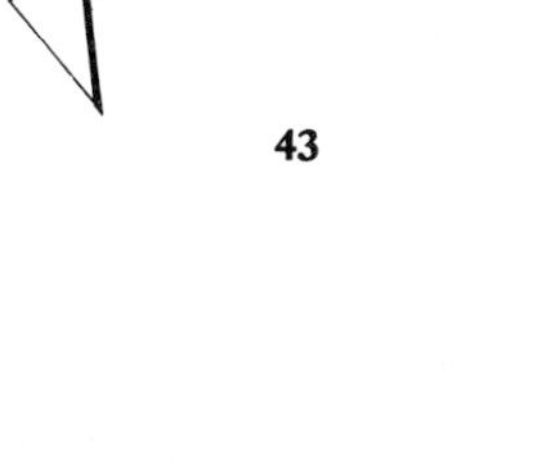

ring C

43

(ring C boat) have been advanced for sparteine. In the case of the mono-perchlorate of the N-oxide of sparteine, a decision in favour of the all-chair form **43** may be made from PMR evidence of intramolecular hydrogen bond formation (also established for the corresponding α-iso-sparteine derivative). Thus the $^{+}$N—H resonance of the N-oxide salt is abnormally low field in $CDCl_3$ (δ 18·52) and is little affected when the solvent is replaced by DMSO-d_6. Monoperchlorates of the parent alkaloids in DMSO-d_6 have NH signals at relatively high field positions near δ 6·5 and close to those of derivatives in which intramolecular hydrogen bonding is not expected such as diperchlorates. It is argued that intramolecular bonding in the mono salts ($^{+}$NH····: N) is disrupted in DMSO-d_6 solution; PMR data upon the salts in $CDCl_3$ are not reported, presumably because of solubility limitations.

References

Bottini, A. T. and O'Rell, M. K. (1967). *Tetrahedron Lett.* 429.

Casu, B., Reggiani, M., Gallo, G. G. and Vigevani, A. (1966). *Tetrahedron* **22**, 3061.

Casy, A. F. (1968). *J. Med. Chem.* **11**, 188.

Demarco, P. V. (1969a). *Tetrahedron Lett.* 383.

Demarco, P. V. (1969b). *J. Antibiot.* **22**, 327.

Demarco, P. V., Farkas, E., Doddrell, D., Mylari, B. L. and Wenkert, E. (1968). *J. Am. Chem. Soc.* **90**, 5480.

Eliel, E. L., Allinger, N. L., Angyal, S. J. and Morrison, G. A. (1965). "Conforma-tional Analysis". Wiley, New York.

Hutchinson, D. W. and Tomlinson, J. A. (1969). *Tetrahedron* **25**, 2531.

Kier, L. B. (1968). *J. Med. Chem.* **11**, 441.

Krueger, P. J. and Skolik, J. (1967). *Tetrahedron* **23**, 1799.

Laszlo, P. (1967). *In* "Progress in Nuclear Magnetic Resonance Spectroscopy", Vol. 3 (J. W. Emsley, J. Feeney and L. H. Sutcliffe, eds.), Pergamon, Oxford.

Murthy, A. S. N. and Rao, C. N. R. (1968). *Applied Spectroscopy Reviews* **2**, 69.

Ouellette, R. J. (1964). *J. Amer. Chem. Soc.* **86**, 4378.

Pimental, G. C. and McClellan, A. L. (1960). "The Hydrogen Bond", p. 142. Freeman, San Francisco.

Pople, J. A., Schneider, W. G. and Bernstein, H. J. (1959). "High Resolution Nuclear Magnetic Resonance", McGraw-Hill, New York.

Ronayne, J. and Williams, D. H. (1969). *In* "Annual Review of NMR Spectros-copy", Vol. 2 (E. F. Mooney, ed.). Academic Press, London.

Stien, M. L., Ottinger, R., Reisse, J. and Chiurdoglu, G. (1968). *Tetrahedron Lett.* 1521.

Wenkert, E. and Mylari, B. L. (1967). *J. Am. Chem. Soc.* **89**, 174.

Whelton, B. D. (1969). Ph.D. Thesis, University of Washington.

Selective Bibliography

GENERAL TEXTS

1. Becker, E. D. (1969). "High Resolution NMR. Theory and Chemical Applications", Academic Press, New York.
2. Bible, R. H., Jr. (1965). "Interpretations of NMR Spectra, An Empirical Approach", Plenum Press, New York.
3. Bible, R. H. Jr. (1967). "Guide to the NMR Empirical Method". Plenum Press, NewYork. A workbook of exercises in the interpretation of NMR spectra.
4. Bovey, F. A. (1969). "Nuclear Magnetic Resonance Spectroscopy". Academic Press, New York. A general account with useful appendices on ring current shielding effects and calculated spectra.
5. Carrington, A. and McLauchlan (1967), "Introduction to Magnetic Resonance with Applications to Chemistry and Chemical Physicals", Harper and Row, New York. An account of fundamental principles.
6. Chapman, D. and Magnus, P. D. (1966). "Introduction to Practical High Resolution Nuclear Magnetic Resonance Spectroscopy", Academic Press, London. Contains a useful collection of chemical shift and coupling constant tables.
7. Emsley, J. W., Feeney, J. and Sutcliffe, L. H. (1965). "High Resolution Nuclear Magnetic Resonance Spectroscopy", Vols. 1 and 2, Pergamon Press, Oxford. An extensive reference book.
8. Hecht, H. G. (1967). "Magnetic Resonance Spectroscopy", Wiley, New York. An account of fundamental principles.
9. Jackman, L. M. and Sternhell, S. (1969). "Applications of Nuclear Magnetic Resonance Spectroscopy in Organic Chemistry", 2nd Ed., Pergamon Press, 1969. A new edition of Jackman's well-known text first published in 1959.
10. Lyndon-Bell, R. M. and Harris, R. K. (1969). "Nuclear Magnetic Resonance Spectroscopy", Nelson, London. A concise account of fundamental principles in paperback form.
11. Mathieson, D. W. (Ed.) (1967). "Nuclear Magnetic Resonance for Organic Chemists", Academic Press, New York. A general text based on Summer Schools in NMR Spectroscopy organized by the Royal Institute of Chemistry.
12. Pople, J. A., Schneider, W. G. and Bernstein, H. J. (1959). "High Resolution Nuclear Magnetic Resonance", McGraw-Hill, New York. The classic text on NMR spectroscopy.

INTRODUCTORY TEXTS. Each contains a useful chapter on NMR spectroscopy.

1. Brand, J. C. D. and Eglington, G. (1965). "Applications of Spectroscopy to Organic Chemistry", Oldbourne, London.
2. Dyer, J. R. (1965). "Applications of Absorption Spectroscopy of Organic Compounds", Foundations of Modern Organic Chemistry Series, Rinehart, K. L. Ed., Prentice Hall, Englewood Cliffs, New Jersey.
3. Fleming, I. and Williams, D. H. (1966). "Spectroscopic Methods in Organic Chemistry", McGraw-Hill, London.

RECENT ADVANCES

1. "Advances in Magnetic Resonance" (Ed. Waugh, J. S.), Vols. I–IV (1965-1970). Academic Press, New York.
2. "Annual Review of NMR Spectroscopy" (Ed. Mooney, E. F.), Vol. I (1968), Vol. II (1969), Vol. III (Annual Reports on NMR Spectroscopy) (1970). Academic Press, London.
3. "Progress in Nuclear Magnetic Resonance Spectroscopy" (Eds. Emsley, J. W., Feeney, J. and Sutcliffe, L. H.), Vols. I–VII (1966–1971). Pergamon Press, Oxford.

JOURNALS

1. *Journal of Magnetic Resonance* (1969–). Academic Press, New York and London.
2. *Organic Magnetic Resonance* Vol. I (1969–). Heyden and Sons, London.

SOURCE BOOKS OF SPECTRAL DATA

1. Bovey, F. A. (1967). "NMR Data Tables for Organic Compounds", Vol. I, Interscience, New York.
2. "High Resolution NMR Spectra Catalog", Vols. 1 (1962) and 2 (1963). Varian Associates.
3. Wiberg, K. B. and Nist, B. J. (1962). "The Interpretation of NMR Spectra", Benjamin: New York. A collection of calculated spectra.
4. Howell, M. G., Kende, A. S. and Webb, J. S. (1966). "Formula Index to NMR Literature Data," Vols. 1 and 2. Plenum Press, New York.
5. Hershenson, H. M. (1965). "NMR and ESR Spectra Index". Academic Press, New York.
6. "Sadtler Standard NMR Spectra" (1967). Sadtler Laboratories, Philadelphia, Pennyslvania.

OTHER REFERENCES

1. Becker, E. D. (1965). *J. Chem. Educ.*, **42**, 591.
2. Garbisch, E. W. Jr. (1968). *J. Chem. Educ.*, **45**, 311, 402, 480.

Author Index

Numbers in *italics* indicate the pages on which the references are listed

A

Abraham, R. J., 122, 123, *132*
Adamson, D. W., 233, *237*
Ahmed, F. R., 211, 215, *238, 239*
Albert, A., 67, *84*
Alexander, T. G., 2, *51*
Alekseeva, L. M., 80, *84*
Allinger, N. L., 86, 128, *133*, 149, 151, 163, *168, 169*, 331, 353, *382*, 386, *402*
Aminoto, T., 29, 31, *52*
Anderson, J. K., 65, *84*
Anderson, W. R., 63, *84*
Anet, F. A. L., 103, 118, 122, 130, *132*, 185, *187*
Angyal, S. J., 86, 128, *133*, 149, 151, *169*, 330, 331, 345, 353, *382*, 386, *402*
Anteunis, M., 123, *133*
Aono, K., 58, *85*
ApSimon, J. W., 95, 119, *133*
Archer, S., 243, *279*
Armstrong, N. A., 39, *51*

B

Babao, H., 7, 9, *51*
Babiec, J. S., 32, 33, *51*
Bak, B., 281, 291, 292, *327*
Baker, R. H., 6, *52*
Baker, R. H. Jr., 362, *383*
Ballantine, J. A., 48, *51*
Barbieux, M., 101, *133*
Barcza, S., 10, *51*
Barfield, M., *133*, 270, *279*
Barlow, R. B., 322, *327*
Barnes, W. H., 250, *280*
Barrante, J. R., 32, 33, *51*
Barrett, P. A., 233, *237*

Barrow, K. D., 263, *279*
Bartle, K. D., 43, 46, *51*
Barton, D. H. R., 268, *279*
Bass, R. J., 28, *51*
Bauld, N. L., 29, *51*
Bavin, P. M. G., 43, 46, *51*
Bayer, E., 283, *329*
Beare, S. D., 180, 182, 183, 184, 185, *187*
Becconsall, J. K., 146, 149, 160, *168*
Beck, B. H., 125, 128, *134*
Becker, E. D., 191, 204, *237, 238*, 304, *329*
Beckett, A. H., 23, 32, *51*, 78, *84*, 151, *169*
Bell, M. R., 243, *279*
Belleau, B., 254, *279*
Bentley, K. W., 189, *237*
Bergmann, F., 80, *84*
Berlin, A. J., 118, 125, *133, 134*
Bernstein, H. J., 16, *52*, 67, *85*, 87, *134*, 216, *239*, 332, *383*, 398, *402*
Best, D. C., 105, 107, *134*
Beyerman, H. C., 249, *280*
Bhacca, N. S., 92, *133*, 227, *239*, 268, 270, 273, 277, *279, 280*, 281, 291, 292, *327*, 333, *382*
Bible, R. H., 217, 221, *237*
Billinghurst, J. W., 233, *237*
Biltonen, R. L., 301, 302, *328*
Bishop, R. J., 164, *169*, 240, 243, *279*
Bommer, P., 186, *187*
Booth, H., 104, 118, 132, *133*, 136, 138, 140, 158, 164, 165, 167, *168*, 196, 212, 223, 224, 225, 227, *237*, 253, *279*, 336, *382*
Bottini, A. T., 242, *280*
Bottomley, W., 96, *133*
Bourn, A. J. R., 103, 118, *132*

405

N

Subject Index

Asterisk denotes Figure (usually a spectrum) or Figure reference

A